D0883367

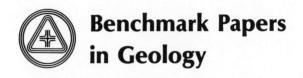

Benchmark Papers
in Geology

Series Editor: Rhodes W. Fairbridge
Columbia University

**Benchmark Papers
in Geology / 37**

A Benchmark® Book Series

STATISTICAL ANALYSIS
IN GEOLOGY

Edited by

JOHN M. CUBITT

Syracuse University

STEPHEN HENLEY

Institute of Geological
Sciences, Nottingham, UK

**Dowden, Hutchinson
& Ross, Inc.**

STROUDSBURG, PENNSYLVANIA

To Cynthia, Lyn, and Cerynne

Copyright © 1978 by **Dowden, Hutchinson & Ross, Inc.**
Benchmark Papers in Geology, Volume 37
Library of Congress Catalog Card Number: 78-17368
ISBN: 0-87933-335-9

80 79 78 1 2 3 4 5
Manufactured in the United States of America.

LIBRARY OF CONGRESS CATALOGING IN PUBLICATION DATA
Main entry under title:
Statistical analysis.
  (Benchmark papers in geology ; v. 37)
  Bibliography: p.
  Includes index.
  1. Geology—Statistical methods—Addresses, essays, lectures.
I. Cubitt, John M. II. Henley, Stephen.
QE33.2.M3S7   550'.1'82   78-17368
ISBN 0-87933-335-9

Distributed world wide by Academic Press,
a subsidiary of Harcourt Brace Jovanovich,
Publishers.

SERIES EDITOR'S FOREWORD

The philosophy behind the "Benchmark Papers in Geology" is one of collection, sifting, and rediffusion. Scientific literature today is so vast, so dispersed, and, in the case of old papers, so inaccessible for readers not in the immediate neighborhood of major libraries that much valuable information has been ignored by default. It has become just so difficult, or so time consuming, to search out the key papers in any basic area of research that one can hardly blame a busy man for skimping on some of his "homework."

This series of volumes has been devised, therefore, to make a practical contribution to this critical problem. The geologist, perhaps even more than any other scientist, often suffers from twin difficulties—isolation from central library resources and immensely diffused sources of material. New colleges and industrial libraries simply cannot afford to purchase complete runs of all the world's earth science literature. Specialists simply cannot locate reprints or copies of all their principal reference materials. So it is that we are now making a concerted effort to gather into single volumes the critical material needed to reconstruct the background of any and every major topic of our discipline.

We are interpreting "geology" in its broadest sense: the fundamental science of the planet Earth, its materials, its history, and its dynamics. Because of training and experience in "earthy" materials, we also take in astrogeology, the corresponding aspect of the planetary sciences. Besides the classical core disciplines such as mineralogy, petrology, structure, geomorphology, paleontology, and stratigraphy, we embrace the newer fields of geophysics and geochemistry, applied also to oceanography, geochronology, and paleoecology. We recognize the work of the mining geologists, the petroleum geologists, the hydrologists, the engineering and environmental geologists. Each specialist needs his working library. We are endeavoring to make his task a little easier.

Each volume in the series contains an Introduction prepared by a specialist (the volume editor)—a "state of the art" opening or a summary of the object and content of the volume. The articles, usually some thirty to fifty reproduced either in their entirety or in significant extracts, are selected in an attempt to cover the field, from the key papers of the last century to fairly recent work. Where the original works are in foreign languages, we have endeavored to locate or commission translations. Geologists, because of their global subject, are often acutely aware of the oneness of our world. The selections cannot, therefore, be restricted to any one country, and whenever possible an attempt is made to scan the world literature.

To each article, or group of kindred articles, some sort of "highlight commentary" is usually supplied by the volume editor. This should serve to bring that article into historical perspective and to emphasize its particular role in the growth of the field. References, or citations, wherever possible, will be reproduced in their entirety—for by this means the observant reader can assess the background material available to that particular author, or, if he wishes, he too can double check the earlier sources.

A "benchmark," in surveyor's terminology, is an established point on the ground, recorded on our maps. It is usually anything that is a vantage point, from a modest hill to a mountain peak. From the historical viewpoint, these benchmarks are the bricks of our scientific edifice.

RHODES W. FAIRBRIDGE

PREFACE

We have attempted in this volume to provide an extensive historical review of the application of statistical methods in geology. It has been very difficult to draw boundaries around the field covered. Methods that are primarily intended for use in analyzing spatially distributed data are excluded from this study, as we think that they would fit more logically into a separate review of spatial techniques. Many of the recent developments in statistics that have dealt with modelling applications (e.g. conditional simulation) have also been regarded as beyond the scope of the compilation. Finally, there are in the study of geology a number of mathematical and other methods, not strictly statistical (largely because they are not based on probabilistic considerations) and these too have mostly been excluded. Some of the papers included, however, show the strong influence exercised by geologists, among other natural scientists, on the development of statistical thought.

Even in the field of "mainstream" statistics that we have attempted to cover, there are difficulties in choosing material. It is obvious that the subject matter is extremely heterogeneous, with varied types of statistical analysis being applied to the wide range of disciplines that are to be found under the geological umbrella. Any bias in the selection of papers, therefore, that are included in this volume may well result from the similar backgrounds of the editors. Both are interested in sedimentary petrology, with secondary fields of interest in igneous and metamorphic rocks and structural geology.

Compiling this volume has been a very enjoyable task, although joint editorship on two sides of the Atlantic Ocean has had its difficulties. We wish to express our thanks to all the authors and publishers for granting us permission to reprint their works, to colleagues in the Natural Environmental Research Council of the United Kingdom and those at Syracuse University, as well as many other friends worldwide, for their comments and advice. In particular, our thanks are extended to Professor D. F. Merriam, Dr. K. G. Jeffery, Dr. R. B. McCammon, Dr. R. Reyment, Dr. J. C. Davis, and Dr. F. Chayes for their helpful criticisms of the manuscript.

JOHN M. CUBITT
STEPHEN HENLEY

CONTENTS

Contents

PART II: STATISTICAL METHODS—A SECOND LOOK

PART III: NEW STATISTICAL METHODS

CONTENTS BY AUTHOR

STATISTICAL ANALYSIS
IN GEOLOGY

INTRODUCTION

Modern statistics is the result of the convergence of three independent lines of activity pursued in separate areas of human knowledge whose common elements were recognized and brought together in the early part of the twentieth century. The first is traditionally related to political science and developed from the Latin word *status* (a political state) as a quantitative description of the affairs of a government or state (hence, the term *statist-ics* or political arithmetic). The word was first coined by Hooper in 1770 who described "... the science that teaches us what is the political arrangement of all the modern states of the known world" as *statistics* in his text *The Elements of Universal Erudition* (Yule and Kendall, 1953; Till, 1974). Although the description was at first mainly verbal, advances in communication and transportation have made it feasible to estimate numerically the resources and needs of a state. The current use and importance of statistics in political arithmetic is attested to by the reams of government statistics published each year by statistics offices throughout the world.

Statistical science was also affected by work on vital statistics (birth, marriage, divorce, sickness, and death) conducted by Willian Pelty (1623–1687) and John Graunt (1620–1674). For example, the book entitled *Natural and Political Observations . . . Made Upon the Bills of Mortality* by John Graunt in 1662 interpreted birth, death, and christening figures for London parishes from 1604–1661 in terms of mass biological and social phenomena. Together with articles published thirty years later by astronomer Edmund Halley, the foundations for modern work on life expectancy and vital statistics were laid.

However, at about the same time, the third root of statistics—the mathematical theory of probability engendered by interest in games of chance among the leisure classes of the time—was also derived. Contributions to the theory were provided by Blaise Pascal (1623–1662) and Pierre de Fermat (1601–1665), but the foundation for modern probability theory was laid by Christiaan Huygens in 1657 with his treatise, *De Ratiociniis in Ludo Aleae (On Reason in Games of Dice)*, and developed by the Bernoulli family, in particular, Jacques Bernoulli with his publication *Ars Conjectandi* in 1713. Early development of probability theory came from famous astronomers and mathematicians of the eighteenth century, such as Pierre Simon de Laplace (1749–1827) and Karl Friedrich Gauss (1777–1855) who introduced the classical theory of errors of observations. Their work initially served scientists who were making optical measurements of great precision and constructing increasingly sophisticated models of natural phenomena.

Karl Pearson, Ronald Fisher, and W. C. Gossett revolutionized statistics at the beginning of the twentieth century by consolidating these diverse fields into a single scientific discipline. The impetus for this achievement was provided by the discovery of the χ^2 and *t*-statistics, which led to the rapid evolution of current statistical techniques. It has also resulted in the modern nontechnical interpretation that statistics ". . . is concerned with the summarization and presentation of large masses of data and with conclusions from these data" (Krumbein and Graybill, 1965, p. 3).

It is chastening, therefore, to realize that geology, now one of the least numerate sciences, was at the forefront of statistical application in 1833 with the publication of Charles Lyell's *Principles of Geology*, a situation most aptly portrayed by Sir Ronald Fisher in his address to the Royal Statistical Society in 1953 (Paper 1).

This auspiscious introduction of fundamental statistical principles to geology, however, did not encourage the profession to adopt a more rigorous approach to science. In fact, geological applications of statistics then disappeared from scientific literature for approximately sixty years until the eminent statistician Karl Pearson published a series of essays entitled "Contributions to the Mathematical Theory of Evolution" in 1897 and in 1901 helped found the journal *Biometrika*. Both these events marked significant breakthroughs for the paleontologist because in many situations, the problems studied were based on data from fossils. As a result, biometrical classification received considerable attention as a statistical science (Brinkmann, 1929; Pearson, 1928; Hersch, 1934; Robb, 1935; Simpson and Roe, 1960). For the reader interested in the early development of paleobiometrics, Schmid (1934) and Sokal and Sneath (1963) provide bibliographies of both European and North American literature.

Paralleling the early application of statistics in paleontology was the quantification of petrology and geochemistry. Following the early work of Reyer (1877), the increasing quantity of data on rock chemistry allowed a number of systematic studies of data distributions (Harker, 1909; Richardson and Sneesby, 1922; Niggli, 1923; Richardson, 1923; Loewinson-Lessing, 1924, 1925, 1930, 1935; Niggli, de Quervain, and Winterhalter, 1930; Kupletsky and Oknova, 1934). The purpose of most of these studies was to establish boundaries between various igneous rock types by using simple population statistics such as mean, variance, and skewness of frequency distributions (see in particular Loewinson-Lessing, 1925, 1930).

However, Clarke (1892, 1924) and later Washington (1920, 1925) statistically analysed large volumes of data on rock chemistry to derive well-reasoned average compositions for the more common igneous and sedimentary rock types—averages that remain in use to this date. Similar studies by Vogt (1921–1923) on granites and quartz porphyries and by Chirvinsky (1909, 1911) on the quantitative mineralogical composition of granite and gneiss led to a better interpretation of the granite eutectic (Loewinson-Lessing, 1936).

In sedimentology, the underrated work of Sorby (1908) and Udden (1898, 1914) on wind-borne deposits and clastic sediments were followed by the routine application of histograms to sedimentary distributions. Similarly, graphical analyses of size frequency distributions developed by Wentworth (1922, 1929) and modified by Krumbein (1934, 1936b) led to standardized statistical analysis of sediment size data. Among other leading exponents of statistics in sedimentology during this period were Ewing (1931), Krumbein (1934, 1936a, b, 1937a, b, c, 1938a, b, 1939), Niggli (1935), and Otto (1937), as well as Eisenhart who published the correct application of the χ^2 test of significance for sedimentary data in 1935.

From the late 1930s to World War II, applications of statistics to the geological sciences became a regular feature in paleontology, sedimentology, geomorphology, petrology, and geochemistry literature. During this period, geophysics also acquired a significant base in statistics (Court, 1952; Claerbout, 1976), and occasional statistically oriented papers appeared in stratigraphy (Korn, 1938) and structural geology (Pincus, 1951, 1952, 1953). However, all of these applications had to remain relatively simple because of the time-consuming nature of the calculations. Sophistication crept into geological data analysis only after the period of "descriptive" applications of statistics was succeeded by stochastic (probabilistic) modeling of geological processes and the introduction of multivariate statistics. The end of the "descriptive" period is recognized in the work of Razumovsky (1940) on logarithmic distribution of weights of chemical elements in ores (Vistelius, 1967).

The second state in the application of probability methods in geology (initiated by Kolmogorov in 1941) could be considered the stage of transition to stochastic problems. During this period, stochastic modeling was introduced, thus revealing the possibility of applying statistics in the geosciences to estimate parameters of populations from sample observations, and specific probability—theoretical subjects in geology—were formulated (Vistelius, 1967). In recent years the number of such works has increased, especially in the French and Russian literature. However, the content indicates specialization in areas of application; Vistelius (1967, 1968) notes the concentration of U.S. and U.K. geologists on analysis of variance methods (e.g., Krumbein and Miller, 1953; Krumbein, 1955; Griffiths, 1967; Cubitt, 1975), of Russian geologists on distribution functions (Vistelius and Sarmanov, 1947; Razumovsky, 1948; Kolmogorov, 1949; Vistelius, 1963, 1968) and correlation theory (Vistelius, 1956; Sarmanov and Vistelius, 1959; Sarmanov, 1961), and of South African and French geologists on probability distributions (De Wijs, 1951, 1953; Sichel, 1952; Krige, 1960, 1966). The post–World War II period from about 1945 to 1960 showed considerable expansion of statistical applications in almost all fields of geology. Summary and review papers that kept specialists abreast of developments appeared frequently; for example, Ingerson (1954) and Shaw and Bankier (1954) for geochemistry, Horton (1945) and Strahler (1952, 1954, 1956, 1964, 1968) for geomorphology, Prentice (1949) and Leitch (1951) for paleontology, Pincus (1951) for aspects of structural geology, Chenowith (1952), Miller (1953), Krumbein (1954, 1955) and Griffiths (1960b) for geology, Chayes (1949) for petrology, Krumbein (1954) for stratigraphy, Allen (1944) and Griffiths (1962) for sedimentology, and Ayler (1963) for mineral exploration.

Nevertheless, despite the advances made in data analysis (Burma, 1949; Reyment, 1961), problems concerning large numbers of samples or measured parameters remained insoluble on a practical scale because of the complexity and time-consuming nature of the analyses. The era of multivariate statistical analysis had to await the arrival of computers (Krumbein and Sloss, 1958). For example, after laboriously working out principal components solutions by hand (Reyment, 1961), paleontologists switched their attention to computer analysis of data, and within a decade extensive advances were made in paleobiometrics (Sokal and Sneath, 1963; Cole 1969; Sokal and Rohlf, 1970; Blackith and Reyment, 1971; Sneath and Sokal, 1973) and paleontology (Imbrie and Kipp, 1971; Reyment, 1971). Similar changes were observed in other branches of geology and resulted in an exponential increase in publications on statistical applications in geology (Harbaugh and Merriam, 1968). Included in this list are, in alphabetical order, the books of Agterberg (1974); Bath (1974); Chayes (1971); Claerbout (1976); Cole and King (1968); David (1977); Davis (1973); Davis and McCullagh (1975); Fenner (1969, 1972); Griffiths (1967); Guarascio, David and Huijbregts (1976); Harbaugh and Bonham-Carter (1970); Har-baugh, Doveton, and Davis (1977); Harbaugh and Merriam (1968); Hazen (1967); Joreskog, Klovan, and Reyment (1976); Koch and Link (1970, 1971); Krumbein and Graybill (1965); Lafitte (1972); Marsal (1967); Matheron (1962, 1963a, 1965, 1967, 1971, 1975); McCammon (1975); Merriam (1969, 1970, 1972, 1976b, c, 1978, in prep.); Romanova and Sarmanov (1970); Schwarzacher (1975); Sharapov (1971); Smith (1966); Thiergartner (1968); Twomey (1977); Vistelius (1967); Weiss (1969); and Whitten (1975), as well as the reviews of Agterberg (1964, 1967); Agterberg and Robinson (1972); Griffiths (1970); Journel (1973, 1975); Krumbein (1960, 1969); Merriam (1976a); Morisawa (1971); Vistelius (1968, 1976a, b); Wilks, (1963); and Whitten (1964).

During this last period, attention focused on the "shotgun" approach to statistical analysis in geology (Reyment, 1974; Brower, 1974)—that is, the application of every available technique to the data in the hope that some interpretable results might be obtained. This approach remains prevalent today although many geologists have criti-cally evaluated the statistical validity of sampling and analysis metho-dology and found the application of some statistics to geological data to be questionable. Through these "second-looks," the geosciences have acquired a new perspective on the application of statistical techniques, particularly multivariate, univariate, and modeling methods (Krige, 1951; Vistelius and Sarmanov, 1961; Chayes and Kruskal, 1966; McCammon, 1966, 1970; Demirmen, 1969; Miesch, 1969; Chayes, 1970, 1971, 1975; Davis, 1970; Tukey, 1970; Blackith and Reyment, 1971; Griffiths, 1971; Zodrow and Sutterlin, 1971; Drapeau, 1973; Till and Colley, 1973; Saha, Bhattacharyya, and Lakshmipathy, 1974; Butler, 1975; Klovan, 1975; Link and Koch, 1975; Zodrow, 1975, 1976).

The significance of this period cannot be understated, for not only have geologists reevaluated basic approaches to sampling and instrumen-tal analysis, but also fundamentally new techniques of statistical analy-sis have been derived. No more clearly can this be demonstrated than in the mining industry where the statistical methodology termed *geo-statistics* was developed by Matheron and his coworkers (Matheron, 1962, 1965, 1967; Huijbregts and Matheron, 1971; Journel, 1975; David, 1977) to solve problems in the analysis of grade distributions in ore bodies, as identified by De Wijs (1951, 1953), Krige (1951, 1964, 1966), and Sichel (1952, 1973). The French school also was responsible for introducing correspondence analysis to geology as a complement to factor analysis (Dagbert and David, 1974; David, Campiglio and Dar-ling, 1974; Melguen, 1974; Teil, 1975, 1976; Teil and Cheminee, 1975).

It is with this background that the editors have arranged the twenty-three published contributions in this volume into three sections:

5

classical methods, a second look, and new approaches. The three periods are disparate in the length of time each occupies in the development of statistical analysis in the geological sciences. However, even though there has been disproportionate number of important developments in the field during the past twenty years, the editors consider that all three periods should be given equal representation in this benchmark volume on the development of statistical analysis in geology. The reader should note that the editors have made no attempt to include any literature on certain geological spatial data handling techniques, such as trend-surface analysis and contouring, since these will form the basis of a future benchmark volume.

Finally, the role played by scientific journals in the development of statistical applications in geology should be noted. In 1953, 1954, and 1966, the *Journal of Geology*, on the initiative of R. L. Miller and R. B. McCammon, advanced the theory, status, and popularity of statistical and mathematical geology by publishing three complete issues on the applications of statistical methods in geology. Subsequently, the journal has continued to publish papers with a statistical bias and has maintained its leading position in this area. Statistically oriented papers also have appeared regularly in the *American Association of Petroleum Geologists Bulletin, Bulletin of the Geological Society of America, Journal of Sedimentary Petrology, Journal of Paleontology, Geochimica Cosmochimica Acta, Doklady Academy Sciences of the USSR, Chemical Geology, Canadian Journal of Earth Sciences,* and *Transactions of the Institute of Mining and Metallurgy.*

However, with the advent of the Special Distribution Series (1963–1966) and *Computer Contributions* (1966–1970) of the Kansas Geological Survey; *Geocom Bulletin* (1968-1976) of Lea Associates, London; *Computer Applications* (1973 to date) of Nottingham University Geography Department; *Science de la terre, informatique, geologique* (1973 to date); in particular, *Mathematical Geology* (1969 to date); and the recent *Computers & Geosciences* (1975 to date), a new phase in the rapid advancement of all branches of statistical, mathematical, and computer geology has been entered. This advance has encouraged the further development of new statistical methods, the reevaluation of classical statistics in geological applications, and the ever-expanding use of statistical techniques in geology.

Readers interested in the philosophical development of statistical analysis in geology and the historical relationship of statistical and mathematical geology should refer to papers by Vistelius (1962, 1963, 1967, 1968, 1969, 1976a, b), Craig (1974), Krumbein (1974), Merriam (1974, 1976b), and Reyment (1974). For those who wish to follow the latest developments in the subject, *Geotimes* publishes an annual review of mathematical geology.

1

Reprinted from *J. Roy. Stat. Soc., Ser A.,* **116**(1):2-3 (1953)

THE EXPANSION OF STATISTICS

Sir Ronald Fisher, F. R. S.

[*Editor's Note:* In the original, material precedes this excerpt.]

Charles Lyell, the geologist, was born in 1797, and in 1830 there appeared the first of the three volumes of his celebrated *Principles of Geology*. The second volume was received by Darwin at Monte Video in 1832 during the outward voyage of the "Beagle", and no book could have been more stimulating for the theories which Darwin later formed. The third volume is dated 1833, three years after the first. The work is, of course, a scientific classic, a masterpiece of lucid and effective literary style, embodying proposals by the author of revolutionary importance, and preserving for readers of our own age the memorials of a scientific environment with its ideas and arguments so remarkably different from our own.

Geologists prior to Lyell had recognized the sequences of strata which we know as Primary and Secondary, using in the first place the regularity of order of superposition in the same locality. They observed, too, that particular components of these formations could be recognized, though far apart, by their characteristic fossils. They could not by these means recognize or establish the order among the Tertiary rocks, for, in the part of the world then accessible, these occur in patches, and not over wide areas overlying one another. Lyell determined the order and assigned to the successive rock masses the names they now bear by a purely statistical argument. A rich group of strata might yield so many as 1,000 recognizable fossil species, mostly marine molluscs. A certain number of these might be still living in the seas of some part of the world, or at least be morphologically indistinguishable from such a living species. It was as though a statistician had a recent census record without recorded ages, and a series of undated records of previous censuses in which some of the same individuals could be recognized. A knowledge of the Life Table would then give him estimates of the dates, and, even without the Life Table, he could set the series in chronological order, merely by comparing the proportion in each record of those who were still living.

With the aid of the eminent French conchologist M. Deshayes, Lyell proceeded to list the identified fossils occurring in one or more strata, and to ascertain the proportions now living. To a Sicilian group with 96 per cent. surviving he gave, later, the name of Pleistocene (mostly recent). Some sub-appenine Italian rocks, and the English Crag with about 40 per cent. of survivors, were called Pliocene (majority recent). Forty per cent. may seem to be a poor sort of majority, but no doubt scrutiny of the identifications continued after the name was first bestowed, and the separation of the Pleistocene must have further lowered the proportion of the remainder. The Miocene, meaning "minority recent", had 18 per cent., and the Eocene, "the dawn of the recent", only 3 or 4 per cent. of living species. Not only did Lyell immortalize these statistical estimates in the names still used for the great divisions of the Tertiary Series, but in an Appendix in his third volume he occupies no less than 56 pages with details of the classification of each particular form, and of the calculations based on the numbers counted. There can be no doubt that, at the time, the whole process, and its results, gave to Lyell the keenest intellectual satisfaction.

Now the point of this little history, its point for statisticians, is that the statistical argument by which this revolution in Geological Science was effected was almost immediately forgotten. In later editions of the *Principles* this great Appendix, in which so much labour had been expended, has disappeared; it survived indeed only two years. It had served its purpose; the ladder by which the height had been scaled could be kicked down. Geologists could quickly recognize a fossiliferous stratum by a few characteristic forms with clear morphological peculiarities. There was no need to wait for extensive collection, or statistical tabulation. Nor does it seem to have been thought that future geologists had anything to learn from the example, or the particular method, of Lyell's discovery. His designations of the Tertiary formations remain as records of a forgotten past, like fossils themselves, less intelligible to the geological students than the casts of sea-shells they extract from the rocks.

The obliteration of the Appendix, so decisive for readers of later editions, was presumably actuated by motives stronger than the desire to save space, and these motives have an importance for us in so far as they are characteristic of statistical inquiry. Among thousands of identifications it would be surprising if at least a hundred were not questionable. The state of the evidence on each of these would constantly change as better preserved or more complete specimens came to be scrutinized. The labour of bringing the list up to date would be great, new editions were frequent, and with the Tertiary succession clearly established, and in practical use by all, this labour would serve no immediate purpose. Moreover, Lyell could not say, as might a later geological evolutionist, "These minor differences do not matter, it is only a question of convention at what level of distinctness a new name is required", for he and his associates firmly refused to admit the possibility of the gradual transformation of one form into another. Any "real" difference must mean to them both extinction and creation, and must be given full weight in estimating the proportion of forms "still living".

[*Editor's Note:* In the original, material follows this excerpt.]

Part I

CLASSICAL STATISTICS

Editors' Comments
on Papers 2 Through 13

13   WATSON
The Statistics of Orientation Data

In the early decades of the twentieth century, one of the leading exponents of studying igneous rock distributions was William Richardson,
an outstanding mineralogist and petrologist who was the first geologist
to realize the implications of silica percentage in differentiating types of
igneous rock. His initial attempts to subdivide igneous rocks centered
on the absolute frequency distribution of silica in various igneous rocks
and involved some of the earliest applications of normal or Gaussian
distributions. In two papers, Richardson and Sneesby (1922) and
Richardson (1923) concluded that the frequency distribution of silica
in igneous rocks could be accounted for by two major populations of
samples, each described by normal distribution curves: acid (granite)
and basic (basaltic) igneous rock groups. An important additional feature of Richardson's contribution was the recognition of variation within normal populations and the application of univariate statistics such
as mean, variance, and standard deviation to describe this variation. Although Richardson's paper contained lengthy consideration of theoretical problems in igneous rock terminology and description, only the
pages relevent to this benchmark volume are included herein (Paper 2).

The application of univariate statistics to geological data by using
this simplistic approach eventually expanded to most branches of
geology. However, one problem consistently encountered in sedimentology was to determine whether one or more samples of geological
data belonged to the same statistical population. In 1935, Churchill
Eisenhart of Princeton University published a paper introducing the chi-
square test to the geological literature (Paper 3). This publication provided a mechanism for testing population samples and ultimately
enabled the development of hypothesis testing as a major tool of the
geologist (Griffiths, 1967). It is unfortunate, therefore, that Eisenhart,
a statistician of great foresight and intelligence, only published two
papers in geology. However, to acknowledge the small, but highly significant contribution Eisenhart has made to the development of statistical analysis in the geological sciences, the editors have included his
paper reporting the chi-square test research in this volume.

During the 1930s the use of univariate statistical methods in sedimentology was restricted by the unequal intervals of grade scales used
in presenting data. Workers such as Wentworth (1922, 1929) soon discovered that regular moment statistics (mean, standard deviation,

skewness, and kurtosis) could not be applied to data in this form and were obliged to develop their own statistics. However, the statistics were difficult to compute, and their interpretation was often complex. Nevertheless, sedimentologists persisted with the arduous and time-consuming analysis of sediment size distribution by using the unequal interval scales until William Krumbein of Northwestern University re-examined the situation in 1936 and concluded that the problem lay with the unequal interval scale and not the statistics (Paper 4). By applying a logarithmic scale (ϕ) to his sediment data, Krumbein discovered that a series of simpler classical statistics based on moments of distribution usually applied in other sciences were available to geologists. The ϕ scale subsequently became standard analytical procedure in sedimentology, and Krumbein's paper has become a benchmark in both sedimentological and statistical applications.

Russian scientists Razumovsky and Kolmogorov discovered that Krumbein's logarithmic scale was also characteristic of chemical distributions and published a number of short papers reporting the nature of the logarithmic distribution and the associated statistics (Razumovsky, 1940 and Paper 5; Kolmogorov, 1941). Simplification of the statistical analysis of geochemical data immediately resulted, because logarithmic transformation of data, particularly trace element values, enabled standard univariate statistical techniques to be applied. It is now common practice in applied geochemistry to transform logarithmically all trace element data before univariate and multivariate statistical analysis is performed. Among the noted exponents of the logarithmic distribution have been Ahrens (1953, 1954a, b, 1957) and Vistelius (1967) who brought the Russians' work to the attention of European and U.S. geologists. Subsequently other types of distribution were recognized in geological data, including Poisson, gamma, binomial, and circularnormal (Krumbein and Miller, 1953; Pincus, 1953; Krumbein, 1955; Chayes, 1956; Mason, 1962; Krumbein and Graybill, 1965; Curl, 1966; Griffiths, 1966; Watson, 1966; Koch and Link, 1970, 1971; Krumbein and Shreve, 1970; Reyment, 1971; Mardia, 1972; Yevjevich, 1972; Till, 1974; Smart, 1976). Multimodal distributions have also been examined by Ghose (1970), Mundry (1972), Clark and Garnett (1974), and Clark (1976). However, heated debate still continues on the nature of many geologic data distributions (e.g., Krumbein, 1938; Sichel, 1947; Chayes, 1954; Miller and Goldberg, 1955; Durovic, 1959; Jizba, 1959; Griffiths, 1960a; Krige, 1960; Vistelius, 1960; Becker and Hazen, 1961; Friedman, 1962; Middleton, 1962; Koch and Link, 1970, 1971; Dapples, 1975; Link and Koch, 1975). A summary of the statistical background to frequency distributions is given by Meyer (1975).

Within the decade of the 1940s, univariate statistics became standard analytical techniques in most branches of geology. In practice, however, geologists retained a limited view of the capabilities of statistics. As will be evident from the state-of-the-art review published in 1944 by Percival Allen of Reading University, most geologists envisaged statistical analysis simply as the analysis of population distributions, sampling patterns, or error variation (Paper 6). However, the first successful attempts in 1949 to study the variation of several geological parameters simultaneously (multivariate statistics) opened numerous new lines of research.

One of the earliest applications of multivariate statistics and a forerunner of modern-day statistical analysis in geology was a multiple regression study of blastoids from Illinois. Benjamin Burma at the University of Nebraska simultaneously compared the variations of eight characters in two fossil assemblages and deduced that the two groups of fossils were significantly different (Paper 7). The procedure described, however, involves many time-consuming operations on an electronic calculator, and analysis of the type pioneered by Burma had to wait for the computer for practical application. Nevertheless, the first steps towards modern-day multivariate statistical applications had been made, even though Burma was, in effect, many years ahead of his time.

While Burma was initiating geologists into the benefits of multivariate analysis, Krumbein continued systematically to develop univariate and bivariate statistical procedures. In 1955 he published the first detailed account of analysis of variance procedures and experimental design in the earth sciences (Paper 8). Griffiths and Krumbein subsequently used controlled experiments to analyze error variation especially in instrumental analysis and sedimentology (see Griffiths, 1967, for details). In 1958 Krumbein, with L. L. Sloss, was also the first to publish a computer program specifically for geological analysis.

Developing from Korn's (1938) work on Devonian and Lower Carboniferous bed thickness, a number of publications appeared in journals during the late 1940s and early 1950s and outlined methods for the analysis of lithostratigraphic successions. In essence, the papers examined trends in rock components through stratigraphic sections; examples can be found in Vistelius (1949, 1952, 1957) and Romanova (1957). The technique usually employed, time-trend analysis, is a method of smoothing measurements of a phenomenon at successive increments in time, such as stratigraphic sequences. The technique was particularly popular in the 1960s as is evident from the number of publications on the subject (Vistelius, 1961; Anderson and Koopmans, 1963, 1968; Fox, 1964, 1967; Schwarzacher, 1964; Fox and Brown, 1965; Weiss et al., 1965), but has been superseded by more sophisticated analytical tools such as time-series analysis, Markov chain analysis, and spectral analysis (Bath, 1974; Krumbein, 1975; Schwarzacher, 1975). It does, however, retain a few advocates including W. Fox (1975) from Williams College.

At the peak of its popularity, Andrei Vistelius of the Academy of

Sciences of the USSR in Leningrad published what is considered a major work on time-trend analysis in geology (Paper 9). His fifteen years of experience in stratigraphic analysis and sedimentological research in USSR provided the background for this lengthy and complex, but at the same time, accurate and extremely practical description of time-trend analysis. The paper now is recognized as a benchmark in the development of analytic techniques for time-depth incremental measurements.

With the advent of computers and development of a technology to handle and manipulate large quantities of data in the late 1950s, new and important avenues were opened to the advancement of statistical analysis in geology. Multivariate analysis became a standard tool for the geological sciences, and numerous methods were adapted for geological data analysis. In 1961 Reyment introduced the method of principal components analysis, and 1962 marked the first geological publication employing factor analysis (Imbrie and Purdy, 1962). Both of these techniques were developed originally by psychologists to study interrelationships among human character attributes, but they can be applied in numerous branches of scientific inquiry, with geology being no exception (Davis, 1973; Agterberg, 1974).

The purpose of principal components analysis is to evaluate the structure of a variance-covariance or correlation-coefficient matrix (a standardized variance-covariance matrix) of a multivariate sample. The structure of the data can be thought of either as a series of vectors in multidimensional space, representing the variance and covariance of each variable, or in the form of an $m \times m$ variance-covariance matrix (m equals the number of variables). The vectors or elements of the matrix can be regarded as defining points lying on an m-dimensional ellipsoid where the eigenvectors of the matrix yield the directions of the principal axes of the ellipsoid and the eigenvalues represent the length of these axes. Because a variance-covariance matrix is symmetrical, these m eigenvectors will be orthogonal. Thus, the method of components analysis essentially involves the rotation of coordinate axes to a new frame of reference in the total variable space—that is, an orthogonal transformation wherein each of the m original variables is described in terms of m new principal components. Factor analytical techniques, such as those described by Davis (1973) and Klovan (1975) are considered to be complementary to principal components analysis. However, "in true factor analysis the factors are defined to account maximally for the intercorrelations of the variables. Thus, components analysis can be said to be variance oriented, whereas true factor analysis is correlation oriented" (Joreskog, Klovan, and Reyment, 1976, p. 59).

Examples of applications are numerous but the reader is referred to the following publications for insight into the statistical methods and data analytic problems embraced by the techniques: Reyment (1961, 1963); Cooley and Lohnes (1962, 1971); Imbrie and Purdy (1962); Imbrie (1963); Krumbein and Imbrie (1963); Imbrie and Newall (1964); Imbrie and Van Andel (1964); Manson and Imbrie (1964); Harbaugh and Demirmen (1964); Middleton (1964); Cattell (1965); Harris (1966); Klovan (1966, 1968, 1975); McCammon (1966, 1968, 1969); Miesch, Chao, and Cuttitta (1966); Nichol, Garrett, and Webb (1966); Webb and Briggs (1966); Harmar (1967); Le Maitre (1968); Garrett and Nichol (1969); Childs (1970); Davis (1970, 1973); Howarth (1970); Koch and Link (1970, 1971); Falconer (1971); Parks (1971); Read and Dean (1972); Drapeau (1973); Mather (1973, 1976); Potenza (1973); Rao, Mann, and Carozzi (1973); Size (1973); Till and Colley (1973); Saager and Sinclair (1974); Symons and De Meuter (1974); Waitr and Stenzel (1974); Jaquet, Froidevaux, and Vernet (1975); and Joreskog, Klovan, and Reyment (1976).

From amongst these, the editors have chosen the papers by John Imbrie and Edward G. Purdy in 1962 (Paper 10) and Richard Reyment in 1961 as landmarks in the historical development of multivariate analysis in geology. However, in the latter case mistakes in the text and calculations detract from the impact of the presentation, and so Reyment's other equally lucid description of the technique has been substituted for it in this volume (Paper 11). Coinciding with the development of principal components and factor analysis in geology was the appearance of numerical taxonomy classification techniques such as cluster analysis. Pioneered by Peter Sneath, F. James Rohlf, and Robert Sokal, taxonomic classification procedures that put the objects into a number of distinct groups, with the objects in each group being more similar to each other than to the objects in all other groups, were a common feature of systematic biology literature as early as 1960. Six years later James Parks of the Union Oil Company of California and subsequently of the Department of Marine and Environmental Sciences at Lehigh University published a description of cluster analysis (Paper 12) and demonstrated how it might be applied to the analysis of geological data—that is, the Bahamian bottom sediment data of Imbrie and Purdy (1962) and Purdy (1963). Parks recognized that cluster analysis is a two-fold process involving:

1. Preparation of a similarity matrix representing relationships between every pair of samples or variables. A number of similarity coefficients are available in geology (Sepkoski, 1974), but in general the correlation coefficient and distance coefficient are employed.

2. Searching the similarity matrix for the variables or samples with the largest similarity coefficient. This pair-by-pair comparison of samples or variables results in a two-dimensional hierarchical representation

of the relationships known as a dendrogram or dendrograph (McCammon and Wenniger, 1970).

Partially as a result of Parks' timely publication and the need for a multivariate statistical technique with easily interpretable results, cluster analysis has become a frequently applied classification method in geology, for example, Purdy (1963), Howd (1964), Kaesler and McElroy (1966), Parks (1966, 1970), McCammon (1968), Kaesler (1969), Rhodes (1969), Kaesler and Taylor (1970), Collyer and Merriam (1973), Joyce (1973), and Denness et al. (1978). Theoretical aspects are covered by Krumbein and Graybill (1964), Harbaugh and Merriam (1968), Tryon and Bailey (1970), Davis (1973), and Everitt (1974).

At this point the editors would like to diverge from the historical review of "mainstream" statistical analysis in the geological sciences that has been presented so far and examine an area of statistics that is almost unique to geology. How to manipulate and analyze data whose basic characteristic is that of directional measurement was a problem for structural geologists, paleoecologists, sedimentologists, and paleomagnetists for many years. Yet no significant progress was noted until the eminent statistician Sir Ronald Fisher (1953) was introduced to the problem. His suggestions laid the foundations for the works of Pincus (1953), Steinmetz (1962), Loudon (1964), and Watson (1966) of which the latter (Paper 13) formed a cornerstone upon which all future research was based; for example, Jones (1968), Jeran and Mashey (1970), Mardia (1972), Schuenemayer, Koch and Link (1972), Till (1974), and Harvey and Ferguson (1976). The contribution of Geoffrey S. Watson, a statistician with Princeton University, to the statistical theory of orientation data is recognized and acknowledged in the final benchmark paper of this section on classical statistics.

2

Reprinted from *Mineral. Mag.* **20**:1–4 (1923)

THE FREQUENCY-DISTRIBUTION OF IGNEOUS ROCKS. PART II. THE LAWS OF DISTRIBUTION IN RELATION TO PETROGENIC THEORIES

W. Alfred Richardson

Lecturer in Petrology, University College, Nottingham

1. *The Law of Distribution.*

IN Part I[1] of this paper the actual frequency-distribution of igneous rocks as revealed by the latest edition of Washington's collection of analyses was examined, and compared with that of earlier records. An attempt will now be made to determine the law of this distribution —a necessary preliminary to any petrogenic application of the results. The silica frequency will be considered since this has been found to give a curve more characteristic of igneous rocks than does any other oxide· Now Dr. Harker[2] pointed out that the silica distribution of records published in 1903 did not obey any simple law such as that of the probability curve, and the same conclusion applies to the collection of 1917. However, when the empirical method of Karl Pearson[3] was

---

[1] W. A. Richardson & G. Sneesby, Min. Mag., 1922, vol. 19, pp. 303–313.

[2] A. Harker, Nat. Hist. of Igneous Rocks, London, 1909, p. 148.

[3] Karl Pearson, Chances of death and other studies in evolution, London and New York, 1897, vol. 1, p. 26.

applied to analyse the curve into component curves of error, fairly simple conclusions resulted.

This attempt at analysis was suggested largely by the close resemblance that the extreme ends of the distribution showed to portions of normal curves of error. In fact the frequency curve appeared divisible into three parts :—(1) A semi-normal curve at the ultrabasic end. (2) A typical U-type distribution between the modes. (3) Another semi-normal curve at the ultra-acid end.

It was discovered that normal curves could be fitted to these extreme portions with exactness. In fig. 1A the chosen equations are plotted and written against the curves. The chain-dotted portion is the result of adding ordinates, and the small circles give the positions of the actual frequencies. It is clear that the distributions mutually modify one another but little, and only over a very narrow range.

Two normal curves of error, situated at the modes of the actual distribution, therefore, reproduce the frequencies below 52 and above 72. Centrally, however, while the combined curve reflects the general character of the actual distribution, the fit is less exact than at the outer edges. Further, the influence of the basic curve at the acid mode is almost negligible; and vice versa.

In order to investigate the deviations in the central region, differences between the ordinates of the combined curve and the observations were taken out. These are plotted in the lower diagram (fig. 1B) which may be called the *curve of residuals*. Its vertical scale is half that of the main diagram, but for ease of comparison it is plotted beneath to the same horizontal scale.

If the fit were perfect the curve of residuals would be a straight line coinciding with the axis of X. In the outer regions, indeed, there are only small oscillations on either side of the zero-line showing that the fit here is a fair one. The deviations from R to S, i. e. between the modes, are more formidable. There is a large negative part close to R, followed by a rather irregular group of large positive residuals, and the meaning of this arrangement becomes a matter of interest. Three possible causes of such a disturbance suggest themselves :—

1. There might be a subordinate mode near 64, with a suite of rocks distributed normally around it. This leaves the negative parts somewhat deepened, and there are other objections to this interpretation which will appear in the sequel.

2. There may be certain special processes affecting this region, which do not involve a normal distribution, and will be considered later.

3. The oscillations in the part *RS*, though of greater amplitude, are essentially of the same nature as those at the ends, and may be due to the same cause, namely, irregularities in the sampling which may ultimately be reduced when the collections become representative. Since Daly's investigation points to low frequencies in this region, it seems

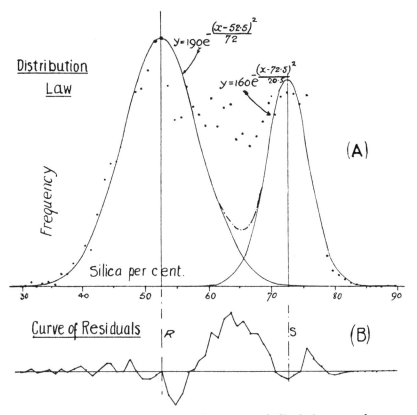

FIG. 1. Analysis of the frequency-distribution of silica in igneous rocks.

likely that this is the true explanation of much of the divergence. On the whole, the matter must be regarded as *sub judice*, and definite conclusions will not be possible until at least one further edition of the collection is available.

Meanwhile, the law of distribution of igneous rocks, deduced from consideration of silica, may be considered as expressed by the sum of two probability equations. It may be stated in the following form, namely, that igneous rocks visible at the earth's surface are grouped about two poles—an acid and a basic—such that their distribution about these conforms more or less closely to two combined curves of error. In other words, *there are two main rock types, one of a basaltic and the other of a granitic character, and all others may be regarded as chance variations about them—meaning by the term ' chance', of course, merely that small departures from type are more probable than large departures.*

Space does not allow of a detailed analysis of the frequencies of the other oxides. They are less characteristic, having smaller dispersions, but it may be stated that they reveal no features antagonistic to the law just stated.

[*Editor's Note:* In the original, material follows this excerpt.]

3

Reprinted from *J. Sed. Petrol.* 5(3):137–145 (1935)

A TEST FOR THE SIGNIFICANCE OF LITHOLOGICAL VARIATIONS

CHURCHILL EISENHART

Princeton University, Princeton, New Jersey

ABSTRACT

It is pointed out that the notion of "homogeneity" is dependent on the classification used. Sampling fluctuations are such that two or more samples are rarely identical for any classification, and, therefore, some criterion is needed to indicate plausibility of homogeneity in a given instance. The Chi-Square test is an appropriate criterion, and it is easily applied to geological data, as is shown in some worked samples.

Statement of the Problem.—Geologists believe "that a sediment must carry with it definite clues of its past history." If this be assumed as a working hypothesis, what are its consequences, and how may its tenability be tested?

One consequence of this hypothesis is that, if it be true, then two or more samples of sediments which are *known* to have had the same past history (i.e., are known to have been drawn from the same parent deposit) ought to show certain similarities. That is, they ought to be "homogeneous" with regard to certain attributes.

In the second place, if two or more samples have been taken from parent deposits which are *known* to be different, then the samples ought to show significant differences with regard to some attribute(s). Moreover, a test designed to indicate whether or not samples are similar with respect to these attributes (i.e., can be considered as coming from the same parent population) should, in the present case, assert dissimilarity on the basis of the information contained in the samples.

If homogeneity is plausible, one can proceed to estimate the true means, medians, measures of dispersion, etc., of the parent deposit in order to make comparisons with other deposits. But one can estimate none of these unless the data are *comparable*, and this implies that group A is composed of homogeneous items, and that group B is also homogeneous—a group composed of dogs and chickens cannot be compared with another composed of cats and ducks and result in a meaningful answer. The notion of homogeneity is fundamental.

Homogeneity.—According to the dictionary a group is *homogeneous* if it consists of parts all of the same kind. Obviously, things that are identical are also homogeneous; but identity is not necessary for homogeneity. To take an example, the group "native white, age 35, of New York City" is homogeneous in respect of color, nationality, age, and place of residence, but the individuals are not all identical as some may be of native and others of foreign parentage. Identity implies homogeneity in all respects, and hence is rarely, if ever, realized in practice. Moreover, homogeneity depends on the classification used—the group above is not, off hand, homogeneous in respect of parentage.

Testing for Homogeneity.—Since homogeneity is dependent on the attribute(s) under consideration, it is of interest to know how to test for homogeneity in a given case.

The simplest case is that of a single sample. In respect to a chosen attribute the sample either *is* homogeneous with regard to the attribute, or *it is not*—there is no half way point. Suppose the lithology of a sample of pebbles is of interest. If they are all limestone pebbles, the sample is homogeneous; if they are mostly limestone, but also some shale pebbles are present, then it is non-homogeneous; it is, however, homogeneous if we are using the coarser classification "igneous—not igneous."

Proceeding to the case of two samples, if both samples contain limestone and shale pebbles and no others, under what conditions can the two samples, considered together, be regarded as lithologically homogeneous? What about the proportions? If they are not identical samples, some criterion must be used to assist one in deciding whether or not the two samples are homogeneous in the sense that they have both

come from the same parent deposit, whose mineral proportions are approximated by the proportions in the samples, the variation being due to sampling fluctuations.[1]

In the absence of knowledge other than that contained in the samples

[1] The need for some mathematical criterion in problems of this sort has been appreciated by Lincoln Dryden and in a recent paper (A statistical method for the comparison of heavy mineral suites; *Am. Jour. Sci.*, (5), Vol. 29, No. 173, pp. 393–408, May 1935) he has proposed using the correlation coefficient, r, as a measure of the similarity of the various *proportions* in the samples to be compared. Unfortunately he was unaware of the pitfalls of the method he selected.

As will become more evident further along in this paper, *it is necessary* to take into account the *sizes* (number of items in the respective samples) of the samples under consideration when testing for similarity. As the size of a sample is increased the proportions of the various minerals in the sample tend to approximate the proportions of the parent deposit with greater and greater stability, that is, sampling fluctuations will account for less and less of observed discrepancies in proportions as the sample size is increased. Dryden's method does not take this into account, as he uses the percentages of the various types of minerals in his samples and makes no mention of the total sizes. A short reference to samples A and B from page 395 of Dryden's paper will emphasize the necessity of utilizing sample sizes: the two samples do not differ greatly as regards the percentages of their mineral contents, and, if these samples are assumed to contain about 300 items each, the probability is .4 that the observed differences are ascribable to sampling fluctuations, but, if it is assumed that the samples contain about 3000 items each, the corresponding probability is .00000 00000 002213 which implies that it is practically impossible to ascribe the dissimilarities to chance.

Furthermore, *the coefficient of correlation*, r, is *not applicable* to samples classified in respect to mineral composition. This coefficient can in general, be used only in connection with attributes which are measurable or quantitative; it can be used with qualitative characteristics only when these can be ordered in a numerical sense (e.g. poor, fair, good excellent). As mineral classification is purely descriptive and unordered, the correlation coefficient is unserviceable, and it is necessary to turn to the theory of probability in order to obtain a criterion to test homogeneity in respect of lithological composition.

themselves, the question of whether or not they are homogeneous in a statistical sense depends on whether or not their differences can be ascribed wholly to chance fluctuations. In drawing inferences from data one must always accept a certain risk of being wrong, the amount of risk acceptable depending on the purpose for which the sample information is to be used and on the rashness of the observer. For example, what risk is assumed if the two samples hereinafter are considered as homogeneous in a statistical sense?

Sample	Limestone	Shale
α	103	794
β	109	781

The table gives the frequencies of these two types of pebbles as found in samples α and β. The size distinction which determines whether or not a particle is a pebble being the same in each case—i.e. the data are comparable.

An appropriate test (the Chi-Square Test, to be discussed hereinafter) asserts that in drawing many pairs of samples at random from the same parent deposit, one would find between the two samples such variations as were observed or worse in approximately sixty-two percent of the cases. In other words, in a single trial such variation or greater variation is very likely—from a sampling point of view it has a probability of 0.62. Hence one does not run any great risk in assuming that α and β are samples drawn from the same parent deposit.

On the other hand, if samples α_1 and β_1 are being examined, samples whose frequencies are

	Limestone	Shale
α_1	180	520
β_1	96	655

the same test asserts that about once in 10^{15} samples would one find such (or a worse) irregularity between samples drawn from the same parent population. In such a case the hypothesis that they are homogeneous is untenable, and should be rejected.

This criterion (the Chi-Square Test) is very generally applicable, and may be applied to a group of several samples with many categories in each sample (the same categories, of course). It has been used, and is being used, with great success in anthropometric, biometric, and econometric studies. Therefore it seems fitting to discuss it in this paper with the view of adopting it as a criterion in certain geological problems involving sampling.

The Chi-Square Test.—For simplicity, let it be assumed that one has only two samples, S_1 and S_2, and that one has classified some attribute into two well-defined categories C_1 and C_2. A table can now be formed

Sample	Attribute		Totals
	C_1	C_2	
S_1	(S_1C_1)	(S_1C_2)	(S_1)
S_2	(S_2C_1)	(S_2C_2)	(S_2)
Totals	(C_1)	(C_2)	N

where ( ) denotes summing: thus (S_1C_1) denotes the total number of items from sample S_1 which possess the attribute C_1; likewise (S_1) denotes the total number of items in sample S_1; N is the total number of items in the sample, $N = (S_1) + (S_2) = (C_1) + (C_2)$.

If one now assumes that both samples are from the same parent universe, one expects to find the same proportion of, say, C_1 in sample S_2 as in sample S_1 (i.e., the proportion should be independent of the sample chosen). In the present notation this is expressed by

$$\frac{(S_1 C_1)}{(S_1)} = \frac{(S_2 C_1)}{(S_2)}. \tag{1}$$

Moreover, since $(S_2 C_1) = (C_1) - (S_1 C_1)$ and $(S_2) = N - (S_1)$, substitution in (1) yields $[N - (S_1)](S_1 C_1) = (S_1)(C_1) - (S_1)(S_1 C_1)$, which reduces to

$$(S_1 C_1) = \frac{(S_1)(C_1)}{N}, \tag{2}$$

a relation which enables one to calculate the frequency of a cell on the assumption of complete independence. Since the total of the first row is to be (S_1), evidently $(S_1 C_2) = \dfrac{(S_1)(C_2)}{N}$

giving $(S_1) = \dfrac{(S_1)[(C_1) + (C_2)]}{N}$. Consequently, if the proportions are to be truly independent the various frequencies must be distributed as in the following table:

Sample	Attribute		Totals
	C_1	C_2	
S_1	$\dfrac{(S_1)(C_1)}{N}$	$\dfrac{(S_1)(C_2)}{N}$	(S_1)
S_2	$\dfrac{(S_2)(C_1)}{N}$	$\dfrac{(S_2)(C_2)}{N}$	(S_2)
Totals	(C_1)	(C_2)	N

The reasoning is perfectly general, and if one has samples S_1, S_2, $\cdots$, S_i, $\cdots$ and categories C_1, C_2, $\cdots$, C_j, $\cdots$ then the frequency appropriate to complete independence for the cell in the jth column of the ith row is

$$(S_i C_j) = \frac{(S_i)(C_j)}{N},$$

the Independence Frequencies. (3)

Denoting the *Observed Frequency* by $(S_i C_j)'$, due to sampling fluctuations $(S_i C_j)' \neq (S_i C_j)$, and the difference $d_{ij} = (S_i C_j)' - (S_i C_j)$ can be formed. (4)

It is evident that some d's will be plus and others minus, and that the sum of all the d's will be zero since

$$\sum_{ij} (S_i C_j)' = \sum_{ij} (S_i C_j) = N,$$

and hence any criterion must depend on the values of the d's without regard to sign.

A mathematical analysis of the problem indicates that an appropriate criterion is the Chi-Square Test, based on

$$\chi^2 = \sum_{ij} \frac{(d_{ij})^2}{(S_i C_j)} \tag{5}$$

in which d_{ij} is defined above in (4), leading to a probability measure of the homogeneity of the data. It is evident that the nearer the values observed are to the independent frequencies, the smaller the χ^2. Consequently the magnitude of χ^2 is an indication of the degree of departure from perfect homogeneity.

It has been shown by R. A. Fisher (Phil. Tran., Vol. 222A, p. 357; Jour. Roy. Stat. Soc., Vol. XXXVII, p. 447) and by others that for *large samples* the probability distribution of χ^2 defined by (5) is approximately the same as that of another χ^2 defined years before by Karl Pearson in connection with Goodness of Fit. Tables are available (Fisher's *Statistical Methods for Research Workers*, and Pearson's *Tables for Biometricians and Statisticians*) which give the probability of obtaining by chance a value of χ^2 as great or greater than that actually observed in sampling from truly homogeneous material. In using these tables of probability one looks in the table under the number of degrees of freedom, n, given by

$$n = (r-1)(c-1) \qquad (6)$$

for a table of data consisting of r rows and c columns.[2] As Pearson's table was originally constructed for another purpose, one must enter his table with $n' = n+1 = (r-1)(c-1) +1$; the present notation is the same as Fisher's. For a fuller discussion, including short cut methods applicable to special cases see Chapter IV of Fisher's book.

Finally, it must be remembered that the tables of probability have been calculated on the assumption of large samples. It has been found that this calculated distribution is not very closely realized for very small class frequencies, and, therefore, to play safe one should try to have the smallest independence frequency greater than 10 (a minimum). Nothing is gained by including nearly vacuous classes; it is far better to throw several classes together and

[2] It was seen above that in a 2×2 table, when one value has been calculated, all the rest are determined, since the marginal totals are to remain the same. In general, as can readily be seen by trial, in a table of r rows and c columns, there are only $(r-1)$ $(c-1)$ frequencies that need be calculated, the rest then being obtained by subtraction from the marginal totals.

enter the probability table with a correspondingly smaller number of degrees of freedom.

Worked Examples (with comments). —(1) The two samples α and β referred to above will afford data to illustrate the application of the Chi-Square technique. In the tabulation immediately following the upper number in a cell is the observed frequency, and the number in ( ) the corresponding independence frequency.

Sample	Attribute		Totals
	Limestone	Shale	
α	103	794	897
	(106.42)	(790.58)	
β	109	781	890
	(105.58)	(784.42)	
Totals	212	1575	1787

In obtaining the independence frequencies only one value had to be calculated, e.g., $\dfrac{212 \times 897}{1787} = 106.42$, the others being obtained at once by subtraction from the marginal totals. Consequently, there is only one degree of freedom, $n = 1$.

In the table which follows, the upper number is the absolute value of the difference between the observed and the expected frequency; the lower number is the square of the difference just mentioned divided by the expected frequency $\left(\text{e.g.,} \dfrac{(103-106.42)^2}{106.42} = \dfrac{(3.42)^2}{106.42} = .109^+ \right)$, and this is the contribution to χ^2 from this cell of the table. By adding the separate contributions it is found that $\chi^2 = .249$. By interpolating between values in Fisher's table, for

Sample	Li	Sh	Totals
α	3.42 .109	3.42 .015	.124
β	3.42 .110	3.42 .015	.125
Totals	.119	.030	$.249 = \chi^2$

$n = 1$, one finds a corresponding probability of .62 which means that sixty-two times in a hundred one ought to obtain an equivalent or larger value of χ^2 in drawing two such samples from a common parent population. One assumes no great risk, therefore, in accepting α and β as samples from the same parent deposit and in asserting that they are homogeneous in that sense.

(2) Proceeding in the same way with samples α_1 and β_1 the contributions of the four cells to χ^2 are found to be:

Sample	Li	Sh	Totals
α_1	25.84	5.53	31.37
β_1	20.93	5.33	26.26
Totals	46.77	10.86	$57.63 = \chi^2$

These individual contributions are so great that it is not in the least surprising to find that the probability of such a χ^2 (or a larger one) is of the order 10^{-15}. In such a case one rejects at once the hypothesis of homogeneity, the probability by chance of such a variation from homogeneity of occurrence being so small.

It may be asked: "Where is the border-line beyond which a hypothesis should be rejected?" There is no absolute answer to this question, but, in general, experience has shown that if P is less than .05 it is wise to reject the hypothesis. Of course, in doing this, one will, once in twenty times, consider as non-homogeneous two samples that really are homogeneous. If extremely desirous not to lose these samples, $P = .01$ can be chosen as the criterion point, but it must be remembered that in so doing one admits to the homogeneous class samples that are really non-homogeneous and which would have been excluded by using $P = .05$. Every experimenter must decide for himself what he will accept.

What if $P = .99$? In this case the hypothesis ought to be rejected, in general, because the probability of choosing samples that agree so well from all possible samples is so small that one suspects that they were not *random*. If, however, it is certain that no bias has entered, then homogeneity can be accepted, but with a grain of salt, remembering that such luck happens only once in a lifetime.

(3). As a further illustration, interesting results follow the application of the χ^2 method to a problem treated by Dryden (loc. cit. p. 402 ff.). While examining a contact between Eocene and Miocene beds in Maryland, Dryden selected samples from about one foot below the contact (in the Eocene) and about a foot above the contact (in the Miocene). Off hand the figures seemed to indicate that some of the Eocene material had been reworked into the base of the Miocene bed, the Eocene and Miocene samples showing certain strong similarities. An application of the χ^2 method to Dryden's first data indicates at once that the Eocene and the Miocene beds are distinctly different in their mineral make up, the value of χ^2 obtained being nearly three times the one per cent value.

Apparently unfamiliar with the χ^2 criterion, and not liking the hypothesis regarding the reworking of Eocene material into the Miocene, Dryden "decided to find in a little more detail just where and how the mineral change from typical Eocene to typical Miocene took place. With a rather hazy idea of what was demanded in the way of samples, collections were made from the top inch of the Eocene (if there had been reworking at this place the material farther down could have had nothing to do with it), in one-inch intervals from the contact to six inches above the contact, and one sample from 6–9 inches above and one from 9–12 inches above the contact." He gives the percentages of various minerals by count in his Table V. These are reproduced below using 4 categories in place of his 9 on account of the paucity of data in certain categories.

The corresponding frequencies (rounded to integers) are given on the assumption of samples in the neighborhood of 300 items (Dryden has informed the writer that this is the general size of his samples). Calculating χ^2 for the whole table one finds $\chi^2 = 96.30$, of which 55.93 is contributed by the Eocene sample. For $n = (9-1)(4-1) = 24$, the one per cent value is 42.9, and since the observed value is far greater than this, one concludes that the Eocene and Miocene samples are not collectively homogeneous in respect to heavy minerals. Since more than half of the total χ^2 comes from the Eocene sample, it is of interest to examine the situation with this sample omitted.

Calculating χ^2 for the complete table less the Eocene sample, one finds $\chi^2 = 32.54$ with $n = (8-1)(4-1) = 21$. From tables it is found that $P(\chi^2) = .05$, and in view of the fact that the data has been rendered more homogeneous by bunching categories together (for by adding together pairs like 5, 9 and 9, 5 one gets 14,14 which is more homogeneous) the evidence suggests that *all* the samples of Miocene material cannot collectively be considered as homogeneous in respect to heavy

	Eocene −1,0		Miocene														
			0,1		1,2		2,3		3,4		4,5		5,6		6,9		9,12	
	%	f	%	f	%	f	%	f	%	f	%	f	%	f	%	f	%	f
Z	50	150	28	84	30	90	30	90	28	84	31	93	26	78	29	87	41	123
St	24	72	36	108	33	99	31	93	33	99	33	99	32	96	29	87	25	75
G	6	18	20	60	18	54	20	60	20	60	18	54	22	66	17	51	19	57
All others	20	60	16	48	19	57	19	57	19	57	18	54	20	60	25	75	15	45

All percentages add to 100, all frequencies to 300.

27

minerals. In addition, an examination of the individual contributions shows that. 17.57 (more than half) of the total χ^2 came from the [9,12] sample. If this sample were removed it is clear that the remainder of the Miocene samples could be regarded as homogeneous.

A further investigation of the [9,12] sample yields:

A Table of Comparisons of the [9, 12] Sample with

Sample	χ^2	n	$P(\chi^2)$
(Eocene) $-1,0$	25.16	3	much less than .01* ;
(Miocene) 0,1	13.47	3	less than .01
(Miocene) 6,9	14.89	3	less than .01

(As a comparison, Miocene [0,1] and [1,2] when compared yield $\chi^2=1.68$, giving $P(\chi^2)=.65$.)

* For $n=3$, $P=.01$ when $\chi^2=11.34$.

These figures suggest that sample [9,12] is

(a) significantly different from the Eocene sample,

(b) significantly different from the Miocene samples,

and

(c) that (b) is not due to a "distance effect."

Considering in connection with this the other evidence of Dryden it appears that there is a sudden change in heavy mineral content in the neighborhood of 9 to 12 inches above the Eocene-Miocene contact. Does this change continue farther up, or does the composition revert to that in accord with samples [0,1] to [6,9]?

Before summing up, there are several warnings that need to be explicitly stated: First, the χ^2-test should not take the place of common sense; simply because two samples from widely separated regions yield a small value of χ^2 it should not be concluded at once that they have had a common geologic history. Secondly, the χ^2-test is not omnipotent; samples may be homogeneous, say, in respect of lithology (as determined by the χ^2-test) and decidedly different in respect of shapes of particles. In such a case it is up to the geologist to decide whether mineral content or particle shape is more important—after all complete similarity can rarely be established. Finally, the χ^2-test *cannot be applied to data where the units are weights, lengths, etc., as it is only applicable to frequencies.* If applied to weights, for example, it would correspond to saying that a given unit of weight, say a gram, at all times retained its identity and never dispersed—i.e., if a gram of material were colored pink, put back in the main pile, the material mixed, then one should expect to find all the pink particles still all together forming one unique indispersible unit. Of course, this difficulty does not arise in the case of pebbles counts, as one pebble at all times retains its identity and cannot fall apart into smaller particles.

SUMMARY

The meaning of "homogeneous" is arbitrary, and in a given case needs to be defined with care, since it will depend on the classification used. An appropriate criterion to test for homogeneity corresponding to a certain classification is the Chi-Square Test. The test can be used to indicate whether a given deposit is homogeneous throughout in a statistical sense, and it will also indicate when applied to several deposits whether they can be considered as mutually homogeneous. Since homogeneity is intimately connected with comparability of data these concepts must be fully understood if conclusions due to noncomparable data are to be avoided.

REFERENCES

FISHER, R. A., *Statistical methods for research workers*, 4th ed., Edinburgh and London, Oliver and Boyd, Chapter 4, especially p. 94, 1932.

————, On the interpretation of χ^2 from contingency tables, and the calculation of *P: Roy. Statistical Soc., Jour.*, vol. 85, pp. 87–94, 1922.

————, Statistical tests of agreement between observation and hypothesis: *Economica*, p. 139, 1923.

————, The conditions under which χ^2 measures the discrepancy between observations and hypothesis: *Roy. Statistical Soc., Jour.*, vol. 87, p. 442, 1924.

YULE, G. UDNY, On the application of the χ^2 method to association and contingency tables, with experimental illustrations: *Roy. Statistical Soc., Jour.*, vol. 85, pp. 95–104, 1922.

————, *An introduction to the theory of statistics*, 10th ed., London, Charles Griffin and Co., Ltd., Chapters III, V, and pp. 370–387. At the end of this book will be found an excellent list of references to statistical literature.

4

Reprinted from *J. Sed. Petrol.* 6(1):35–47 (1936)

APPLICATION OF LOGARITHMIC MOMENTS TO SIZE FREQUENCY DISTRIBUTIONS OF SEDIMENTS

W. C. KRUMBEIN
University of Chicago, Chicago, Illinois

ABSTRACT

A symbol for the negative logarithm of grain diameters to the base 2 is introduced as a device for statistical computation. The advantages of the symbol are that it simplifies geometric grade scales, it permits the application of common statistical procedures to the data, and yet it yields directly a series of values which express the logarithmic properties of the size frequency curve. By means of a simple chart the logarithmic values may be converted to their "diameter equivalents" it desired. An example is included of the application of the symbol to a statistical technique based on the moments of the distributions, and the significance of the measures thus obtained is discussed.

In relatively recent years emphasis in sedimentary petrology has shifted from isolated rock samples to suites of samples, collected with a view of obtaining more complete knowledge of the formation being studied. This change has brought with it a growing realization of the need for statistical analysis of the resulting data, as numerous papers on the subject demonstrate. Unfortunately, the use of common statistical measures has been somewhat limited by the unequal intervals of the grade scales used in presenting the data. Thus workers in sedimentary petrology have had to develop their own statistical techniques. The net effects of this situation have been two, in the main. First, measures have been adopted that lend themselves directly to data arranged in unequal classes, such as the median and the quartiles, and second, graphic methods have largely been used because of their relatively greater convenience.

Sedimentary petrologists at present use a wide range of statistical devices, including numerous empirical methods of characterizing the sediments, but among common methods is one conspicuous by its absence. This is the method of moments.[1] Wentworth (1)* developed a technique based on moments, but because of the complexity of his computations, and the unsatisfactory manner in which his theory was presented, little use has been made of it. Hatch and Choate, (2) at about

[1] The moments of a frequency distribution are parameters or constants that describe the properties of the distribution. In ordinary statistical practice the first moment is the *arithmetic mean*, and the second moment is the basis of the *standard deviation*. The moments as used in this paper will be discussed more fully below.

* Numbers in parentheses refer to bibliography and comments at the end of this paper.

the same time, defined a series of logarithmic measures and developed graphic methods for frequency data that plotted as straight lines on logarithmic probability paper. Other techniques have been developed, but have not been widely used by geologists.

The thesis of the present writer is that as long as the uses of statistical measures are mainly descriptive and comparative, as they are in sedimentary petrology at present, the widest possible choice of methods should be available, so that any situation that arises may be attacked in the manner best suited to the data themselves, or to the problem at hand. In this connection a method of moments, as applied to the frequency curves of sediments, offers an important and significant method of attack. The moments of the distributions have been more extensively studied, and they have been related more closely to the probabilities underlying the distributions, than have any other statistical measures. Thus there is available a large body of statistical technique which may be drawn upon if a satisfactory method of moments may be applied to sediments. The chief difficulty in the past has been due largely to the unequal intervals, but this difficulty is not insurmountable.

From an understanding of the underlying problem presented by the unequal intervals, and by a proper choice of logarithm, not only may the curves be greatly simplified during plotting, but the data themselves may be treated by arithmetic meth-ods no more tedious than the usual procedure of computing the moments in common statistical practice. Moreover, the use of logarithms in this manner results in the computation of *logarithmic moments*, which it will be shown are directly applicable to sedimentary data. Thus the statistical devices used by sedimentary petrologists at present can be supplemented by an entirely different method of approach, which opens other possibilities in the direct use of conventional statistical procedures with the data.

It is the purpose of the present paper to describe a method of moments as applied to sedimentary data, and to stress the significance of the resulting measures in evaluating the size frequency distributions of sediments.

THE SYMBOL ϕ

The discussion in the following text will proceed from first principles, so that the steps involved in the method may be developed in logical sequence. In most mechanical analyses the data are assembled in classes according to the Wentworth grade scale (3) in which each succeeding grade is half as large as its predecessor. Thus the class 1/2–1/4 mm. is followed on the right by the class 1/4–1/8 mm., an interval half its size. If the grade scale is plotted on ordinary arithmetic graph paper, the grades rapidly decrease in width as the scale is followed to the right, due to the convention of plotting the coarser grades at the left. Similarly a histogram or frequency curve plotted on such a scale will be quite unsymmetrical, with much of the material assembled at one end or the other. (See Fig. 1.) In this direct plotting of the diameter along the horizontal axis, and frequency along the vertical axis, the diameter is chosen as the *independent variable;* i.e., the particular frequencies found depend upon the presence of diameters of such a size, rather than the reverse.

It has long been common practice to plot the classes equal in width both to simplify the diagram and to give equal significance to each grade. The increased symmetry of the resulting frequency curve is ample justification for the procedure. Actually, of course, the logarithms of the diameters are being plotted when the classes are drawn equal in width, which means that for the independ-

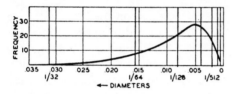

FIG. 1. Frequency curve of a Pennsylvanian underclay, showing frequency (*dependent variable*) plotted against diameters in mm. (*independent variable*). The heavy ordinates represent the Wentworth class limits. Note the lack of symmetry of the curve.

ent variable *diameter* is being substituted the independent variable *log diameter*. This change of independent variables thus serves an immediate and easily recognized purpose. (See Fig. 2.)

That the intervals become equal when logs are used is easily shown.

Thus $\log_{10} 2 = +0.301$; $\log_{10} 1 = 0$; $\log_{10}(1/2) = -0.301$. Here each class is equal in width (interval $= 0.301$), with the origin at 1 mm. Note that the logs of numbers smaller than 1 are negative, and further, that if the base 10 is used, the class limits are not marked by integers. This sug-

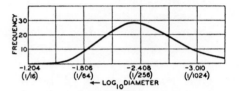

FIG. 2. Frequency curve of the same sediment as fig. 1, with *frequency* plotted against *log₁₀diameter*. The Wentworth class limits are now equally spaced, and the curve is much more symmetrical.

gests that logs be taken to such a base that each class limit is an integer, and this is accomplished by using the base 2. Then $\log_2 2 = +1$; $\log_2 1 = 0$; $\log_2(1/2) = -1$, etc. Notice, however, that negative values still apply to diameters smaller than 1 mm. Now, there probably are more fine-grained sediments (sands, silts, clays) than there are gravels, so that most analyses will lie in the range below 1 mm. This suggests that negative logs be used, both to avoid negative numbers in this important range, and also to convert the grade scale to one which increases to the right, as most ordinary scales do. By using negative logs, then, the scale is simply reversed, and the computations are in no wise made more difficult; indeed, they are often much simplified as the sequel will show.

For computational reasons it is advisable to adopt some name for these negative logs of the grade scale, and the writer suggested in another connection (4) that the symbol ϕ be used for this purpose. The substitution involved is:

$$\phi = -\log_2 \xi \qquad (1)$$

where ξ is the numerical value of the grain diameter (or class limit) in mm., and ϕ is defined by the writer as the *Wentworth exponent*, from the exponential form of equation (1), $\xi = (1/2)^\phi$, and from its application to the Wentworth grade scale. With values of ξ equal to 2, 1, 1/2, ξ equals -1, 0, $+1$, respectively. Thus ϕ increases with decreasing grain diameters, and the Wentworth scale becomes arithmetic, with $\phi = 0$, ($\xi = 1$ mm.), as the origin. It is possible to interpolate arithmetically on this scale, and to express all grain diameters in ϕ-terms directly. The conversion of one symbol to the other, by means of a simple chart, will be discussed below. For purposes of reference, Table 1 shows the ϕ-

TABLE I. *Conversion of Wentworth grade limits from ξ to ϕ*

Grade limit in mm. (ξ)	ϕ	Grade limit in mm. (ξ)	ϕ
32	-5	1/8	$+ 3$
16	-4	1/16	$+ 4$
8	-3	1/32	$+ 5$
4	-2	1 64	$+ 6$
2	-1	1/128	$+ 7$
1	0	1/256	$+ 8$
1/2	$+1$	1/512	$+ 9$
1/4	$+2$	1/1024	$+10$

values of the Wentworth grade limits from 32 mm. to 1/1024 mm. One immediately apparent advantage accruing from the ϕ-scale is the elim-

ination of unwieldy fractions or decimals, such as 1/1024 mm. (0.00098 mm.). These inconvenient values become simply $+10$ on the new scale. The use of the symbol adapts itself equally well to the grade scale based on $\sqrt{2}$. The successive points along that scale are half-units on the ϕ-scale.

The substitution of ϕ for ξ, then, merely involves a change in the independent variable, such that the intervals between class limits become equal; each class is one unit in width; and the scale of values increases to the right. Hence it is possible to use the mechanical analysis data arranged according to the ϕ-scale for the computation of a whole series of statistical measures, many of which may be adopted directly from standard statistics texts, with no changes in procedure (5). The resulting measures, furthermore, may be used directly to characterize the frequency curve, they may be used in connection with Gram-Charlier series, they may be used in estimating areas under the curve—in short, many aspects of conventional statistical analysis become available. On the other hand, the measures may be converted to their "diameter equivalents" if desired.

It should be emphasized at this point that the use of any statistical measure does not depend upon the fact that the independent variable has an immediately comprehensible significance; it depends only on the fact that a curve is given. This is equally true of frequency curves plotted on the basis of ξ, $\log_{10} \xi$, or ϕ. Furthermore, it is not necessary that there be simple relations between the measures in terms of one independent variable, with those based on some other independent variable. True, it is often desirable to have such simple relations, and in the present case the mean value of the ϕ distribution bears quite a simple relation to a mean value of the ξ distribution. Similarly the second moments are not unduly complex in their relations, but this is probably not true for higher moments. These relations will be discussed below; the remarks are included here both to establish the approach and to avoid possible misunderstandings.

APPLICATION OF ϕ TO THE METHOD OF MOMENTS

An interesting feature of the ϕ-notation, which renders it a favorable choice for the computation of the moments of the frequency distributions of sediments, is that in any frequency distribution the arithmetic mean of the ϕ-values is a logarithm of the geometric mean of the ξ-values,[2] Thus it is only necessary to find the arithmetic mean of the ϕ-distribution to obtain a log geometric mean that serves well as a central point about which the moments of the curve may be computed. Furthermore, the higher moments of the distribution, in terms of ϕ, may be found by following standardized statistical procedures, thus obviating the necessity of devoloping new techniques of computation. In this manner a series of measures corresponding to the *mean size*, the *standard deviation, skewness*, and *kurtosis* may be obtained, with, however, this important distinction: *the values obtained by the computations with ϕ will be logarithmic measures in terms of the log geometric mean instead of arith-*

[2] This relation is proved in the Appendix to this paper, which also discusses several other theoretical points.

metic measures in terms of the arithmetic mean. Once these measures have been found they may be used directly in describing the characteristics of the curve. To develop these points, it may be well to furnish an actual example of the computations.

In computing the statistical values of a sediment in ϕ-terms, one may follow directly the standardized procedures given in statistics texts for finding the arithmetic mean, the standard deviation, (and skewness and kurtosis, if these last values are desired) (6). The use of the ϕ-notation in these standard procedures yields logarithmic moments directly, as has been pointed out. The following example follows the standard practice directly, except that a few notational changes have been made, so that the interested reader may trace the relations of the moments to one another. A tabular series of values is set up, as shown in Table 2. For simplicity the data (7) are arranged in Wentworth classes, although classes based on $\sqrt{2}$ could just as well be used (8). Column (1) shows the grade limits in diameters (ξ), column (2) the same limits in ϕ-terms, and column (3) lists the weight percentage frequencies (f) in each grade. In column (4) the maximum grade has been chosen as the arbitrary zeropoint of a deviation or d-scale, and the grades above and below are numbered in sequence as negative and positive, respectively. In column (5) the f of each grade in column (3) has been multiplied by the corresponding d from column (4), to yield a series of values called fd. These are added algebraically, and the final sum written below. In this case it is +31.5. Next the d's are squared and entered in column (6). These are then multiplied by the frequencies from column (3), to yield the fd^2 values of column (7). Again a total is struck, here 41.7. Column (8) contains the cubes of the d's, and the fd^3 values of column (9) are found by multiplying

tribution, called M_ϕ, is found by adding n_1 to the mid-point of the d-scale, here ·1.5 (in ϕ-units). This yields $M_\phi = 1.500 + 0.315 = 1.815$.

The *standard deviation* of the ϕ-distribution, σ_ϕ, is $\sigma_\phi = \sqrt{n_2 - (n_1)^2}$. Now $n_1 = .315$, and hence $(n_1)^2 = 0.099$, so that $\sigma_\phi = \sqrt{.417 - .099} = \sqrt{.318} = 0.563$.

TABLE II. *Computation of the moments of the size distribution of beach sand from southern shore of Lake Michigan*

(1) Grade limits in mm. (ξ)	(2) Grade limits in ϕ	(3) Per cent by weight (f)	(4) d	(5) fd	(6) d²	(7) fd²	(8) d³	(9) fd³
1–1/2	0–1	4.9	−1	−4.9	1	+4.9	−1	−4.9
1/2–1/4	1–2	58.9	0	0	0	0	0	0
1/4–1/8	2–3	36.0	+1	+36.0	1	36.0	+1	+36.0
1/8–1/16	3–4	0.2	+2	+ 0.4	4	0.8	+8	+ 1.6
Totals		100.0		+31.5		41.7		+32.7

the values of column (3) by those of column (8). The algebraic total, +32.7, is written below.

These several values are now each divided by the total frequency (100), and the results written to one side, as follows. The value 31.5/100 = 0.315 is called n_1, the first moment about the origin of the d-scale. From column (7) is obtained 41.7/100 = 0.417, called n_2, the second moment about the d-origin. Finally, from column (9), 32.7/100 = 0.327, called n_3, the third moment about the d-origin. From here on standard formulas are used in determining the statistical values; the theoretical derivation of these formulas may be found in statistics texts (Camp) (6).

The *arithmetic mean* of the ϕ-dis-

The *skewness* is more complicated. The third moment about the d-origin, n_3, must first be converted to m_3, the third moment about the mean, by using the standard equation $m_3 = n_3 - 3n_2 n_1 + 2(n_1)^3$. This yields $m_3 = .327 - 3(.417)(.315) + 2(.315)^3 = .327 - .393 + .062 = -0.004$. From this point on one has a choice of several formulas for the skewness. We shall use the one based on $\alpha_3 = m_3/\sigma^3$. Skewness is then $Sk_\phi = \alpha_3/2$. In this example σ_ϕ is .563, so that $2\sigma_\phi^3 = .356$, and $Sk_\phi = -.004/.356 = -0.011$ (Camp) (6).

The *kurtosis* of the curve will not be computed here. It is based on the fourth moment about the mean, and involves a somewhat tedious computation. Interested readers will find

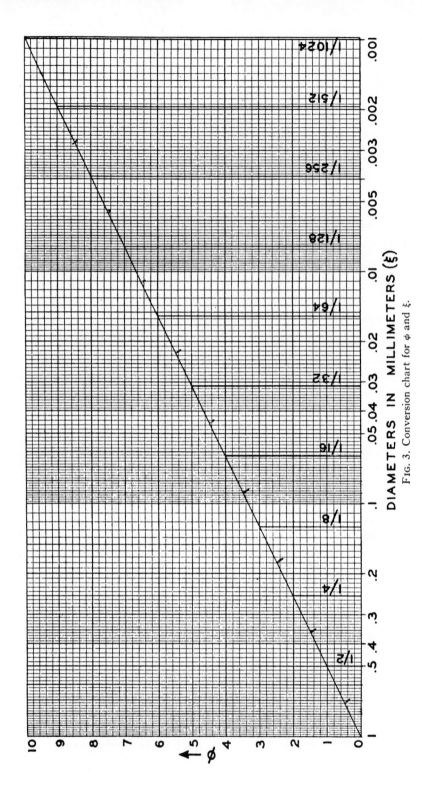

F<small>IG</small>. 3. Conversion chart for ϕ and ξ.

the necessary equations in Camp's book (see reference 6).

As a result of the computations described, three statistical values have been obtained for the beach sand:

$$M_\phi = 1.81$$
$$\sigma_\phi = 0.56$$
$$Sk_\phi = -0.01$$

These values may be used directly in describing and comparing the frequency curves, without converting them back to their ξ-equivalents. However, for purposes of comparing the ξ-equivalents with the above ϕ-values, a simple method of converting one symbol to the other may be described.

GRAPHIC CONVERSION OF ϕ TO ξ

For the conversion of fractional or decimal values of ϕ or ξ to one another, a simple conversion chart may be made with semi-logarithmic paper. By using Dietzgen 3-cycle paper #340–L310, a range of ten units along each axis may be had. Starting with the diameter $\xi = 1$ mm., at the left, the successive Wentworth scale points are laid off along the horizontal logarithmic scale. The values decrease to the right until 1/1024 (0.00098) is reached just a shade beyond the 0.001 point. The scale points will be about 2.4 cm. apart. Along the vertical arithmetic scale the point $\phi = 0$ is set equal to $\xi = 1$ mm., and each major division of the scale (about 1.2 cm. on the paper) is called a ϕ-unit. At the corresponding ξ-points ordinates are erected to the ϕ-values given in Table 1, and a diagonal line is drawn across the tops of the ordinates from (1, 0) to (0.00098, 10). The line makes an angle of about 28° with the horizontal. Figure 3 illustrates such a chart somewhat reduced, so that as shown it is too small for accurate work, but does yield approximate values.

The conversion of one symbol to the other thus becomes simply the reading of a straight line graph, and the charts may be made to represent any range of sizes desired. If it is preferred, however, the symbols may be transformed to one another by logarithmic computations.[3]

By referring back to the three values given above, we may illustrate the conversion of ϕ to ξ by Figure 3. The value of M_ϕ is 1.81; this point is located on the vertical axis, and by moving to the right until the diagonal line is met, the corresponding ξ- value is found to be 0.286 mm. Similarly the other two values are found, yielding the following numbers, where GM_ξ is the geometric mean of the diameters, σ_ξ is the "diameter equivalent" of σ_ϕ, and Sk_ξ is the "diameter equivalent" of Sk_ϕ. Specifically, the relations are $M_\phi = -\log_2 GM_\xi$, $\sigma_\phi = -\log_2 \sigma_\xi$, and $Sk_\phi = -\log_2 Sk_\xi$:

$$GM_\xi = 0.286 \text{ mm.}$$
$$\sigma_\xi = 0.676$$
$$Sk_\xi = 1.05$$

[3] To convert ϕ to ξ logarithmically, recall that $\log_{10}\xi = \log_{10}2 \, \log_2\xi$, and that $\log_{10}2 = 0.301$. Hence $\log_{10}\xi = 0.301 \log_2\xi$. But $\phi = -\log_2\xi$, so that $-0.301\phi = \log_{10}\xi$. When the resulting value is negative, it should be added to $10.000 - 10$ to obtain a positive mantissa, before locating the antilog in common log tables.

It will be advantageous to examine the two sets of statistical values obtained from the given example, in terms of the frequency curve of the sediment. For this discussion the mean value and the standard deviation will be treated in detail; the skewness will be mentioned later. Figure 4 is the frequency curve of the beach sand of Table 2, obtained by graphic differentiation of the corresponding cumulative curve (7). The horizontal scale shows ϕ as the independent variable, and the area under the curve represents the total frequency. At the point $\phi = 1.81$ an ordinate has been erected. This is

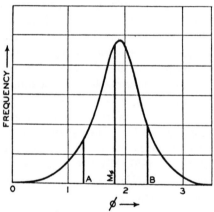

FIG. 4. Frequency curve of beach sand (Table 2) with ϕ as the independent variable. Each unit on the horizontal scale is one Wentworth grade. The distance between ordinates A and B measures the spread of the curve.

M_ϕ, the arithmetic mean of the ϕ-distribution, and it passes through the center of gravity of the distribution. Hence it serves as a measure of the average value of the distribution.

The relation of the standard deviation, σ_ϕ, to this mean may readily be seen from Figure 4. Since σ_ϕ is expressed in ϕ-units, it measures the distance directly in class intervals from the mean to the center of gyration of each half of the curve. Hence

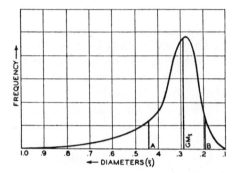

FIG. 5. Frequency curve of the same sediment as fig. 4, with diameters in mm. (ξ) as the independent variable.

it tells directly how many Wentworth grades are included in the central part of the curve on either side of the mean. The value $\sigma_\phi = 0.56$ means that in the distance $M_\phi - \sigma_\phi$ to $M_\phi + \sigma_\phi$ there are 1.12 Wentworth grades, entirely independently of whether the sediment is coarse or fine. Thus σ_ϕ is a measure of the degree of scatter about the mean value. Consequently it serves as a measure of the "sorting" of the sediment, at least in a descriptive sense.

To locate the ordinates marked A and B in Figure 4 it is only necessary to move 0.56 ϕ-units (grade sizes) to the left and right of the mean value. Point A is $M_\phi - \sigma_\phi = 1.81 - 0.56 = 1.25$, and B is $M_\phi + \sigma_\phi = 1.81 + 0.56 = 2.37$. Between these two ordinates, then, is the central portion of the distribution.[4]

We may now turn to the "diameter equivalents" of the measures just discussed. To evaluate these in terms of the frequency curve, the curve itself may be drawn with the diameters (ξ) as independent variable, as in Figure 5. At the point $\xi = .286$ an ordinate is erected. This is the geometric mean of the distribution. To locate ordinate A, divide GM_ξ by σ_ξ: $.286/.676 = 0.421$; to locate B, multiply GM_ξ by σ_ξ: $.286 \times .676 = 0.193$.[5] These two ordinates are shown in the figure. Notice that just as the frequency curve itself is drawn out to the left when ξ is used as the independent variable, so the spread between A and B is drawn to the left with respect to the geometric mean. In fact, GM_ξ is itself the geometric mean of the ξ-values at A and B.[6]

A comparison of Figures 4 and 5 suggests that σ_ϕ is a more readily visualized measure of the characteristics of the sediment than σ_ξ, although in particular cases the latter may have distinct advantages. It is a point that requires further attention.

The skewness of the curves has not been included in the above discussion for several reasons. One of these is that unless one is certain he has a single homogeneous population in the sample, the single curve may represent a mixture of two distributions (9). This is of decided significance in the study of sediments because it is often not possible to determine whether the sample is homogeneous, or whether during its collection two or more distributions may not have been included. Further, diagenic changes may conceivably affect the normality of the curve, and such factors cannot at present be evaluated. Nevertheless, the skewness may afford important data. Sudden changes in its value in a series of related sediments may indicate sampling difficulties or even genetic changes. In this respect the kurtosis of the curves may also be significant.

In a geometrical sense the skewness of the curve may be interpreted as a measure of the relative spread of the data on one side or the other of the mean. Thus it is a measure of the departure of the curve from normality or symmetry, and the sign before the value indicates the direction in which this departure occurs. In the curve of Figure 4 the skewness is negative, and hence toward the negative ϕ-scale. An easy way to visualize skewness is to recall that in a normal curve the mean, median, and mode all coincide, whereas with increasing skewness these values spread apart. The extent of their spread, except in extreme cases, may be used as a measure of the skewness (6, p. 166).

Entirely aside from skewness as a descriptive and comparative measure for sediments, the third moment (on which the skewness is based) is im-

---

[4] In a perfectly normal distribution, symmetrical about the mean, the interval between the ordinates A and B would include about 68 per cent of the distribution.

[5] These manipulations follow from the fact that when M_ϕ and σ_ϕ are added, GM_ξ and σ_ξ must be multiplied, and when σ_ϕ is subtracted from M_ϕ, GM_ξ is divided by σ_ξ.

[6] Thus, $GM_\xi = \sqrt{(GM_\xi/\sigma_\xi)(GM_\xi\sigma_\xi)}$.

portant in fitting an equation to the curve and in smoothing the data. When the third moment is so small that it may be neglected, the data may be assumed to approximate normality, with the result that curve-fitting is tremendously simplified. The third moment of the curve in Figure 4 is -0.004, which is practically negligible, while the third moment of Figure 5, computed about the *arithmetic mean* of the ξ-values, is found to be approximately 0.24, a value not negligible, as the curve itself indicates. Now in all statistical computations, they are simple in proportion as the higher moments may be neglected. Thus the use of ϕ as the independent variable has greatly simplified the statistical approach in this instance, and it is the resulting simplification that "justifies" the substitution of ϕ for ξ in sedimentary analysis.

SUMMARY

The principal points stressed in this paper have been that the disadvantages arising from the use of unequal class intervals in sedimentary analysis may be offset by the substitution of a new independent variable for the usual diameter values. The use of ϕ as the new independent variable serves this purpose, and makes available a series of statistical measures based on the moments of the distributions. In the present instance the definitions of the parameters involve weight percentage frequencies, inasmuch as the method is applied to conventional mechanical analysis data. The question of other frequencies, such as number frequencies, introduces additional factors into the discussion. Obviously a new series of ϕ-measures may equally readily be defined in terms of number frequency, volume frequency, and so on. Measures on such different bases need not necessarily bear simple relations to each other.

The whole subject of statistical analysis as applied to sediments is still largely an open field, and there are practical difficulties in predicting which set of measures best adapt themselves to a genetic study of sediments. The obvious implication is that workers should express their results in more than one manner, so that the data be available as and when needed. At the present stage of development it seems desirable to use several techniques rather than any single approach which may by its limitations mask significant points points about the sediment. The method of moments, in this connection, affords a means of supplementing the quartile measures commonly used by sedimentary petrologists.

The writer takes pleasure in acknowledging his indebtedness to Dr. Carl Eckart of the Department of Physics of the University of Chicago for his friendly criticism and many helpful suggestions during the preparation of this paper.

APPENDIX

The choice of the arithmetic mean of the ϕ-distribution, M_ϕ, and the standard deviation in ϕ-terms, σ_ϕ, as the parameters of the size frequency distribution, permits an expression of the normal distribution in the ϕ-notation as

$$y = \frac{1}{\sigma_\phi \sqrt{2\pi}} e^{-(\phi - M_\phi)^2 / 2\sigma_\phi^2}$$

where ϕ is the value of the independent variable at any point. The analogy to the normal probability law is apparent. The relations between the normal law in ϕ-terms and in ξ-terms have not been fully examined by the writer, but they are doubtlessly complex. However, there is no necessity for simple relations, inasmuch as it is not necessary to convert the ϕ-distribution to its ξ-equivalent in order to describe the distribution.

It happens that the choice of independent variable in the present case affords a fairly simple relation between the arithmetic mean of the ϕ-distribution and the geometric mean of the ξ-distribution. Thus, by definition,

$$M_\phi = \frac{\phi_1 + \phi_2 + \cdots + \phi_n}{n}, \text{ where } \phi_1, \phi_2 \text{ are in-}$$

dividual values of the variable. However, $\phi = -\log_2 \xi$, so that $M_\phi = -\log_2 M_\xi = $

$$\frac{\log_2 \xi_1 + \cdots + \log_2 \xi_n}{n} \text{ where } M_\xi \text{ is some mean}$$

of the ξ's. The antilog of this expression is $M_\xi = \sqrt[n]{\xi_1 \xi_2 \cdots \xi_n}$, so that M_ξ proves to be identical with the usual definition of the geometric mean of the diameters, called GM_ξ in the text. Higher moments of the ϕ-distribution increase in complexity when expressed in ξ-terms; the geometrical relations between

the second moments was pointed out in the text in conjunction with Figures 4 and 5.

One of the advantages of a choice of independent variable such as the present is that when the sedimentary data plot as a symmetrical curve in ϕ terms, the third and higher moments may be neglected, and the data may be used in conjunction with tables of the probability function, both for determining areas under the curve and for smoothing the data; in addition it yields a simple equation for the curve, the parameters of which (M_ϕ and σ_ϕ) uniquely determine the frequency distribution. However, even when the curve is not symmetrical, the higher moments of the ϕ-distribution may be used to build up a form of the Gram-Charlier series, of which the normal ϕ-curve is the first term. Thus in any except extreme cases of asymmetry, detailed statistical analysis is possible by conventional procedures. Further, such measures as the Chi-square test may be used to determine the goodness of fit of such smoothed curves. The writer has in preparation a paper on these aspects of the method, which will appear in an early issue of this *Journal*.

It is interesting, in connection with the ϕ-notation, that the method of moments developed by Wentworth (see footnote 2, text) yields values practically identical with those of the ϕ-method. As far as the writer can trace the steps in Wentworth's procedure, he actually computes moments with respect to a log of the geometric mean, although in his discussion he appears to confuse this with the arithmetic mean.

REFERENCES

1. WENTWORTH, C. K., Method of computing mechanical composition types of sediments: *Geol. Soc. America, Bull.*, vol. 40, pp. 771–790, 1929.
2. HATCH, T., and CHOATE, S., Statistical description of the size properties of non-uniform particulate substances: *Franklin Inst., Jour.*, vol. 207, pp. 369–387, 1929.
3. WENTWORTH, C. K., A scale of grade and class terms for clastic sediments: *Jour. Geol.*, vol. 30, pp. 377–392, 1922.
4. KRUMBEIN, W. C., Size frequency distributions of sediments: *Jour. Sed. Petrology*, vol. 4, pp. 65–77, 1934.
5. Since ϕ is a logarithm, operations involving multiplication, division, or the extraction of roots in ξ-terms reduce merely to addition, subtraction, and simple division in ϕ-terms. Thus, such measures as Trask's sorting coefficient and skewness (Trask, P. D., *Origin and environment of source sediments of petroleum*, 1932, pp. 68–76) may be very simply computed, since not even a slide rule is needed.
6. MILLS, F. C., *Statistical Methods*, New York, 1924, describes the method of finding the arithmetic mean on p. 118, and the standard deviation on p. 156, in the first 6 columns of the table. Skewness and kurtosis are not included, although the former will be included in

the example below. For the necessary conversion equations needed in computing skewness and kurtosis, see Camp, B. H., *The mathematical part of elementary statistics*, New York, 1931, chapter 2.

7. The data used are the mechanical analysis of a beach sand. See Krumbein, W. C., The probable error of sampling sediments for mechanical analysis, *Am. Jour. Sci.*, vol. 27, 1934, p. 208, sample 7a.

8. When the data are arranged in Wentworth grades, the statistical values are expressed directly in the Wentworth grades as units. When the $\sqrt{2}$ or $\sqrt[4]{2}$ scales are used, constants depending on the new class sizes must be used to obtain comparable values. See Mills, (6) in this connection. Other things being equal, of course, the use of smaller classes will yield closer approximations to the true values of the statistical measures.

9. PEARSON (*Phil. Trans. Royal Soc.*, vol. 185A, 1894), has shown that a skew curve may be broken down into two normal curves, even if the original curve is unimodal. For a possible genetic significance of skewness, see Gripenberg, Stina, A study of the sediments of the North Baltic and adjoining seas, *Fennia*, vol. 60, no. 3, 1934, p. 203.

5

Reprinted from *Comptes Rendus (Doklady) de l' Acad. de Sciences de l'*
URSS 33(1):48-49 (1941)

ON THE RÔLE OF THE LOGARITHMICALLY NORMAL LAW OF FREQUENCY DISTRIBUTION IN PETROLOGY AND GEOCHEMISTRY

By N. K. RAZUMOVSKY

In petrology all quantitative analyses of rocks represent arithmetical means of a certain, usually small number of analyses for a given rock type. This is supposed to be the way to obtain the most probable values for the content of components characterizing the composition of the rock as a whole. It is in this way that have been derived the average rock compositions given by Dana, Osanne, Rosenbuch, Loewinson-Lessing, Zavaritsky.

An analysis of the fluctuation of values of the content of any component in a rock makes one, however, conclude that the distribution deviates widely from the law of the arithmetical mean, displaying sharp asymmetry. The asymmetry vanishes when the content intervals of a given component are arranged in a geometrical progression. Hence it follows that the logarithms of contents will obey Gauss' law, or, in other words, that it will be the arithmetical mean from the logarithms that will give the most probable average value for the content of the component concerned. But since the sum of logarithms is equal to the logarithm of the product of the magnitudes given, the most probable content is a geometrical mean, and equidistant values to either side of it will be equiprobable.

Let us illustrate the above said by an example. A. N. Zavaritsky ([1]) gives in his book 151 analyses of all types of effusive rocks (after Osanne). We shall take component a as corresponding to the content of the total of alkali in the shape of the compound $(K, Na)AlO_2$. Within the interval $a = 4—6$ there are 6 analyses; $6—8 = 13$ analyses; $8—10 = 16$; $10—12 = 23$; $12—14 = 21$; $14—16 = 19$; $16—18 = 16$; $18—20 = 10$; $20—22 = 12$; $22—24 = 4$; $24—26 = 5$; $26—28 = 3$; $28—30$, $30—32$, $32—34$ 1 analysis in each, which makes a total of 151 analyses. Let us represent these data graphically, plotting along the abscissae the intervals, and along the ordinates the frequencies. We obtain an asymmetrical curve (Fig. 1, *I*) with a longer right branch and having its arithmetical mean ($a = 14.4$) outside the most frequent value, or mode, which for our case is about 11. If we arrange the intervals in a geometrical progression, and compute the number of analyses in each according to the same table, the results will be different: a in the interval $4—5.65 = 4$ analyses; $5.65—8 = 15$; $8—11.32 = 32$; $11.32—16 = 47$; $16—22.63 = 40$; $22.63—32 = 12$; $32—45.3 = 1$. A total of 151 analyses. Here the mean (geometrical) is equal to 13.5 ($M_0 = \lg a = 1.131$) and corresponds with the mode within the accuracy of the figures obtained. Standard deviation $\sigma = 0.204$. The mean probable error, expressed as σ in decimal logarithms, equals $\rho = 0.1376$. It has a very simple physical meaning: it measures the fluctuation range of a; σ is the logarithm of the number by which the geometrical mean should be multiplied or divided, in order to obtain the interval embracing a half of all the values known by us. For our case $A_1 = \dfrac{13.5}{1.33} = 10.15$ and $A_2 = 13.5 \times$

43

$\times 1.33 = 17.97$, and the number of analyses within the range $A_1 - A_2$ is equal to 75 out of 151, i. e. almost precisely 50%. The curve constructed in a logarithmical scale (Fig. 1, curve II) is rather symmetrical. Theoretically, the frequencies for our intervals are equal to 3—13—33—48—35—16—3, and very closely approach the empirical frequencies 4—15—32—48—40—12—1 = 151.

From the above said the following conclusions are to be drawn:

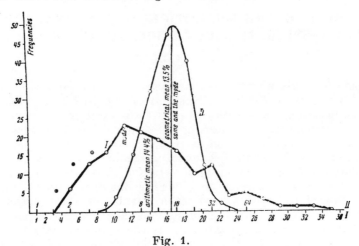

Fig. 1.

1) The frequency distribution of the content of chemical elements in rocks obeys a logarithmically normal law.

2) Grouping the analyses within intervals arranged in geometrical progression, we obtain a symmetrical frequency distribution. The constants of this distribution are: M_0—the arithmetical mean of the logarithms of content and σ—standard deviation in the same logarithms. One can also employ $\rho = 0.674\ \sigma$—probable error.

3) All the average analyses made by the method of the arithmetical mean should be revised. It is difficult to predict whether or not the revision will result in any material correction, but the figures will unquestionably be changed, and the constants replaced by others.

4) The geometrical table of the average composition of the earth crust, and the per cent of the element content made by Clarke and Washington on the basis of arithmetical means computed for almost every analysis of rocks, also calls for critical revision. An investigation in this line will be most essential, since Washington and Clarke's numbers lie at the basis of many important and general conclusions bearing on the occurrence and content of rare and precious elements in the earth crust.

5) Vernadsky's decades of elements are not only a convenient method to evaluate the range of distribution of a given element, but they present also natural classes, according to which the elements must be distributed at a regular frequency. A critical revision of the original material will probably serve to disclose these new regularities.

Geological Institute.
Academy of Sciences of the USSR. Moscow.

Received
15. VIII. 1941.

REFERENCES

[1] А. Н. Заварицкий, Пересчет химич. анализов изверженных горных пород (1933).

Translated by T. Rogalina.

44

6

Reprinted from *Nature* 153(3872):71–74 (1944)

STATISTICS IN SEDIMENTARY PETROLOGY

By Dr. P. ALLEN
University of Reading

RECENT publications show that at long last there is a flood-movement in British geological research towards the accumulation of increasingly precise quantitative data. The transition from the qualitative phase, painfully slow at first, has lately been accelerated in most branches of the science. This limited progress is especially marked in sedimentary petrology, where Fleet, Butterfield, Smithson and Walder have raised a banner which all must surely follow.

Without a doubt this vigorous movement marks only a beginning. Already some of its protagonists, wallowing in a mire of seemingly endless figures, are beginning to question the wisdom, and indeed the reliability, of their laborious endeavours. Attempts at extrication by crude mathematical methods (concerning the reduction and not the reliability of data) have met with little success. Other petrographers, less desirous of reaching immediate conclusions, are content merely to amass information of unqualified precision in the hope that one day it may prove useful.

All the present difficulties and discrepancies have one common underlying cause. They are due entirely to the widespread inability of geologists first to recognize, and secondly to deal with, the various types of population commonly encountered in petrological research. No matter if these populations are finite or infinite, and irrespective of which sedimentary variates they concern, their behaviour and its repercussions remain uninvestigated and almost completely ignored. Yet, to quote the most fundamental example, all petrological work, even that concerned with the smallest finite populations, relies on sampling for its practicability. Consequently, though the distinction between some of the statistics actually used, and the parameters estimated, may often be recognized, no allowance is ever made for it—the penumbræ of error darkening our petrological morass are left well alone ! Yet it is idle merely to record the useless fact that zircon reaches a mean size of "50 μ" in a certain locality ; statement of the statistic in conjunction with its (qualified) standard error—as "50 μ ± 3 μ"—conveys information of considerable geological value.

In consequence of the shortcomings outlined above, interpretations drawn from quantitative petrographical data usually either do not justify the laborious methods used, or (especially when 'conclusive') are largely unwarranted. The former situation involves waste of data and both involve waste of time. Indeed, I know of no geological work (published in Britain) that contains one conclusion stated as an honest mathematical probability. I am also unaware of any paper devoted mainly to quantitative information which is enlightened by a really comprehensive account of the sampling and analytical techniques employed. Truly, appraisal by the sceptic is normally quite impossible. Consistently with this state of affairs, investigations planned to achieve desired degrees of precision—indeed degrees of precision themselves—are practically unknown in British geology. The present difficulties are all the more surprising when we consider that the mathematical techniques necessary for their removal have existed for some time. One can only conclude that for three decades most geologists have worked in complete oblivion of the progress of statistical science.

During the past five years, much work in the Department of Geology at the University of Reading has been devoted to the application of statistical methods to geological problems, especially those of a petrological and stratigraphical nature. In the sphere of sedimentary petrology the mathematical techniques have already more than justified their adoption.

Selection of Fields of Study

Statistical control exerts its first influence in the selection of fields of study. The preferable fields are those which permit detailed investigation of (1) the areal petrographical characters of widespread but thin horizons (less than 6 in. thick), and (2) the vertical petrographical successions in single localities. By the first approach an attempt is made to treat certain selected geological 'moments' separately throughout their accessible extents, and so to elucidate the *spatial* distributions of their petrographical characters. By the second, the distributions *in time* of petrographical characters are studied at selected points. In such a manner the invalidating effects of the two types of distribution upon one another may be minimized each in turn. Unfortunately, the nature and number of actual and potential exposures very often leave the sedimentary petrologist no choice but to concentrate largely upon one or other of the two viewpoints. When quarries and other sites are normally small but very numerous (as in the Wealden rocks of south-east England), horizontal studies often yield the more valuable results ; when they are extensive but few (as in the Tertiaries of the Isle of Wight, now being studied by my colleague Miss P. S. Walder) investigation of the petrographical sequence in time frequently appears to be the more promising. I recently completed an areal study of the Top Ashdown Pebble Bed[1] in the Weald of Kent, Surrey and Sussex, using an arbitrary 2-in. horizon situated 4 in. down in the underlying sandstone as 'control'. Since the efficacy of statistical methods has been most fully tried out on this sediment, particular reference thereto is made in the present outline.

If a horizontal study has been decided upon, the petrologist should next ensure that the sampling on which it is to be based will be as nearly random as possible. The distribution of satisfactory exposures usually limits him in this respect, because, being controlled by factors other than chance, it is not necessarily a random one. Secondary sampling from them at random is seldom possible either, for the exposures at any one widespread horizon less than 6 in. thick are usually all too few at the start. The sandpits and openable sites at the Top Ashdown Pebble Bed horizon were apparently distributed according to factors quite unrelated to the petrographical characters of the bed. The risks inherent to an assumption of randomness, being therefore considered minimal, were ignored, and attention was merely focused upon ensuring random sampling within the sites located.

Choice of Statistics

Preliminary analysis of a few field samples (by the usual methods of gravity separation, and counting of *entire* residues) indicates the range of statistical values likely to be necessary for the subsequent sample and horizonal characterizations. In general, the constituents of sediments are found to be most easily dealt with through separate consideration

of two or three fractions, namely, (1) the allogenic pebble suite (when present), (2) the allogenic light and heavy grain suites, and (3) the authigenic light and heavy suites. The minimum quantitative requirement is that in every field sample each of these five suites shall be fully characterized by statistics relating to frequency, abrasion, grade size and sorting. The frequencies (expressed as percentages) of all pebble and mineral grain species, varieties and other distinctive types, should be determined, together with their 'errors'. The degrees of abrasion are best expressed as the percentages of euhedral, subangular, angular and rounded individuals (suitably defined), average size as their arithmetic mean sizes (based on measurements of intermediate axes), and degrees of statistical sorting as the coefficients of variation derived from the data yielding their means. Calculation of errors is of course necessary in all cases.

During the study of the Top Ashdown Pebble Bed, the determination of the grade sizes and degrees of sorting of all the species, varieties, etc., in every subsample was found to be too time-consuming. This difficulty was overcome in the case of the allogenic heavy suite by using the values pertaining to zircon (probably the most stable mineral) as indexes typifying the whole suite. The validity of the method was confirmed by replicate analysis and by detailed comparisons with other species. The allogenic heavy grain suite of each Ashdown field sample was thus finally characterized by at least (1) complete frequency analysis, (2) the 'zircon abrasion index' (equals per cent of euhedra among the zircons), (3) the 'zircon size index' (equals the mean size of zircon), and (4) the 'zircon statistical sorting index' (equals the coefficient of variation of the zircon grain-size distribution).

Estimation of Minimum Subsample Size and Standard Errors

The field sampling and characterization designs having been settled, it is necessary to estimate the minimum number of pebbles or grains in the subsamples pertaining to each suite which will be necessary to achieve sufficient accuracy for studying the variations anticipated in the various populations of the sediment. Analysis of a few subsamples of arbitrary sizes from widely different sedimentary grades usually suffices for a provisional estimate. In the case of the Top Ashdown Pebble Bed, the subsample sizes were standardized at greater than 200 for pebbles and greater than 1,000 for grains.

In order to establish the degrees of significance of the chosen statistics and to recognize when significant differences exist between them (within the aims of the investigation), the petrologist must be able to estimate their associated 'errors'. Because these (best expressed in the form of standard errors) necessarily always embody components due to random sampling (SE_r) and experimental treatment (SE_e) they may be termed 'total standard errors'. Their magnitude will vary with the project and the consequent unit of comparison—normally either the subsample, the 'patch' or the 'locality'.

1. *Standard Errors of the Subsample (SE_s).* During the preliminary work it is convenient first to investigate the total standard errors (SE_s) of statistics referring to the adopted subsample alone. Values for these, in terms of their component sampling errors, may be obtained from series of replicate analyses of homogenized material drawn at random from the known grade-size range of the sediment. For pur-

poses of improvements in laboratory technique, SE_e may also be calculated for each species from the data. In the case of frequencies, significant differences between the total and the sampling errors within series are best tested by χ^2, using Brandt and Snedecor's formula[2]. Graphed against frequency, the values (k_e) of the ratio SE_e/SE_r determined for each species may be used in the subsequent assay as a basis for estimating the total errors from the sampling errors of single values. In this way the otherwise inevitable replication of each sample in the main investigation is obviated. SE_e is most necessary for use in the tests of significance involved during the study by concentrated sampling, of local small-scale petrographical variation, or 'patchiness'.

2. *Standard Errors of Patchiness* (SE_p). Combined with the thin sampling normally forced upon petrologists, SE_e is not, unfortunately, necessarily large enough to deal with the establishment of more widespread changes, such as the recognition of minor petrographic 'regions'. Estimation of a sufficiently comprehensive error, the 'standard error of patchiness' (SE_p), allowing for small-scale variation within the sediment, may be conducted in the same general way as that of SE_e, but using instead the ratios (k_p) derived from random series of replicate samples of *unmodified* material, the latter being taken at random from small horizontal areas of constant size. For the Top Ashdown Pebble Bed these areas were fixed at $\frac{1}{4}$ sq. ft.

3. *Standard Errors of Locality* (SE_l). When horizonal sampling is thin, and broad petrographic regions are to be established, the unit of comparison strictly becomes the locality, and an error even more comprehensive than SE_p is needed. One such, the 'standard error of locality' (SE_l), may be estimated as before from its component sampling error by means of specific factors (k_l) obtained from preliminary replicate sampling within 'localities'. During the study of the Top Ashdown Pebble Bed, a 'locality' was defined as a quarry of a certain size. Unexpectedly, however, analysis of variance showed that none of the values obtained for k_l significantly exceeded the corresponding k_p's, and SE_l was consequently taken as approximately equal to SE_p.

Manipulation of Standard Errors

The three standard errors may be manipulated during the subsequent work in the usual way, the particular error used and the level of significance chosen varying with the aims in view and the sampling concentrations achieved. The assumption that the distributions of the variates concerned are normal must necessarily be provisional until they are more extensively investigated.

Much arithmetical labour may be avoided in the large numbers of frequency comparisons afterwards carried out, by preparing beforehand curves of significant differences based on the various k-factors and the chosen level of significance. The level of significance adopted during the Ashdown work was $P < 0.05$, and since all grain counts either equalled or exceeded 1,000 individuals, a frequency p_1 per cent of a mineral was considered to be significantly different from that $(p_2$ per cent) in another sample only when

$$p_2 > \frac{50(5p_1+k^2)+k\sqrt{2500k^2+100(k^2+500)p_1-(k^2+500)p_1^2}}{250+k^2},$$

where k was the value of k_e, k_p or k_l appropriate to the mean of p_1 and p_2. Curves were therefore prepared for the above relation as an identity (graphing p_2 against p_1) when $k = 1, 1.5, 2, 2.5, \ldots, 6.5, 7$. From these, the status of a large majority of differences could be read off at sight.

The pebble counts were treated in the same general way.

Mapping of Statistics

The kinds of statistic relating to the several species, varieties, etc., obtained during the main investigation, are best plotted on to separate maps as they are obtained, and their frequency distributions finally inserted thereon. The latter may be examined by moment-analysis.

During the insertion of 'contours' designed to emphasize the main trends of change and similarity, rigid statistical control is vitally necessary if they are to have any real meaning. By far the best scheme is to make the contour intervals equivalent to minimum significant differences; for example, for $P = 0.05$ when the total counts always exceed the intended minimum size.

When the collection of data concerning the horizon is completed, statistical reduction and analysis of variation become guides to interpretation rather than policemen to methods. The analysis and inter-horizonal comparison of correlations between variates, neatly carried out in terms of the correlation coefficient (r), is illustrative in this respect. Several significant correlations (up to $+0.71 \pm 0.087$) were established between mineral frequency and sedimentary grade size, stratigraphical variates and grade size, etc., in the Top Ashdown Pebble Bed. These in turn led to further investigations which showed, for instance in the first example, that they were usually direct results of the interplay of former hydrodynamical conditions and the relative abundance of source materials. Post-depositional alteration was shown to have taken a very minor role.

Statistical Characterization of the Sediment as a Whole

The final statistical characterization of the entire horizon is most conveniently given by the arithmetic means, standard deviations and standard errors of the frequencies, abrasion indexes, size indexes and sorting indexes of each mineral and pebble species, variety, form, etc., together with the significant correlations (qualified by their standard errors) established between certain of these variates. This should be regarded merely as a development of Butterfield's pregnant creations—his "characteristic formula" and "range formula"[3].

The problem of comparing and contrasting sediments with sediments and other rock types by means of characteristic statistics is still under investigation. Though the possibilities of using r and χ^2 appear promising in this respect[4,5], the difficulties of translating the statistical into geological correlations remain unsurmounted. Until, by their combined efforts, petrologists amass a fairly large sample of all such statistical relations, we can have no basis for making direct petrogenetical probability statements. The frequency suites of the arbitrary Ashdown horizon previously mentioned and of certain Yoredale Sandstones[2] yield (at present) $r = +0.89 \pm 0.093$, but our ignorance of the geological status of this value forbids us to conclude that it necessarily implies a close genetical relationship.

A full exposition of the statistical methods outlined above will be given elsewhere in due course.

[1] Allen, P., *Proc. Geol. Assoc. Lond.*, **52**, Fig. 57A (1941).
[2] Fisher, R. A., "Statistical Methods for Research Workers", 8th edit., 85 (1941).
[3] Butterfield, J. A., *Trans. Leeds Geol. Assoc.*, **5** (1940).
[4] Dryden, L., *Amer. J. Sci.*, **29**, 393 (1935).
[5] Eisenhart, C., *J. Sediment. Petrol.*, **5**, 137 (1935).

Reprinted from *J. Paleo.* **23**(1):95–103 (1949)

STUDIES IN QUANTITATIVE PALEONTOLOGY

II. MULTIVARIATE ANALYSIS—A NEW ANALYTICAL TOOL FOR PALEONTOLOGY AND GEOLOGY

BENJAMIN H. BURMA

ABSTRACT—Multivariate analysis is the simultaneous comparison of three or more variables in order to determine the significance of their likenesses and differences. The technique has been in use for some thirteen years, but no simple account of the method has hitherto been available. An illustrative example from invertebrate paleontology is here used, but the method is applicable to any problem in geology involving several quantitatively known variables.

In recent years, increasing attention has been paid to quantitative methods in both geology and paleontology. One result of this trend has been an increased interest in statistical analysis, particularly that part of the discipline which deals with tests for the significance of the likeness and differences between two samples (or collections). These tests have, however, been rather slow in gaining acceptance and for a variety of reasons. Most of these reasons are not germane to the issue, but there is one, rather poorly defined, which seems to have been effective. There seems to have been an almost subconscious realization that the usual tests of significance involve only one, or at the most two, variables, whereas geological problems usually involve several variables. A more extended discussion of this phase of the problem will be found in the paper on quantitative methods by Burma (1948). A short summary will serve here.

A comparison of a single variable between two samples is readily effected by comparing their means (Simpson and Roe, 1939; pp. 192–197, 210–212). For certain simple problems, such a comparison is adequate. A better comparison of two samples may be made on the basis of two variables, provided that these variables show a notable amount of correlation as will be found to be the case in the great majority of problems. This can readily be accomplished by computing the regression line of one variable on the other for each sample, and then comparing the two regression lines (Simpson and Roe, 1939; pp. 262–280).

Both the above methods ignore the fact that much geological data, and particularly paleontological data, involves several variables and that the characteristics of a given specimen depend not on characters A, and B, and C, but on characters A, B and C considered together and simultaneously. We are usually interested in comparing objects as a whole, and not characters as such. Since the methods discussed above will not make a comparison of more than two characters simultaneously, it is obvious that they will be inadequate in such cases. Such a comparison would have to involve a comparison of the mutual regressions of several characters simultaneously.

The problem of handling multiple regressions was successfully solved by Hotelling in 1931 (Hotelling, 1931). The first person to use this method in the solution of a practical problem seems to have been M. M. Barnard, who applied it to a study of four series of Egyptian skulls (Barnard, 1935). It has been used by others since that time, but still is not widely known. Simpson and Roé's "Quantitative Zoology" which treats other statistical methods for biological data with some thoroughness, does not mention this method. Multivariate analysis is treated in a number of more advanced works on statistics (e.g., Fisher, 1946; pp. 285–289) but in these books the method is presented in a way which is not usable to a person without advanced mathematical background.

In essence, the method may be pictured thus: If we are dealing with one variable, we may use the value of this variable as a coordinate and plot the variable onto a line, using a point for each specimen, and then study the distribution of these points. If we have two variables, we have two coordinates for our points and plot them into a plane in two dimensions. If we have three variables, our points would be plotted in a "solid" of three dimensions. If we have n variables, we use n-dimensional space. We would then study the dispersion of the points in this n-dimensional cluster, doing this for each of the two samples we wish to compare. Next we would combine our two samples and study the dispersion of the combined sample. The relative dispersions of the original and combined samples will enable us to decide whether the two samples differ significantly or whether they do not.

Fortunately, the actual computations involved in this method are all very simple. These operations are, however, rather lengthy and ordinarily it will be found that long-hand computation is much too time-consuming except possibly for a comparison of three variables. A slide rule *cannot* be used because of the low order of accuracy it allows. A calculating machine is thus a necessity in the use of this method. It should be an electrically driven model with at least ten rows of keys and preferably with automatic division and multiplication. Such a machine is suitable for calculations involving up to eight variables or so. If it is necessary to handle more than eight variables it would probably be well to use an International Business Machines calculator, feeding the data on punched cards. Ordinarily, it will be found that the number of characters to be compared can be held to a reasonable figure.

METHOD OF CALCULATION

To illustrate the calculations of multivariate analysis, we may use an example taken from the blastoids. These were collected from the Paint Creek formation near Floraville, Illinois. They are of the group commonly referred to *Pentremites godoni*, but it was soon discovered that they were susceptible to division into two groups, one of which had a somewhat stellate horizontal section associated with a rather low deltoid-standard radial ratio and another with a more rounded section and a markedly higher deltoid-standard radial ratio. The first of

these we shall call *P. godoni* alpha and the second *P. godoni* beta. Comparisons of regression lines of two characters indicated that the two did not differ significantly. In spite of this, it was felt that there was a real difference between the two, so it was decided to try multivariate analysis. For this purpose, eight characters were measured on each specimen: the standard radial, length of the deltoid, the base of the radial, height of the azygous basal, length of the ambulacrum, the total height, the thickness, and the number of side plates per ambulacrum. (For an explanation of this terminology, see Burma, 1948.) In all, 20 usable specimens of *P. godoni* alpha and 20 of *P. godoni* beta were available. These two samples will hereafter be referred to as "alpha" and "beta" respectively. The original measurements are shown in Table 1. All measurements are in millimeters. In this and subsequent tables, the column headed by (1) will be the data for the standard radial; (2) the deltoid; (3) the base of the radial; (4) the azygous basal; (5) the ambulacrum; (6) the height; (7) the thickness; and (8) the number of side plates. All measurements in any one horizontal row are of one single specimen. Column (1) should always be the character on which the other characters will regress, usually the "time" character.

Since we will be dealing with regression formulae which are accurate only in straight-line relationships, we should be sure that our data fulfills this requirement. If the present data is plotted arithmetically, it will be found to show a definitely curvilinear trend. Knowing, however, that growth is commonly governed by an exponential function, we will find that the logarithms of these data will yield a plot whose trend approximates a straight line.

For that reason, we will in this case deal with the logarithms of the data rather than the original measurements. In order to deal with whole numbers and digits as low as possible, the following procedure has been adopted. Three place logarithms have been used as sufficiently accurate and each has been multiplied by 1000 to get rid of the decimal point. The size of the digits was then reduced by subtracting a number from the logarithms for each character, the same number being subtracted, for example, from character (1) in both alpha and beta. This is important. From character (1), 700 was subtracted; from (2), 600; (3), 500; (4) 300; (5) 1000; (6), 1000; (7), 1000; and (8), 1500. This computation is then checked. This step

first step is to obtain for each set of data, for alpha and beta, the sum of each column ($\Sigma(X)$), the sum of the square of each member of each column ($\Sigma(X^2)$), and the sum of twice the product of corresponding members

TABLE 1.—ORIGINAL DATA

		P. godoni alpha									*P. godoni* beta					
(1)	(2)	(3)	(4)	(5)	(6)	(7)	(8)	(1)	(2)	(3)	(4)	(5)	(6)	(7)	(8)
6.4	4.7	4.1	3.0	10.3	12.3	11.5	33	5.5	4.9	4.0	2.5	10.0	12.0	10.6	33
7.4	6.6	4.0	2.5	13.0	14.6	12.0	41	6.2	4.7	4.0	2.6	10.1	12.4	11.1	34
7.5	5.4	4.1	2.5	11.5	14.2	11.5	37	6.6	6.5	4.0	2.7	12.5	14.4	12.0	41
7.7	5.4	4.8	3.5	11.6	14.5	12.4	37	7.0	5.9	3.9	2.4	12.2	14.4	11.3	38
8.8	7.4	5.9	3.5	14.5	17.2	16.1	47	7.4	6.4	5.0	3.6	12.2	15.1	12.9	38
9.0	7.5	5.6	3.4	15.1	17.7	16.0	43	7.5	9.0	4.8	3.3	16.0	16.5	15.9	50
9.3	8.4	6.5	4.4	16.9	20.0	19.0	50	7.5	7.7	4.6	3.2	14.9	17.0	14.5	49
9.4	8.2	5.5	4.1	16.6	19.4	17.5	47	7.9	7.0	4.9	3.3	14.2	16.6	15.1	43
9.5	7.9	4.6	3.5	16.0	18.8	15.3	47	8.3	7.4	4.9	3.0	14.9	16.4	15.4	49
9.6	8.6	5.1	3.5	17.0	19.0	17.5	53	8.4	7.4	5.4	4.0	14.0	16.5	15.5	42
9.8	8.6	5.5	3.2	16.9	19.7	16.4	52	8.4	8.2	5.3	3.6	16.2	19.2	16.2	48
9.5	8.0	6.0	4.4	16.4	19.5	17.5	47	8.5	7.8	4.6	3.5	14.5	17.6	14.7	46
10.0	9.6	6.0	4.3	19.0	21.0	18.8	52	8.5	8.6	4.4	3.3	16.8	18.6	15.3	53
10.1	9.0	5.4	3.7	18.5	19.8	17.6	52	8.5	7.9	5.0	3.6	15.9	17.5	16.8	48
10.1	8.9	6.1	4.5	18.0	19.7	19.5	54	8.5	9.6	4.9	3.4	17.7	19.0	16.4	50
10.4	8.7	5.2	3.2	17.5	19.0	16.8	53	8.6	8.7	4.5	3.5	16.3	18.0	15.0	49
10.5	8.2	6.0	4.4	17.2	19.5	17.5	54	8.6	8.1	5.5	4.0	15.8	18.6	16.2	46
10.7	8.5	4.7	3.5	16.9	19.5	16.5	54	8.6	9.5	5.3	3.9	18.0	19.5	17.0	57
12.4	9.2	6.6	5.0	19.9	23.6	18.7	52	9.5	8.9	5.3	4.0	16.5	18.4	17.5	53
8.3	6.5	5.0	3.5	13.9	15.3	14.5	42	10.8	9.9	5.5	4.1	19.5	21.9	19.4	58

TABLE 2.—ADJUSTED LOGARITHMS OF DATA

		P. godoni alpha									*P. godoni* beta					
(1)	(2)	(3)	(4)	(5)	(6)	(7)	(8)	(1)	(2)	(3)	(4)	(5)	(6)	(7)	(8)
106	072	113	177	013	090	061	019	040	090	102	098	000	079	025	019
169	220	102	098	114	164	079	113	092	072	102	115	004	093	045	031
175	132	113	098	061	152	061	068	120	213	102	131	097	158	079	113
186	132	181	244	064	161	093	068	145	171	091	080	086	158	053	080
244	269	271	244	161	236	207	172	169	206	199	256	086	179	111	080
254	275	248	231	179	248	204	133	175	354	181	219	204	217	201	199
268	324	313	343	228	301	279	199	175	286	163	205	173	230	161	190
273	314	240	313	220	288	243	172	198	245	190	219	152	220	179	133
278	298	163	244	204	274	185	172	219	269	190	177	173	215	188	190
282	334	208	244	230	279	243	224	224	269	232	302	146	217	190	123
291	334	240	205	228	294	215	216	224	314	224	256	210	283	210	181
278	303	278	343	215	290	243	172	229	292	163	244	161	246	167	163
300	382	278	333	279	322	274	216	229	334	143	219	225	270	185	224
304	354	232	268	267	297	246	216	229	298	199	256	201	243	225	181
304	349	285	353	255	294	290	232	229	382	190	231	248	279	215	199
317	340	216	205	243	279	225	224	234	340	153	244	212	255	176	190
321	314	278	343	236	290	243	232	234	308	240	302	199	270	210	163
329	329	172	244	228	290	217	232	234	378	224	291	255	290	230	256
393	364	320	399	299	373	272	216	278	349	224	302	217	265	243	224
219	213	199	244	143	185	161	123	333	396	240	313	290	340	288	263

gives the data as recorded in Table 2. From this point on, these data will be used entirely. If the original data have an essentially straight-line trend, such a step as this could be omitted.

The actual computation now begins. The

of columns (1) and (2), (1) and (3), (1) and (4), and so on ($\Sigma(XY)$). To explain this last part of the computation further, in alpha, one would multiply 106 by 072, 169 by 220, 175 by 132, and so on, and then sum these products. With a calculating machine, $\Sigma(X)$,

$\Sigma(X^2)$, and $\Sigma(XY)$, may be obtained simultaneously for each pair of columns, greatly reducing the work involved. The result for the data in hand is given in Table 3. Do *not* "round off" any results of computations to less than ten figures on any place except where it is done in this sample computation. Such random dropping of figures can have a very adverse effect on the accuracy of the method.

This computation is checked by the following method: For each pair of columns

other checks mentioned *must* be done. A single, small, undetected error early in the computation will mean hours of work wasted to obtain an erroneous result. The frequent checks which are recommended will prevent this.

The next step is to obtain $\Sigma(X)$, $\Sigma(X^2)$, and $\Sigma(XY)$ for the combined data, that is, by combining both samples or collections into one. This is very easily done by adding the results already obtained for $\Sigma(X)$,

TABLE 3
P. godoni alpha

	(1)	(2)	(3)	(4)	(5)	(6)	(7)	(8)
$\Sigma(X)$	5,291	5,652	4,450	5,173	3,867	5,107	4,041	3,419
$\Sigma(X^2)$	1,483,209	1,737,058	1,071,792	1,462,603	862,787	1,395,943	918,595	661,581
$\Sigma(XY)$		3,187,580	2,472,956	2,878,762	2,231,676	2,870,010	2,298,640	1,955,956
Check		6,407,847	5,027,957	6,824,574	4,577,672	5,749,162	4,700,444	4,100,746

P. godoni beta

	(1)	(2)	(3)	(4)	(5)	(6)	(7)	(8)
$\Sigma(X)$	4,010	5,566	3,552	4,460	3,339	4,507	3,381	3,202
$\Sigma(X^2)$	885,022	1,704,298	676,304	1,088,940	673,421	1,096,807	666,861	600,928
$\Sigma(XY)$		2,426,908	1,519,658	1,935,988	1,513,338	1,960,018	1,519,196	1,428,814
Check		5,016,228	3,080,984	3,909,940	3,071,781	3,941,847	3,071,079	2,914,764

TABLE 4
Combined data

	(1)	(2)	(3)	(4)	(5)	(6)	(7)	(8)
$\Sigma(X)$	9,301	11,218	8,002	9,633	7,206	9,614	7,422	6,621
$\Sigma(X^2)$	2,368,231	3,441,356	1,748,096	2,551,543	1,536,208	2,492,750	1,585,456	1,262,509
$\Sigma(2XY)$		2,807,244	1,996,307	2,407,375	1,872,507	2,415,014	1,908,918	1,692,385
Check		11,424,075	8,108,941	9,734,524	7,649,453	9,691,009	7,771,523	7,015,510
a		55.548	46.582	51.3554	−42.663	37.2389	−21.6328	−7.39789
b		0.967 2170	0.660 0068	0.814 835	0.958 231	0.873 502	0.891 013	0.743 674

lowing method: For each pair of columns used in computing $\Sigma(XY)$, add the corresponding figures in each row, square this sum, and add the squares for the entire columns $(\Sigma(X+Y)^2)$. For example in alpha, columns (1) and (2), $106+072=178$, $(178)^2=31,684$; $169+220=389$, $(389)^2 =151,321$; and so on, these successive squares being summed. The following relation should then hold true,

$$\Sigma(X+Y)^2 = \Sigma(X^2) + 2\Sigma(XY) + \Sigma(Y^2).$$

The right hand terms have already been calculated and the individual terms need only be added. If the two sides of the equation are not equal, the computation for those two columns must be redone. This and all

$\Sigma(X^2)$, $\Sigma(XY)$ and $(\Sigma(X+Y)^2)$ for alpha and beta. This has been done in Table 4. This addition is checked as before,

$$\Sigma(X+Y)^2 = \Sigma(X^2) + 2\Sigma(XY) + \Sigma(Y^2).$$

The calculation of the regression equation is next. This equation has the form $Y = a + bX$ in which we must calculate the coefficients a and b.

$$a = \frac{\Sigma(Y)\Sigma(X^2) - \Sigma(X)\Sigma(XY)}{N\Sigma(X^2) - [\Sigma(X)]^2},$$

$$b = \frac{N\Sigma(XY) - \Sigma(X)\Sigma(Y)}{N\Sigma(X^2) - [\Sigma(X)]^2}.$$

In these equations, X is the value for column (1) and Y for the other columns. N is the total number of specimens in the combined sample, in this case 40. *a* and *b* are calculated for each pair of columns, (1) and (2), (1) and (3), (1) and (4), and so on. The calculated value of these coefficients is shown on Table 4.

A sample calculation for the coefficient "*a*" for columns (1) and (2) is as follows:

$$a = \frac{(11218)(2368231) - (9301)(2807244)}{(40)(2368231) - (9301)(9301)}$$

$$= \frac{456638941}{8220639} = 55.548.$$

These coefficients should be checked by re-

filled in by subtracting a given computed value of Table 5 from its corresponding observed value of Table 2. For example, in column (2), the first item will be $072 - 158 = -82$, the second is $220 - 219 = 1$, and so on. Be sure to keep proper account of plus and minus signs.

After Table 6 is completed, the sum of each column, and the sum of the squares of each item for each column $[\Sigma(X), \Sigma(X^2)]$ is computed as is shown in Table 7. These are computed exactly as in Table 3. In obtaining the sums of the columns, due account must be taken of plus and minus signs. To check,

TABLE 5

			P. godoni alpha									*P. godoni* beta			
(1)	(2)	(3)	(4)	(5)	(6)	(7)	(8)	(1)	(2)	(3)	(4)	(5)	(6)	(7)	(8)
106	158	117	138	059	130	73	071	040	094	073	084	−004	072	014	022
169	219	158	189	119	185	129	118	092	145	107	126	045	118	060	061
175	225	162	194	125	190	134	123	120	172	126	149	072	142	085	082
186	235	169	203	136	200	144	131	145	196	142	170	096	164	108	100
244	292	208	250	191	250	196	174	169	219	158	189	119	185	129	118
254	301	214	258	201	259	205	181	175	225	162	194	125	190	134	123
268	315	223	270	214	271	217	192	175	225	162	194	125	190	134	123
273	320	227	274	219	276	222	196	198	247	177	213	147	210	155	140
278	324	230	278	224	280	226	199	219	267	191	230	167	229	173	155
282	328	233	281	228	284	230	202	224	272	194	234	172	233	178	159
291	337	239	288	236	291	238	209	224	272	194	234	172	233	178	159
278	324	230	278	224	280	226	199	229	277	198	238	177	237	182	163
300	346	245	295	245	299	246	216	229	277	198	238	177	237	182	163
304	350	247	298	249	303	249	219	229	277	198	238	177	237	182	163
304	350	247	298	249	303	249	219	229	277	198	238	177	237	182	163
317	362	256	308	261	314	261	228	234	282	201	242	182	242	187	167
321	366	258	312	265	318	264	231	234	282	201	242	182	242	187	167
329	374	264	318	273	325	272	237	234	282	201	242	182	242	187	167
393	436	306	370	334	381	329	285	278	324	230	278	224	280	226	199
219	267	191	230	167	229	173	155	333	378	266	323	276	328	275	240

calculation, a matter of a few moments using a calculating machine.

For the next step (Table 5), column (1) (our *X* column) is copied directly from Table 2 for both alpha and beta. Columns (2) to (8) are filled in with the *calculated* value (*Y*) of that particular character which would correspond to the *observed* value of *X*, column (1), using the appropriate regression equation. For example, the equation for calculating the values for column (2) of alpha would be $Y = 55.548 + 0.9672170X$ and the first value in column (2) would be $55.548 + (0.9672170)(160) = 158.0730$. This is rounded off to 158. In this manner, all of Table 5 is filled in.

After this is done, Table 6 is computed. In this Table, column (1) is dropped and is not used hereafter. The other columns are

add corresponding columns of $\Sigma(X)$ of alpha and beta together. The answer should be close to zero and never greater than $\pm \frac{1}{2}N$, in this case, ± 20.

The next step is to obtain the sum of the cross products of each pair of columns as shown in Table 8, that is, $\Sigma(XY)$ for columns (2) and (3), (2) and (4), (2) and (5) ..., (3) and (4), (3) and (5), and so on. In Table 8 these sums of cross products (ΣXY) are shown in the corresponding row and column, for example, the item in row (4) and column (6) is the $\Sigma(XY)$ for columns (4) and (6). These sums of products are calculated just as the $\Sigma(XY)$ in Table 3. The diagonal items are the sum of the squares of the corresponding columns. These calculations are checked by calculating $\Sigma(X + Y)^2$ for each pair of columns of Table 8 as shown

TABLE 6

| | P. godoni alpha | | | | | | | | P. godoni beta | | | | | |
(2)	(3)	(4)	(5)	(6)	(7)	(8)	(2)	(3)	(4)	(5)	(6)	(7)	(8)
−86	−4	39	−46	−40	−12	−52	−4	29	14	4	7	11	−3
1	−56	−91	−5	−21	−50	−5	−73	−5	−11	−41	−25	−15	−30
−93	−49	−96	−64	−38	−73	−55	41	−24	−18	25	16	−6	31
−103	12	41	−72	−39	−51	−63	−25	−51	−90	−10	−6	−55	−20
−23	63	−6	−30	−14	11	−2	−13	41	67	−33	−6	−19	−38
−26	34	−27	−22	−11	−1	−48	129	19	25	79	27	67	76
9	90	73	14	30	62	7	61	1	11	48	40	27	67
−6	13	39	1	12	21	−24	−2	13	6	5	10	24	−7
−26	−67	−34	−20	−6	−41	−27	2	−1	−52	6	−14	15	35
6	−25	−37	2	−5	13	22	−3	38	68	−26	−16	12	−36
−3	1	−83	−8	3	−23	7	42	30	22	38	50	32	22
−21	48	65	−9	10	17	−27	15	−35	6	−16	9	−15	0
36	33	37	34	23	28	0	57	−55	−19	48	33	3	61
4	−15	−31	18	−6	−3	−3	21	1	18	24	6	43	18
−1	38	54	6	−9	41	13	105	−8	−7	71	42	33	36
−22	−40	−105	−18	−35	−36	−4	58	−48	2	30	13	−11	23
−52	20	30	−29	−28	−21	1	26	39	60	17	28	23	−4
−45	−92	−75	−45	−35	−55	−5	96	23	49	73	48	43	89
−72	14	27	−35	−8	−57	−69	25	−6	24	−7	−15	17	25
−54	8	14	−24	−44	−12	−32	18	−26	−10	14	12	13	23

TABLE 7
P. godoni alpha

	(2)	(3)	(4)	(5)	(6)	(7)	(8)
$\Sigma(X)$	−577	26	−166	−352	−261	−242	−366
$\Sigma(X^2)$	43,765	39,992	65,674	20,058	12,417	28,278	20,652
$\Sigma(X)$	576	−25	164	349	259	242	368
$\Sigma(X^2)$	59,108	17,981	29,380	29,177	12,979	17,084	31,994
alpha minus beta	−1,153	51	−330	−701	−520	−484	−734
Combined data (X^2)	102,873	57,973	95,054	49,235	25,396	45,362	52,646
(X)	−1	1	−2	−3	−2	0	2

TABLE 8
P. godoni alpha

	(2)	(3)	(4)	(5)	(6)	(7)	(8)
(2)	43,765	6,219	4,648	28,481	19,249	24,151	25,884
(3)		39,992	38,085	6,899	9,443	23,253	914
(4)			65,674	8,511	11,408	27,143	−916
(5)				20,058	13,248	17,763	15,909
(6)					12,417	13,594	9,161
(7)						28,278	14,088
(8)							20,652

P. godoni beta

	(2)	(3)	(4)	(5)	(6)	(7)	(8)
(2)	59,108	−1,022	11,336	39,198	23,187	23,068	37,922
(3)		17,981	16,487	−1,406	483	7,690	−4,128
(4)			29,380	3,305	4,971	11,084	1,210
(5)				29,177	17,255	16,411	27,796
(6)					12,979	9,201	15,414
(7)						17,084	15,934
(8)							31,994

in Table 9. This sum, $\Sigma(X+Y)^2$, will equal $\Sigma X^2 + 2\Sigma(XY)\Sigma Y^2$, these last items having been already been calculated.

We are now ready to set up a determinate which is shown in Table 10. The diagonal terms (column (2), row (2); column (3), row (2), etc.) in this determinant are the sums of the squares of the corresponding columns of the combined data. The other items are the sums of the corresponding items of alpha and beta in Table 8, e.g., column (3), row (2) of the determinant equals 6219

TABLE 9

(3)	(4)	(5)	(6)	(7)	(8)
96,195	118,735	120,785	94,680	120,345	116,185
	181,836	73,848	71,295	114,776	62,472
		102,754	100,907	148,238	84,494
			58,971	83,862	72,528
				67,883	51,391
					77,106

P. godoni beta (Check)

(3)	(4)	(5)	(6)	(7)	(8)
75,045	111,160	166,681	118,461	122,328	166,946
	80,335	44,346	31,926	50,445	41,791
		65,167	52,301	68,632	63,794
			76,666	79,083	116,763
				48,465	5,801
					80,946

TABLE 10

	(2)	(3)	(4)	(5)	(6)	(7)	(8)	Difference
(2)	102,873	5,197	15,984	67,679	42,436	47,219	63,806	57.65
(3)		57,973	54,572	5,493	9,926	30,943	−3,214	−2.55
(4)			95,054	11,816	16,379	38,227	294	16.50
(5)				49,235	30,503	34,174	43,705	35.05
(6)					25,396	22,795	24,575	26.00
(7)						45,362	30,022	24.20
(8)							52,646	36.70

TABLE 11

	(2)	(3)	(4)	(5)	(6)	(7)	(8)	Difference
(2)	320.73[1] 82110	16.203 24563	49.835 03524	211.01 00945	132.30 72795	147.21 97528	198.93 48260	0.1797 4160
(3)		240.23 .0005	223.80 43107	8.6331 9138	32.394 75706	118.87 59172	−26.796 77691	−0.0227 38198
(4)			206.11 1862	−3.0655 2440	12.300 96980	20.791 41076	−17.576 32891	0.0612 84574
(5)				68.013 31209	34.447 18632	31.561 52825	28.012 23093	−0.0366 56345
(6)					74.185 18053	−25.303 72986	−21.920 64849	0.0466 98337
(7)						86.534 20859	32.900 38940	0.0174 02250
(8)							98.469 81712	0.0293 38193

[1] To conserve space, 320.7382110 is written 320.73 82110

$+(-1022)=5197$. The rest of the determinant is filled in in this manner. The figures for the difference column are obtained from $\Sigma(X)$ of alpha minus $\Sigma(X)$ of beta, as shown on Table 7, and divided by $\frac{1}{2}N$, in this case, 20. Thus the first item of the difference column is $-1153/20=57.65$. The calculations involved should be checked by recalculation.

There are several ways to solve a determinant set up in this manner. The one chosen here is probably the simplest and shortest method. First an auxiliary determinant is set up as shown in Table 11. This is accomplished in the following manner. In giving these directions, which are rather complex, we will refer to the "determinant" with numbered and lettered squares as shown in Table 12. In this table, the diag-

TABLE 12

A	1	2	4	7	11	16	I
	B	3	5	8	12	17	II
		C	6	9	13	18	III
			D	10	14	19	IV
				E	15	20	V
					F	21	VI
						G	VII

onal terms are given letters and other terms, arabic numerals. The terms of the difference column are shown by roman numerals.

The diagonal term A of the auxiliary is the square root of the original term ($\sqrt{102873}=320.7382110$). Term "1" of the auxiliary is the original term divided by auxiliary A ($5197/320.7382110=16.20324563$). Auxiliary B is equal to square root of the original B minus the square of auxiliary term 1. Auxiliary C and the rest of the diagonal terms equal the square root of the original diagonal terms minus the sum of the squares of the auxiliary terms in the column above it. Terms 2, 4, 7, 11, 16 and I of the auxiliary equal the original term divided by auxiliary A. Other terms are calculated in the following manner; all terms used in these calculations use auxiliary terms calculated as shown above.

$$\text{Auxiliary} \quad 6=\frac{(2)(4)+(3)(5)}{C},$$

$$\text{Auxiliary} \quad 19=\frac{(4)(16)+(5)(17)+(6)(18)}{D},$$

$$\text{Auxiliary} \quad V=\frac{(7)(\text{I})+(8)(\text{II})+(9)(\text{III})+(10)(\text{IV})}{E}.$$

Again, all figures in these calculations are those of the *auxiliary* determinant. All the values of the auxiliary determinant are checked by recalculation.

The solution of this auxiliary determinant for B values is next accomplished (not to be confused with square B of Table 12). In this solution, we begin at the bottom of the determinant and work up. Again we will use the square lettering and numbering system of Table 12 and referring entirely to the terms of the auxiliary determinant. B_8, the first B to be calculated, equals

$$\text{VII}/G, \left(\frac{0.029338193}{98.46981712}=0.000297940973\right).$$

$$B_7=\frac{\text{VI}-(21)(B_8)}{F},$$

$$B_6=\frac{\text{V}-(20)(B_8)-(15)(B_7)}{E}.$$

The rest of the B's are calculated by the same system. Note that 21, 20, and 15 in the calculation of B_7 and B_6 refer to the terms in squares numbered in that manner in Table 12. The values of these B terms for our problem are shown in Table 13. These B's

TABLE 13

$B_2=0.000719136629$
$B_3=0.000402709216$
$B_4=0.000253196212$
$B_5=0.001081004484$
$B_6=0.000747476560$
$B_7=0.000087825110$
$B_8=0.000297940973$
$\Sigma BD=0.04126785739$

are then multiplied by the corresponding value in the difference column (D) of the *riginal* determinant

$$B_2D_2=(0.000719136629)(57.65)$$

and these successive products are added ($\Sigma(BD)$). This sum is shown in Table 13 also.

The calculation of R^2 is then possible, which equals

$$\frac{N_1N_2}{N_1+N_2}\Sigma BD.$$

N_1 and N_2 are the numbers of specimens in the respective samples. The value of R^2 turns out to be, in this case, 0.4126785739. F is next calculated according to the formula

$$F=\frac{n-p+1}{p}\cdot\frac{R^2}{1-R^2}$$

(n is the number of degrees of freedom; in this type of problem it will be $N_1 + N_2 - 2$; p is the number of characters originally used minus 1. In this particular problem $n = 20 + 20 - 2$ or 38 and $p = 8 - 1$ or 7). Thus F equals 3.212.

The final step in the process is to look up the corresponding F in a table of variance ratios. That of Fisher and Yates in "Statistical Tables for Biological, Agricultural and Medical Research," p. 33, is eminently satisfactory. This table has columns numbered n_1 and rows numbered n_2 ($n_1 = p$, $n_2 = n - p + 1$, both as in the paragraph above). The 1% table should be used for almost all problems. If we look for our particular values in this table ($n_1 = 7$, $n_2 = 38$) we find the consequent to be, by interpolation, 3.284. If F is greater than the value in the table, it means that the two samples would differ as much as they do by chance, less than once in a hundred times. In this case, the F is slightly smaller than the table value, but not greatly. We may in this case, then, conclude that the differences observed in these two samples are probably significant.

Although the example above is taken from paleontology, the method is applicable to any two series of samples of several variables. For example, the method could be used to tell whether the difference in proportions of several different kinds of heavy minerals in two formations was significant or not. Although the method is laborious, it does have the advantage of reducing large masses of data into a usable form.

ACKNOWLEDGEMENTS

The writer is greatly indebted to Dr. K. J. Arnold for his patient help in guiding him through the steps of this statistical technique.

BIBLIOGRAPHY

BARNARD, M. M., 1935, The secular variations of skull characters in four series of Egyptian skulls: Annals of Eugenics, vol. 6, pp. 352–371.

BURMA, B. H., 1948, Studies in quantitative paleontology, I. Some aspects of the theory and practice of quantitative invertebrate paleontology: Jour. Paleontology vol. 22, pp. 725–761.

FISHER, R. A., 1946, Statistical methods for research workers: Tenth edition, Oliver and Boyd, London, pp. 354.

HOTELLING, H., 1931, The generalization of Student's ratio: Annals of Math. Statistics, vol. 2, pp. 360–378.

SIMPSON, G. G., and ROE, ANNE, 1939, Quantitative zoology: McGraw-Hill, New York.

8

EXPERIMENTAL DESIGN IN THE EARTH SCIENCES

W. C. Krumbein

Abstract--Many Earth phenomena involve simultaneous change of several variables, usually not directly under man's control. Isolation of certain phenomena permits controlled laboratory experimentation, but there is also need for study of simultaneous natural variation in several factors, which in this context supplements rather than competes with laboratory experimentation. Statistical design provides a technique adapted to both laboratory experimentation and the observation of the effects of natural variations.

The present paper reviews the application of some statistical methods to the Earth sciences, with special reference to the design of experiments for analysis of variance. Consideration is given to the nature of distributions encountered in Earth science data, to sampling problems, and to some principles of experimental design. Geological examples are used, but parallels are drawn from other Earth science fields. The use of probabilistic models in developing Earth process theories is touched upon briefly.

Introduction--Statistics has a long history of applications in the Earth sciences. On a descriptive level, these include collection of observations on air temperatures, rainfall intensities, stream discharges, sedimentary particle sizes, and many others from practically all fields. The records are condensed in tables and graphs, and by computation of averages and other summarizing numbers. The summarized data serve many purposes in classification and comparison.

Summarizing operations constitute only part of the treatment of observational data. Statistical analysis affords many additional techniques helpful in sharpening subject-matter decisions. A difference between description and analysis is that whereas the observed numbers are used directly in descriptive statistics, in analytical statistics they are considered as a sample drawn from some larger population. Hence, it is appropriate to consider the kind of population involved and to demonstrate that the sample was in fact drawn from such a population. This point is emphasized by COURT [1952]. In analytic statistics, one attempts to make inferences about the population from a study of the sample drawn from it, while in descriptive statistics the aim is to give a concise description of the sample.

Laboratory experiments and field experiments--Application of statistical design in the study of Earth phenomena may be illustrated by comparing a conventional laboratory experiment with a corresponding field statistical experiment. It is known, for example, that average particle size of beach sand varies along and across beaches because of selective effects of waves and currents. This 'sorting process' may be studied in a laboratory wave tank with waves approaching an artificial sand beach at a fixed angle. The sand is selected for its known initial particle size and mineral composition. As the waves wash up on the beach, they select certain particles in terms of size, shape, or density, and move them preferentially along and across the beach. The period and height of the waves are varied in a systematic manner, and the changing properties of the sand along the beach are studied for each stage of the experiment.

The experimental data may be used to set up empirical relations showing the dependence of sand sorting on wave characteristics for the fixed angle of approach. Alternatively, a theoretical relationship based on wave energy or other considerations may be tested by the experiment. Scale-model theory may be involved.

The corresponding field experiment would be conducted on a straight segment of natural sand beach (Fig. 1), recognizing that natural waves vary in height and period, that the angle of approach changes seasonally or during storm cycles, and that the sand itself may have varying initial properties along the beach. These field conditions give rise to a variety of 'errors' which may seem to cast doubt on any results obtained. However, by designing an experiment in which the same variables (wave height, period, direction of approach) are the three main factors, with changing sand properties along the beach as dependent variable, it is possible statistically to separate the variability introduced by each main factor from the total variability, and to compare the individual variations with a combined 'error' term. The method for separating the variations is analysis of variance. Extension of the method to analysis of variance components permits evaluation of the

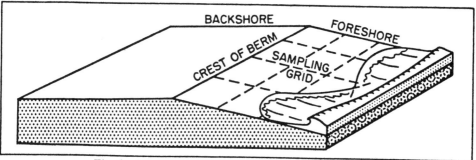

Fig. 1--Diagram of beach segment showing backshore,
berm, and foreshore slope with sampling grid

relative contribution of each main factor to the sorting process. In fact, a mathematical statistical model of the beach, corresponding to the theoretical physical model, could be devised.

Comparison of the two methods for studying beach sorting indicates that they are complementary rather than competitive. The laboratory experiment requires rigid control of certain factors, which in turn requires simplifying assumptions when the theory is applied to natural beaches. The statistical model allows for the natural variations, but runs into the danger that the factors selected may not be the most important among all the factors introduced by nature, so that the somewhat simplified experiment may not be fully decisive. To the extent that the factors selected for study are important, however, a certain measure of predictability for beach processes becomes available. In the ideal case the experiments would permit better prediction of the effects of changes in natural sorting conditions imposed by engineering structures erected on the beach, for example.

The writer is not aware that a rigorously designed beach experiment of this type has been conducted. Sampling design is an essential part of the study, and this in part awaits fuller application of presently growing knowledge of beach particle size, wave, and wave-angle populations.

Although a knowledge of the 'population' or frequency distribution of a variate does not in itself give information on the laws governing a phenomenon, such knowledge is helpful in setting up sampling plans, in selecting experimental models, and in establishing confidence limits for population parameters. As an introduction to later discussion of the effects of different distributions on statistical analysis, some common distribution functions encountered in geological data are considered here.

Geological populations--Table 1 lists six common types of frequency distributions observed in geological data, with examples. The following brief paragraphs indicate the nature of the distributions and furnish references on each type.

The normal (Gaussian) distribution is commonly expressed as

$$y = (1/\sigma \sqrt{2\pi}) \exp\left[-(x-\mu)^2/2\sigma^2\right] \dots\dots\dots\dots\dots\dots (1)$$

where μ is the population mean and σ is the population standard deviation. DIXON and MASSEY [1951] provide an introduction to the properties of the normal distribution.

If the distribution of the variate x is such that ratios instead of differences are the important consideration [HERDAN, 1953, p. 113], the transformation $z = \log x$ may be made. Then if z is normally distributed, its distribution function is

$$y = (1/\sigma_z \sqrt{2\pi}) \exp\left[-(z-\mu_z)^2/2\sigma_z^2\right] \dots\dots\dots\dots\dots (2)$$

and the original variable x is log normally distributed.

The geometry of the normal curve is preserved in (2), which may be referred to as a log-Gaussian distribution. This geometry assures independence of the mean and variance, and permits complete specification of the curve by two z-parameters, which may be expressed in x-terms by taking antilogs.

Table 1--Examples of some distributions observed in geological data

Types of distribution	Example
Normal (or normal approximation)	Natural moisture content of beach sands Beach firmness Particle roundness and sphericity in some sediments Porosity of sediments Particle orientations over narrow range Percentages of abundant minerals in rocks Topographic relief in local areas Angle of slope along valley walls [STRAHLER, 1954]
Log normal	Particle size distribution in many sediments Lengths of first-order streams in drainage basins [STRAHLER, 1954] Thickness of varves [PETTIJOHN, 1949] Permeability of sedimentary rocks [LAW, 1944]
Circular normal	Particle orientations over wide range Orientation of cross-bedding in sands Rock fracture orientation [PINCUS, 1953]
Binomial	Abundant minerals in rocks (number of grains in subsamples of fixed size) Number of frosted grains in some dune sands (in subsamples of fixed size)
Poisson	Rare minerals in rocks (number of grains in subsamples of fixed size)
Gamma distributions[a]	Thickness of bedding in some sediments Percentage of organic content in some sediments Percentage of rarer chemical elements in rocks Sand-shale and clastic ratio in some facies data Particle-roundness and sphericity in some sediments

[a]Gamma distributions are derivable from Pearson Type III distributions by transformation.

The circular normal distribution is defined as

$$\psi (a, a_g, k) = \exp [k \cos (a - a_g)]/2 \pi \, I_0 (k) \quad \dots \dots \dots \dots \dots \quad (3)$$

where a is the angle variate, a_g is the angular co-ordinate of the center of gravity, k is a measure of concentration about the mean, and $I_0 (k)$ is a type of Bessel function. GUMBEL, GREENWOOD, and DURAND [1953] give the theory and furnish tables.

The binomial distribution may be written as

$$P (x) = C_x^n \, p^x (1 - p)^{n-x} \quad \dots \dots \dots \dots \dots \dots \dots \dots \quad (4)$$

where $P (x)$ is the probability of x successes in n trials of an event for which p is the probability of success in a single trial and C_x^n is the combination of n things taken x at a time. The binomial distribution applies to discrete data, but it may be treated by the normal approximation when p does not depart too widely from 0.50.

The Poisson distribution also applies to discrete data when the probability of occurrence is very small. It may be written as

$$y = (m^x e^{-m})/x! \quad \dots \dots \dots \dots \dots \dots \dots \dots \quad (5)$$

where m is the mean value. One characteristic of the Poisson distribution is that its mean and variance are equal. DIXON and MASSEY [1951] describe the binomial and Poisson distributions.

Gamma distributions appear to apply to continuous variates in which the asymmetry is greater than can be normalized by the log transformation. The Gamma function may be expressed in a number of ways; a one-parameter Gamma distribution may be written as

$$y = [1/\Gamma(P + 1)]\, x^P\, e^{-x} \quad \dots\dots\dots\dots\dots\dots\dots (6)$$

where $(P + 1)$ is the mean value, and $x > 0$. In this form the mean and variance are equal, and when $x = (P + 1) = m$, the function has the same value as the corresponding Poisson distribution.

The one-parameter Gamma distribution is skewed, and if $P = 0$, it reduces to the negative exponential $y = e^{-x}$. These are varieties of a more general class of Gamma functions forming a two-parameter family of distributions [MOOD, 1950, p. 112]. In Gamma distributions the mean and variance can be expressed as combinations of both parameters. This is in contrast to the Gaussian distribution where mean and variance are independent.

Other Earth-science distributions--It is instructive to compare the distributions in other Earth sciences with those observed in geology. The Transactions of AGU furnish abundant examples, and Table 2 is based largely on an examination of current and earlier volumes. (To save space in the reference list at the end, the full references are given directly in the Table.)

Table 2--Examples of some distributions from other Earth sciences (references to AGU papers give volume, page, year; others in bibliography)

Types of distribution	Example
Normal	Velocity distribution [KALINSKE, v. 26, p. 261, 1945] Water levels in wells [HUFF, pt. 2, p. 573, 1943] Geothermal gradients [LANDSBERG, v. 27, p. 549, 1946] Cloud drop sizes [suggested by HORTON, v. 29, p. 624, 1948]

Log normal	Hydrologic data [many references; see KIMBALL, pt. 1,
	    p. 460, 1938]
	 Some properties of [KALINSKE, v. 27, p. 709, 1946; also
	    CHOW, v. 33, p. 278, 1952]

Circular normal	Theory [GUMBEL, GREENWOOD, and DURAND, 1953] Angle of wave approach; theory [COURT, 1952]
Binomial	
Poisson	Number of alpha particles emitted per unit time from radioactive rocks
Gamma distributions	Hail-damage frequency [DECKER, v. 33, p. 204, 1952] Height and period of ocean waves [PUTZ, v. 33, p. 685, 1952] Size of raindrops [HORTON, v. 29, p. 624, 1948] Hydrologic data [TODD, v. 34, p. 897, 1953, used log Pearson Type III]
Cube-root-normal	Rainfall data [STIDD, v. 34, p. 31, 1953]
Extreme values	Hydrologic data [GUMBEL, pt. 3, p. 836, 1941] Temperature data [COURT, 1952] . Rainfall data [CHOW, 1953] Ocean wave amplitudes [LONGUET-HIGGINS, 1952]
Rayleigh distribution	Geomagnetic data [BARTELS, 1953] Ocean wave amplitudes [LONGUET-HIGGINS, 1952] Prediction of ocean waves [PIERSON, 1952]

Table 2 includes three distributions not listed in Table 1. The cube-root-normal law is apparently expressible in a cube-root-Gaussian form by expressing the mean and variance in terms of the variate $\sqrt[3]{X}$. The writer knows of no applications of this function in geology. The extreme value distribution is a type of double exponential which can be expressed in linear form by taking a double logarithm of the exponential form: $y = -\log[-\log F(X)]$. The theory was first applied to hydrologic data by Gumbel as indicated in Table 2. It is currently being extended to numerous geophysical fields. The writer is not aware of applications in geology, although the distribution offers possibilities in the study of unusually large boulders sometimes found in gravel deposits, to mention only one example.

The Rayleigh distribution is of basic importance in the study of periodic phenomena. It is a function of the type $y = (2/M^2) r \exp[-(r^2/M^2)]$, where r is a distance and M is a measure related to the spread of the curve. This function apparently has not been applied in geology, but it will be returned to in the closing summary of this paper.

The several frequency distributions described in this section show that many distributions are common to many fields. If a generalization is permitted, it would appear that the several Earth sciences can profit much by mutual discussion of their distribution functions. The need for co-operation with and guidance by mathematical statisticians is also explicit in this generalization.

Analysis of variance--The preceding two sections may seem to be digressions from the main theme of experimental design. As was stated, however, a knowledge of the kind of population from which samples are drawn is helpful in further analysis of the data. Some features of analysis of variance are included here for readers not familiar with the method, as a guide for further reading in such texts as SNEDECOR [1946], COCHRAN and COX [1950], MOOD [1950], DIXON and MASSEY [1951], GOULDEN [1952], and KEMPTHORNE [1952].

In its simplest form analysis of variance compares the variability between groups of data with the variability within the groups. As an example, suppose one were interested in determining whether the firmness of the beach in Figure 1 varied along the crest of the berm. One approach is to randomize a position on the crest, set a stake, and lay off additional stakes at, say, 100-foot intervals. Four or five penetrometer readings randomized about each stake would then supply equally spaced groups of data along the beach crest. Table 3 shows six groups spaced 120 ft apart along the beach at Wilmette, Illinois. Each group has five penetrometer readings.

Table 3--Variation in beach firmness along crest of beach at Wilmette, Illinois (smaller numbers mean greater penetrability), and analysis of variance

Penetration in station						Analysis of variance				
I	II	III	IV	V	VI	Source	Sum of squares	Degrees of freedom	Mean square	F^a
21	31	30	47	52	38	Between groups	788	5	157.60	2.28 NS
52	42	27	38	44	40	Within groups	1660	24	69.17	.....
29	37	30	41	52	25	Total	2448	29	.....	.....
20	51	42	32	35	31					
30	44	46	41	48	39					
152	205	175	199	231	173					

[a]NS - Not significant.

In preparing to analyze these data, a null hypothesis is set up to the effect that there is no significant difference between the group means at the five per cent significance level. The experiment is accordingly designed to test for the occurrence of fixed relations among the group means.

The analysis of variance is given at the right of the data, computed in a manner described in DIXON and MASSEY [1951, p. 121]. The F ratio is computed by taking the ratio of the between-groups mean square to the within-groups mean square. The value 2.28 is less than the critical value of 2.62 for (5,24) degrees of freedom at the five per cent significance level [DIXON and MASSEY, 1951, p. 310] and hence the null hypothesis is not rejected.

One may conclude from this experiment that the beach firmness is homogeneous over the distance sampled, that is, all of the observations could have been drawn from a single firmness population. The mean value of all the observations 37.8 may be taken as an estimate of the population mean, and the within-group variance 69.17 may be taken as an estimate of the population variance. Inasmuch as the penetrometer readings are normally distributed (Table 1), confidence intervals for mean and variance may be readily computed [DIXON and MASSEY, 1951, pp. 74 and 108]. The 95 pct confidence limits for the mean are 34.7 and 40.9. The 90 pct confidence limits for the variance are 45.5 and 119.9.

Table 4--Analysis of variance, single factor basic form
for k groups of n items each[a]

Source	Sum of squares	Degrees of freedom	Mean square
Between groups	$S(C_i^2/n) - CT$	$k - 1$	$SS_b/(k - 1)$
Within groups	By difference	$k(n - 1)$	$SS_w/k(n - 1)$
Total	$S(X_i)^2 - CT$	$kn - 1$	

[a]X_i is a single observation, C_i is a group total, S represents summation, CT is G^2/kn, where G is the grand total, and SS_b and SS_w refer to the between and within sums of squares.

The design used in this experiment is a single factor basic form. Its structure is indicated in Table 4; the notation is in terms of totals instead of means to indicate calculating machine computational procedures. The operation of separating the variability into two parts can be performed upon any sets of numbers, inasmuch as the method is based on algebraic identities [EISENHART, 1947]. The real question is, what use is to be made of the data after the variabilities have been separated? It is here that the concept of analysis of variance models comes in. In the paper referred to, Eisenhart discussed the assumptions underlying analysis of variance. His treatment is used in the following remarks.

The purpose of the firmness experiment was to determine whether there is any significant difference between the means of the firmness populations on the particular beach studied. The parameters thus were population means, and the model belongs in Eisenhart's Class I, used when the purpose is to detect the existence of fixed relations among population means.

In Class I models the estimates of population mean and variance are a form of the method of least squares, inasmuch as analysis of variance solutions are least square solutions. The main contribution of the method here is, according to Eisenhart, the standard form of Table 4, which simplifies the arithmetic and expresses the results clearly.

In contrast to a study of fixed relations among population means, analysis of variance may also be used to detect the existence of components of variance. In the firmness experiment the main interest was in a single beach, yet this beach is only a sample of the much larger population of all beaches. If interest had been focussed on detecting and estimating components of random variation in this larger composite population, emphasis would have to shift from consideration of fixed relations among means to consideration of random deviations in the characteristics of the Wilmette beach from the mean value of these characteristics in the larger population of all beaches.

This second model of the single factor form belongs to Eisenhart's Class II. It requires somewhat different mathematical assumptions, although the arithmetic is the same as in Class I models. A design of somewhat wider applicability, which can also be used for Class I or Class II purposes, is the row-column (two-factor basic) design. It is included here to extend the discussion to the use of grids as illustrated on the foreshore of Figure 1. EISENHART [1947] also uses it as his main example.

Table 5 illustrates the design. The X_{ij} represent single observations in each cell, and the row and column totals are indicated at the margins, a slight modification of Eisenhart's notation. In using this design for Class I problems, the X_{ij} are taken as random variables distributed about true mean values that are fixed constants. This is, if repeated observations are taken from the X_{ij} cell, they would vary at random about some fixed mean m_{ij}, which is the parameter that characterizes the X_{ij}. In using the design for Class II problems, on the other hand, the X_{ij} are taken as random variables distributed about a common mean value $m..$, which is a fixed constant.

The mathematical relations of these models are not discussed here, except to point out that Eisenhart cites certain assumptions for each class of model. The population variance is assumed to be additive (that is, no significant interactions occur); correlations among the variables are

Table 5--Analysis of variance, row-column design
for r rows and c columns, single entry form

| Rows | Columns | | | | | | Row totals |
	1	2		j		c	
1	X_{11}	X_{12}	...	X_{1j}	...	X_{1c}	R_1
2	X_{21}	X_{22}	...	X_{2j}	...	X_{2c}	R_2
	...	...	...	...	...	...	...
i	X_{i1}	X_{i2}	...	X_{ij}	...	X_{ic}	R_i
	...	...	...	...	...	...	...
r	X_{r1}	X_{r2}	...	X_{rj}	...	X_{rc}	R_r
Column totals	C_1	C_2	...	C_j	...	C_c	G

Source	Sum of squares	Degrees of freedom	Mean square[a]	Estimate of
Between rows	$S(R_i^2/c) - CT$	$r - 1$	$SS_R/(r-1)$	$\sigma^2 + c\sigma_R^2$
Between columns	$S(C_j^2/r) - CT$	$c - 1$	$SS_C/(c-1)$	$\sigma^2 + r\sigma_C^2$
Residual	By difference	$(r-1)(c-1)$	$SS_e/(r-1)(c-1)$	σ^2
Total	$SS(X_{ij}^2) - CT$	$rc - 1$		

[a]SS_R, SS_C, and SS_e refer to the row, column, and residual sums of squares, respectively. X_{ij} is a single observation, R_i and C_j are row and column totals, S or SS represents summation, and CT is G^2/rc, where G is the grand total.

assumed absent, and the variances are assumed to be equal (Class I) or homogeneous (Class II). Finally, if the variables as defined for the two models are normally distributed, then all the analysis of variance procedures for testing means and estimating components of variance are strictly valid.

The lower part of Table 5 shows the computations for the row-column design, with the components of variance in the last column. In Class I use, the F ratios are obtained by dividing the row and column mean squares by the residual, and testing for significance. In Class II use, the components of variance may be estimated from the last column. For the row variance component, for example, the residual mean square is subtracted from the row mean square, and the difference is divided by the number of columns.

The analysis of variance designs presented here represent only two of a wide variety available for field and laboratory experiments. The reader is referred to the texts mentioned at the start of this section for further details.

Transformation of data for analysis of variance--Although normalcy of the data has some advantages in analysis of variance, the method itself is 'robust' enough to apply to fairly skewed distributions. COCHRAN [1947] discussed the consequences of failures to satisfy the assumptions of analysis of variance, and BARTLETT [1947] discussed ways of overcoming unsatisfied assumptions. When the data are distributed in certain non-normal ways, certain transformations are of value either for normalizing the data, or for stabilizing the variance. In addition, certain models with multiple cell entries can be used to separate out interaction effects (non-additivity) and to test them for significance.

In order to round out the discussion of distributions, Table 6 lists the six distribution functions of Table 1, and indicates some recommended transformations [BARTLETT, 1947]. Further discussion of the implications of transformations is given in SNEDECOR [1946], KEMPTHORNE [1952], and other texts referred to earlier. The writer is not certain about the need for transformations in circular normal distributions; and for Gamma distributions it may be that a square root transformation is of value for one-parameter cases as against a log transformation for two-parameter cases.

Table 6--Transformations used in analysis of variance

Distribution	Transformation	Remarks
Normal	None	
Log normal	Log X	Normalizes data
Circular normal	None (?)	
Binomial	Arc sine	Stabilizes variance
Poisson	$\sqrt{X}$	Stabilizes variance
Gamma	$\sqrt{X}$ or log X	Stabilizes variance (?)

Sampling designs--Inasmuch as statistical analysis treats a set of observations as a sample from some larger population, the sampling process itself becomes an essential part of experimental design. Only two aspects are touched upon here to continue the main thread of the paper.

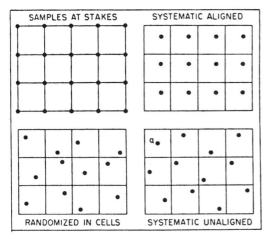

Fig. 2--Four ways of arranging samples in grid, adapted in part from QUENOUILLE [1949]

Figure 2 shows several sample layouts for a row-column grid of the type shown on the foreshore of Figure 1. QUENOUILLE [1949] describes these and other designs and discusses their relative efficiency under different conditions. The upper two diagrams have aligned systematic samples. In the upper left diagram the samples are collected at grid intersections, and in the upper right diagram the sample position was randomized in the first cell. The lower left diagram has its samples randomized in each cell, and in the lower right the samples are systematic but unaligned. For this pattern sample point a is randomized, succeeded by randomization of vertical co-ordinates in the remaining upper row cells, and randomization of horizontal co-ordinates in the remaining first column cells. COCHRAN [1953, p. 183] describes the process. Multiple samples from each cell may be had by repetition of the process.

If the foreshore is homogeneous over the grid, the randomized or aligned systematic samples are satisfactory, but if a population gradient is present, the unaligned systematic samples appear to bring out the gradient more effectively, as Quenouille points out.

The choice of grid cell size, the number of samples per cell, and selection of the randomization process depend to some extent upon the nature of the population being studied. Exploratory sampling plans, designed to bring out the main population characteristics, may be needed. With such exploratory data at hand, or with a background of qualitative knowledge, principles of stratification or systematization can be applied to the sampling problem.

A second approach to sampling, applicable to regional studies, is a hierarchical method in which large cells are successively divided into smaller units, in a manner illustrated in Figure 3, used by POTTER and SIEVER [in press]. The largest squares consist of nine townships each, and each township has 36 sections. Two townships are selected at random from each 'supertownship,' and in each of these two sections are selected at random. Finally, two wells are selected from each of the two selected sections. This method supplies a set of samples in a hierarchy, that is, such that the variance contribution of each level of sampling may be estimated. From such designs the proper spacing of samples for a particular objective (say very local versus broad regional studies) may be set up.

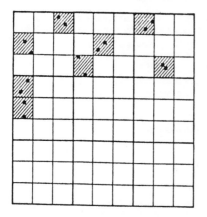

Fig. 3--Hierarchical sampling grid, partially developed; the large cells each consist of nine townships; the small black squares are sections; adapted from POTTER and SIEVER [in press]

A form of hierarchical method was used by YOUDEN and MEHLICH [1937] in a study of soil sampling. Stations were selected a mile apart, substations 1000 ft apart, sampling areas 100 ft apart, and sample points ten feet apart. The data permitted evaluation of the relative efficiency of various sample spacings. It was found that samples ten or 100 ft apart were too close together to constitute an effective method for sampling.

Earth science problems may require collection of data through time, over an area, or both. An example of the wide range of sampling designs available is given by WILM [1943], who described an ingenious design for sampling rainfall with portable equipment. Wilm used a randomized block design which involved stratification of the time period into sub-periods, and stratification of the area into sub-areas. For each time stratum he took two observations at random in each areal stratum with two portable instruments. These were moved about according to plan so that all areal strata were sampled at the end of each time stratum. This experiment gave average values within a few per cent of the values obtained by using ten times as many observations from gages at fixed locations. Wilm's plan has many applications where expensive equipment may be involved, or where it is desirable to allow flexibility in the sampling locations.

Design of experiments--The foregoing discussion of distributions, analysis of variance designs, transformations, and sampling are all related to the broader subject of experimental design. Although these topics were only touched upon, it is hoped that they serve to illustrate the theme of this paper, which is that there can be a sequence of steps in organizing experiments in fields where the number of variables is large, and where relatively few of them can be completely controlled.

Experimental design was first developed in agriculture and biology where control of the variables is limited. As the subject grew it was realized that the same principles apply to controlled laboratory experimentation. Perhaps the main contribution of experimental design was to free experimentation in any field from the somewhat rigid framework of classical laboratory investigation. YOUDEN [1951], in a book addressed to chemists, provides numerous examples of experiments which are improved by explicit recognition of the design elements inherent in them.

The subject of experimental design has received much attention, and the following remarks are based mainly on FISHER [1949], JOHNSON [1949], COCHRAN and COX [1950], YOUDEN [1951], and GOULDEN [1952]. Most writers agree that at least three related principles are involved. First, randomization must be an integral part of the design, to avoid bias. Randomization may be woven through the experiment at several stages, as in sampling, in selection of samples for particular tests, in choice of analysts for part of the work, etc. Many methods are available, from simple coin-tossing to tables of random numbers [DIXON and MASSEY, 1951, p. 290].

Replication assures that the experiment is self-contained, that is, that it provides adequate data for evaluation of its own results. YOUDEN [1951] discusses the importance of replication in some detail.

In many instances the scope of an experiment can be enlarged by permitting simultaneous variation of two or more factors. An example of this is the simultaneous testing of several samples by several analysts using several techniques. Selection of a proper design permits estimation of the variance ascribable to samples, operators, and techniques in a single experiment.

Advances in experimental design, especially in simultaneous variation of several factors, open additional opportunities for parallel field and laboratory investigations in the Earth sciences. These may be particularly appropriate when certain variables can be isolated for laboratory studies while other less controllable factors are concurrently investigated in the field. Examples could be drawn from the study of beach characteristics as developed earlier, in which field experiments including several natural factors supplement detailed laboratory studies of specific controllable variables.

WILM [1952] has recently integrated the principles of design into a pattern of scientific inquiry for applied research, illustrated with problems from forestry. The method is general and has definite applications in other Earth sciences. In essence the pattern includes logical analysis of the variables concerned; exploration of the problem through literature and other sources; experience represented by personal observation; and formal experimentation based on principles of design. The steps follow in sequence in complex problems, but in some inquiries the solution may be obtained at an intermediate stage.

Concluding remarks--The main thread of this paper led to design of experiments based on analysis of variance. This aspect of design is largely concerned with problems of estimation and hypothesis testing. The general method represents one of the most successful branches of analytical statistics.

Although analysis of variance, as applied to the beach experiments, would permit estimation of the relative contributions of several factors in beach processes, the method itself would not directly yield a dynamic theory of the processes. Such a theory would have to be developed by the methods of classical mechanics, or by establishing a statistical analogue of the physical model.

The basis for the statistical approach is that many natural phenomena are subject to random fluctuations, such that no strictly deterministic model can completely predict future events. The statistical model assigns appropriate probabilities to the events and thus makes its predictions on a probabilistic basis, within some selected confidence interval. The investigation of random processes from this point of view has come to be known as the study of stochastic processes.

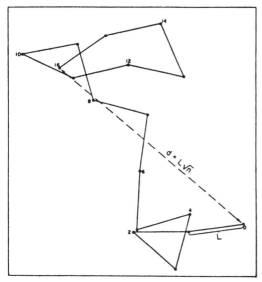

The probabilistic analysis represents an application of the theory of the 'random walk,' which underlies the Rayleigh distribution mentioned earlier. Figure 4 is a diagram of such a random walk, somewhat analogous to Brownian movement. It was prepared from a table of random numbers, as described by BARTELS [1935]. If many such diagrams were drawn, the mean distance of the objects from their initial position would approach the length of a step times the square root of the number of steps.

An example of such an approach to diffusion processes is given by McEWEN [1950]. The displacement of small objects in water from an initial point was regarded as the resultant of a large number of small positive and negative movements following probability laws. The frequency of the resultant displacement and the variance of the positions of the objects were obtained from a limiting form of the binomial series. The underlying theory for the general method is also developed by FELLER [1950].

Fig. 4--Random walk with 16 steps; indicated distance d is expected value when many objects start from same origin

McEwen's statistical development is analogous to the solution of the differential equation for diffusion. The constants of the diffusion equation are equated to the parameters of the statistical model, which thus provides a statistical solution of the physical problem.

The writer is not prepared to discuss stochastic processes in detail. The interested reader is referred to papers by BARTELS [1935], LONGUET-HIGGINS [1952], and PIERSON [1952] for additional examples. Bartels discusses the random walk in some detail and applies the theory to geomagnetic data. Longuet-Higgins used the Rayleigh distribution as a model for ocean wave amplitudes. Pierson applied Gaussian processes to the study of ocean waves, based on a model developed by TUKEY and HAMMING [1949] for analysis of noise in electronic circuits. By using samples of short duration, the assumptions of independent phases during steady state conditions of the sea surface were satisfied, and the analysis became analogous to the random walk problem.

These examples of stochastic processes are touched upon here because they will unquestionably gain in importance in the Earth sciences. The large number of variables in even the simplest Earth process, combined with the apparently random fluctuations which they display, provide a 'natural laboratory' for development of probabilistic models. Analysis of variance, in its usefulness for estimating components of variance and in testing hypotheses, provides an important step toward the ultimate goal of relating observations in nature to their underlying dynamic models of behavior.

References

BARTELS, J., Random fluctuations, persistence, and quasi-persistence in geophysical and cosmical periodicities, Terr. Mag. and Atmos. Elec., v. 40, pp. 1-60, 1935.

BARTLETT, M. S., The use of transformations, Biometrics, v. 3, pp. 39-52, 1947.

CHOW, VEN TE, Frequency analysis of hydrologic data with special application to rainfall intensities, Univ. Ill. Eng. Exp. Sta., Bull. 414, 1953.

COCHRAN, W. G., Some consequences when the assumptions for the analysis of variance are not satisfied, Biometrics, v. 3, pp. 22-38, 1947.

COCHRAN, W. G., Sampling techniques, John Wiley and Sons, 1953.

COCHRAN, W. G., and G. M. COX, Experimental designs, John Wiley and Sons, 1950.

COURT, A., Some new statistical techniques in geophysics, in Advances in Geophysics, v. 1, pp. 75-85, 1952.

DIXON, W. J., and F. J. MASSEY, Introduction to statistical analysis, McGraw-Hill Book Co., 1951.

EISENHART, C., The assumptions underlying the analysis of variance, Biometrics, v. 3, pp. 1-21, 1947.

FELLER, W., An introduction to probability theory and its applications, John Wiley and Sons, v. 1, 1950.

FISHER, R. A., The design of experiments, Oliver and Boyd, London, 5th ed., 1949.

GOULDEN, C. H., Methods of statistical analysis, John Wiley and Sons, 2nd ed., 1952.

GUMBEL, E. J., J. A. GREENWOOD, and D. DURAND, The circular normal distribution, J. Amer. Statist. Assn., J., v. 48, pp. 131-152, 1953.

HERDAN, G., Small particle statistics, Elsevier Pub. Co., Houston, 1953.

JOHNSON, P. O., Statistical methods in research, Prentice-Hall, New York, 1949.

KEMPTHORNE, O., The design and analysis of experiments, John Wiley and Sons, 1952.

LAW, J., A statistical approach to the interstitial heterogeneity of sand reservoirs, Trans. Amer. Inst. Min. Metal.Eng., v. 155, pp. 202-222, 1944.

LONGUET-HIGGINS, J. S., On the statistical distribution of the heights of sea waves, J. Marine Res., v. 11, pp. 245-266, 1952.

McEWEN, G. F., A statistical model of instantaneous point and disk sources with applications to oceanographic observations, Trans. Amer. Geophys. Union, v. 31, pp. 33-46, 1950.

MOOD, A. M., Introduction to the theory of statistics, McGraw-Hill Book Co., 1950.

PETTIJOHN, F. J., Sedimentary rocks, Harper and Brothers, New York, 1949.

PIERSON, W. J., A unified theory for the analysis, propagation, and refraction of storm-generated ocean surface waves, pt. 1, New York Univ., Dept. Met., 1952.

PINCUS, H. J., The analysis of aggregrates of orientation data in the Earth sciences, J. Geology, v. 61, pp. 482-509, 1953.

POTTER, P. E., and R. SIEVER, A comparative study of upper Chester and lower Pennsylvanian stratigraphic variability, J. Geology (in press).

QUENOUILLE, M. H., Problems in plane sampling, An. Math. Statistics, v. 20, pp. 355-375, 1949.

SCHEFFE, H., A method for judging all contrasts in the analysis of variance, Biometrika, v. 40, pp. 87-104, 1953.

SNEDECOR, G. W., Statistical methods, Iowa State College Press, Ames, Iowa, 4th ed., 1946.

STRAHLER, A. N., Statistical analysis in geomorphic research, J. Geology, v. 62, pp. 1-25, 1954.

TUKEY, J. W., and R. W. HAMMING, Measuring noise color 1, Bell Telephone Lab., 1949.

WILM, H. G., Efficient sampling of climatic and related environmental factors, Trans. Amer. Geophys. Union, pt. 1, pp. 208-212, 1943.

WILM, H. G., A pattern of scientific inquiry for applied research, J. Forestry, v. 50, pp. 120-125, 1952.

YOUDEN, W. J., Statistical methods for chemists, John Wiley and Sons, 1951.

YOUDEN, W. J., and A. MEHLICH, Selection of efficient methods for soil sampling, Boyce Thompson Inst., Contributions, v. 9, pp. 59-70, 1937.

Reprinted from *J. Geol.* **69**(6):703–728 (1961)

SEDIMENTATION TIME TREND FUNCTIONS AND THEIR APPLICATION FOR CORRELATION OF SEDIMENTARY DEPOSITS[1]

ANDREW B. VISTELIUS

Laboratory of Aeromethods, Academy of Sciences of the U.S.S.R., Leningrad

ABSTRACT

The method of accentuation of the time trend[2] of sedimentation has been described. This method is based on the arithmetized bed compositions with range numbers and smoothing of these range-number successions by means of a special process. The method can be used for correlation of non-fossiliferous finely stratified sedimentary deposits. The stability of trend function and its peculiarities in different facial environments have been investigated. Trial has been made to give comprehensive examples of the applications of this method for the correlation of sedimentary sections investigated in the Crimea, Caucasus, Turkmenia, and Ural (Orenburg district) in different geological situations. The description and examples are accompanied by computational schemes and practical advice permitting the use of the paper as an instrument for investigation.

INTRODUCTION

The problems studied in connection with (1) generalization of the description of bed sequences; (2) geological mapping; and (3) investigation of sequences of beds exposed by boring which require the creation of methods for the correlation of strata of different exposures and bore holes, as well as a simple method of generalization of the detailed descriptions of bed sequences when series are composed of interchange of thick beds without fossils or typical mineral associations.

Fifteen years of the author's experience show that for the aim mentioned the function approximating the time trend of sedimentation can be used in some cases. This function is very complex and can be applied in two different ways:

a) By accentuation of the local systematic component of compound function and

b) By separation of secular variation in the trend of sedimentation for the approximate regional correlation of sections.

In this paper we shall deal with the method of accentuation of the local peculiarities of the sedimentation trend. The paper embraces principles of the method, calculation

schemes, and some examples of application of the method under different geological conditions.

The data for the paper have been gathered by the author since 1946 and cover several regions of the U.S.S.R. shown on figure 1.

The idea of accentuation of the systematic component of sedimentation was, as far as we know, used first by H. Korn (1938). He smoothed the thicknesses of beds in Upper Devonian–Lower Carboniferous of Thuringia before searching for periodicity by means of Fourier analyses. In 1942 the author applied smoothing in the investigation of porosity distribution in Upper Permian beds of the Buguruslan region (Vistelius, 1944). In this investigation he used the method proposed in a paper by N. N. Michailov on the correction of gravimetrical measurements in the Arctic regions (Michailov, 1940).[3]

A few short papers by A. B. Vistelius on the porosity distribution in the Upper Paleozoic of the Volga River, Tartarian, Bashkirian, and Orenburg regions were published in 1945–51 (Vistelius, 1945, 1946, 1947, 1948, 1948a, 1951). These papers contain results of systematic application of smoothing and approximation of observations by means of trigonometrical series.

---

[1] Manuscript received October 28, 1960.

[2] In compliance with *Webster's New International Dictionary* (2d ed.), we use the word "trend" in the connotation of "general direction taken by something changing or subject to change."

[3] N. N. Michailov's kind help in the work on this paper (Vistelius, 1944) was very valuable.

In 1946 the author began an investigation of the Middle Pliocene of Azerbaidjan using widely autocorrection analyses of stratification, smoothing, and trigonometrical approximation. Some fragments of these investigations were published in the course of 1949–57 (Vistelius, 1949, 1949a, 1950, 1952, 1956, 1957).

M. A. Romanova (1957) published a paper on lithostratigraphy of the Tcheleken red-bed using the methods of smoothing proposed by the present author as a means of mapping of red-bed deposits.

Fig. 1.—Scheme of localization of the sections investigated.

An approximation of the trend in the composition of recent sediments within off-shore regions was made by R. L. Miller (1956) with the aid of orthogonal polynomials. The paper contains valuable examples and general schemes of investigation; its idea corresponds to the idea of secular component separation, which was pointed out above.

The idea used by us can be traced to Dawson's works in Nova Scotia when he wrote about intercalations of sedimentary beds: "I believe

that in each locality these changes succeeded each other in a similar manner and that the great alterations between terrestrial growth and marine deposition extended over very wide areas" (1855, p. 180). The generalized idea of cyclical sedimentation was described by Newberry (1873) in America and Hull (1862) in Britain. The regular character of some cretaceous sections was shown by Gilbert (1895) and discussed in detail for Paleozoic data by Ulrich (1911) in his paper which is still of interest. The papers by Chamberlin (1898, 1909) were close to the subject. All these investigations pointed out the regular character of sedimentation during long periods without concrete determination of bed successions for definite packets of beds. As far as we know the first investigation of a concrete short succession of beds was done by Udden (1912) and Noinsky (1924). These investigations, in subsequent development, led to excellent works by Weller, which were begun in 1926 (Weller, 1931) and still continue. In the U.S.S.R. similar works were produced by Broons (1935) and Forsh (1935, 1940). Many of Weller's papers (Weller, 1930, 1931, 1931a, 1956) and those of his followers (Wanless and Weller, 1932; Jewett, 1933; Wanless, 1939, 1946; Weller, Wanless, Cline, and Stookey, 1942; and many others) shown that there are some rigid successions of beds of definite composition in definite position relative to the beginning of these successions. The appearance of a bed of other composition in these successions can be explained only as an accidental circumstance of sedimentation. Weller's investigations and all conception of cyclical sedimentation incited the idea that any succession of beds has some component of sedimentation connected with a regular process and some component connected with accidental circumstances. In order to check this idea we proceed with the fundamental lines of this paper.

FUNDAMENTAL TOPICS OF THE PROPOSED METHOD

There are many observations on the existence of the trend of sediment composition in sedimentary deposits that can be expressed by means of the equation

$$y = u(t) + \epsilon(t), \qquad (1)$$

where y is the bed composition in the section of sedimentary strata, t is time or thickness, $u(t)$ is the systematic trend component of sedimentation, and $\epsilon(t)$ is the random component of this process.

The mathematical treatment of the problem indicates that $\epsilon(t)$ can be reduced and $u(t)$ may be emphasized. In such a case we shall obtain the values of $u'(t)$ that are close in some sense to $u(t)$. Since $u(t)$ is the systematic trend of sedimentation, it can be used for the reconstruction of the sedimentation history and for numerous other applications.

The separation of component $u(t)$ of equation (1) from the data observed can be achieved by numerous methods. The simplest of these methods is smoothing out, which we shall use for the solution of the problems under consideration in the paper.

The idea of smoothing out is as follows. Let $u(t)$ be the analytical function and $\epsilon(t)$ a random independent component. In such a case $u(t)$ has a smooth succession of finite differences of a definite order connected with the type of function $u(t)$. If we have a succession of observations and as a result of treatment we want to reduce $\epsilon(t)$ and to emphasize $u(t)$, it is possible to find the function $u'(t)$, which will have a smooth succession of finite differences of a necessary order, the values of which are close to the values of $u(t)$. This function $u'(t)$ will approximate the component $u(t)$ in equation (1) and will reduce the component $\epsilon(t)$.

The process of transformation of a series of observations (y) with irregular finite differences into $u'(t)$ with regular ones is termed smoothing or grading. The fit of $u'(t)$ to y is achieved by means of one of the numerous formulae which may be found in many special handbooks (Whitteker and Robinson, 1926; Milne, 1949).

The smoothing requires numeric data and can be applied to the data of analyses, such as determinations of porosity, mineralogical or chemical deposit composition, and so on. If we have a description of a sedimentary section where the rocks are determined in a purely qualitative manner (sand, siltstone, marl, etc.) we must "arithmetize" these rock characteristics. It can be done by ranging the composition of rocks with definite numbers. For example, we can designate the sand by the number 2, the silt by number 1, the clay by 0, and so on. These range numbers may be chosen arbitrarily. When we want to interpret the results of section ranging, it is better to connect the rock range numbers with a natural series of rock types in an ordinary sedimentation environment. Thus, conglomerates are the result of the most extensive denudation process. The conglomerates can be ranged by the biggest number. In such a case sands will have the next range number, siltstone the one following, and so on.

Naturally questions concerning the geological meaning of such "arithmetization" arise. If we are concerned only with the problem of correlation of sections, this permits the use of any numbers for arithmetization. In such a case we receive some functions which reflect some properties of the bed's composition. If these functions can be correlated, our problem is solved independently of the geological sense of the functions themselves. It is important, in such a case, to determine the composition of rocks as precisely as it is possible and to have one and the same number in each case for one and the same rock composition.

In most cases we want not only to correlate sections but to have the possibility of revealing some geological peculiarities from the investigation of sections. In such cases we cannot use arbitrary range numbers but must select them with some geological mean. This problem is very difficult as we do not know the specific values of functions of rock composition which reflect environment of sedimentation. So clastic rocks reflect environment through their granulometric composition, fabric, packing, and so on. When we compare gypsums, carbonaceous rocks, and terrigenous deposits, we lose concrete characteristics. Nevertheless, rocks investigated in the field show us the general character of the dynamics of their accumulation. If we designate the dynamics of rock-building process by y with $m < y < n$ we can build up two series:

$$x_1, x_2, \ldots, x_N$$

and

$$y_1, y_2, \ldots, y_N,$$

where x is the distance from the bottom of the section and y is some value which reflects variation of conditions of sedimentation during accumulation of sediments of the investigated section. These values are principally unknown, and we can define them only from qualitative characteristics.

Let us build up a new series:

$$\varphi(y_1), \varphi(y_2), \ldots, \varphi(y_N)$$

from unknown y. Since we know the qualitative characteristics of y, we can range $\varphi(y)$ in compliance with the qualitative sense of y. So the results of sedimentation under more active conditions will obtain the bigger value of $\varphi(y)$, the less active conditions of sedimentation, the smaller the value of $\varphi(y)$. Thus we obtain monotone function $\varphi(y)$ with

$$\varphi_i(y) \leq \varphi_{i+j}(y)$$

for every i and j.

So, our succession of range numbers is a succession of values of precisely unknown function y after its transformation by means of monotonous function $\varphi(y)$. But monotonous transformation keeps a position of maximum and minimum points of transformed function y (although the values of this function remain unknown). Curve $\varphi(y)$ will have points of maxima and minima with clear geological sense, but intermediate values of $\varphi(y)$ between maxima and minima are distorted by errors of determination of our range numbers. Such a function has some geological meaning and permits its geological interpretation. Of course, results of arithmetical procedures with values of $\varphi(y)$ must be interpreted with caution. Recalling the pure auxiliary mean of $\varphi(y)$, we can work with it within restricted boundaries for incomplete information obtained from observations of rocks.

After the ranging is finished, we draw a curve from co-ordinates (the thickness from the foot of the section [x]; the range number of the rock bed [y]).

The smoothing of (x, y) curves with their transformation to curves in the (x, u') plane, as stated above, can be produced in different ways. The smoothing of curves of sedimentary sections, as we know from many section studies, requires a stable function $u'(t)$, the values of which are influenced by the smallest reflection of the change of smoothing beginning, as well as alteration of the interval length between observations being smoothed. All the methods of smoothing lead to the loss of some observations at the beginning and at the end of the succession being smoothed. When we work on a sedimentary section, the loss of observations, as a rule, is not so important as the stability of the function $u'(t)$. Thus we must take the formulae which warrant the sufficiently stable function $u'(t)$ and accept the fact of the loss of some observations at the beginning and at the end of a series of range numbers of the sedimentary section present.

Practical applications also require a convenient computing scheme for the use of selected formulae.

In our work we have used for some years Spencer's formula:

$$\begin{aligned}
u'_0 = \tfrac{1}{350} [&60u_0 + 57(u_{-1} + u_{+1}) \\
&+ 47(u_{-2} + u_{+2}) + 33(u_{-3} + u_{+3}) \\
&+ 18(u_{-4} + u_{+4}) + 6(u_{-5} + u_{+5}) \\
&- 2(u_{-6} + u_{+6}) - 5(u_{-7} + u_{+7}) \\
&- 5(u_{-8} + u_{+8}) - 3(u_{-9} + u_{+9}) \\
&- (u_{-10} + u_{+10})],
\end{aligned} \tag{2}$$

involving 21 terms. Thus, in applying equation (2), we lose ten observations at the beginning and at the end of the series under investigation.

The loss of ten observations at the beginning and at the end of an investigated series can be accepted when a length of interval of smoothing (x_i, x_{i+1}) is short compared with the length of a whole series. Such a situation is typical for well-exposed mountain outcrops. Investigation of short series (small outcrops) makes use of other smoothing formulae. In our previous investigations of porosity distributions (Vistelius, 1944, 1945, 1946, 1947, 1948), we applied Sheppard's formula (Whitteker and Robinson, 1926):

$$u = \tfrac{1}{35}\,[\,17u_0 + 12\,(u_1 + u_{-1})$$
$$- 3\,(u_2 + u_{-2})\,]\,,$$

which gave rather good results. There are some formulas without loss of observations (Whitteker and Robinson, 1926), but these have not been checked with geological experience yet. As range numbers have a big variation (for instance figures in column 1, table 2), the advantage of formula (2) is obvious.

The application of (2) will be understood better by examining an example.

Example: Let us investigate a bed succession taken from the tertiary red-beds of the Tcheleken Peninsula (the Caspian Sea) and listed in table 1.

Table 1 shows that rock-ranging was done as shown in the accompaning tabulation.

Clays.......................	1.4
Arenaceous clays............	5.6
Argillaceous silt.............	8.4
Silt........................	15.4
Argillaceous sand............	20.3
Fine-grained sand............	22.4
Small-grained sand...........	29.4
Middle-grained sand.........	35.0
Coarse-grained sand.........	42.0
Conglomerates..............	49.0

The range numbers were chosen in compliance with the granulometric composition of rocks or the general type of sedimentation activity. So, the expanded Tcheleken red-beds sands had median diameter about 0.09 mm.; the most coarse sands ⌣ 0.12; silts about 0.04 mm.; and clays about 0.003 mm. There are only intraformational conglomerates with argillaceous pebbles buried in sands being incorporated from underlying beds and deposited nearly contemporaneously with the original beds. We believe that the activity of the sedimentation process forming these conglomerates was not much more than activity of sand sedimentation.

Now let us take the median diameter of clays as a unit and divide all other diameters by it. We will obtain figures as follow: 1 for clays, 13 for silts, 30 for common sands, 40 for the coarsest sands of the investigated section. From computational considerations let us select also figures easily divided by seven. Thus we obtain the scale of range numbers listed above. Intermediate range numbers, we find, take into consideration specific features of suitable rocks.

The range number of our conglomerates is determined as rather more than the range number of the coarsest sands and easily divisible by seven, as is required by the computational scheme of u'_x listed in table 2.

It is necessary to point out that our scale of coarseness of sands is not usual but specially fitted to the investigated sediments.

The calculations of u'_x by means of formula (2) and computational scheme of table 2 are as follows:

The first column (without numbers) is occupied by the number of the points of ob-

TABLE 1

Description of Part of Bed Section from Tcheleken Peninsula*

No. Beds from the Foot	Bed Compositions	Thickness of Beds (Cm.)	No. Observation Points	Range Numbers of Beds
281........	Brick-red clay	103	276–79	1.4
282........	Silt	15	276–80	15.4
283........	Brick-red clay	12	276–81	1.4
284........	Argillaceous silt	39	276–82	8.4
285........	Brick-red clay	33	276–83	1.4
286........	Fine-grained sand	45	284–85	22.4
287........	Brick-red clay	58	286–88	1.4
288........	Arenaceous clay	26	286–89	5.6
289........	Fine-grained sand	67	290–91	22.4
290........	Brick-red clay	44	292–93	1.4
291........	Fine-grained sand	51	294–95	22.4
292........	Brick-red clay	308	296–307	1.4

* Points of observation are selected at 0.25 m. intervals.

servations (*x*). They are taken over equal intervals of the thickness of a section from the bottom of the section. Column 1 contains the range numbers of the composition of beds in the points of observations. It is raw material of observation which must be treated by Spencer's process. The lithological sense of the figures in this column was given above. Columns with numbers from 3 up to 9 show what arithmetical procedure we must follow to obtain smoothed values

column 1 by smooth curve of the function u'_x in column 10. It is important that we were able to make such substitution without any *restriction of a type of smoothing function*, excluding some conditions for finite differences.

STABILITY OF TREND FUNCTION

Application of trend method requires stability of u'_x function. In other words, if we smooth one and the same range number succession many times through different inter-

TABLE 2

AN EXAMPLE OF THE WORKING PROCESS OF SPENCER'S 21-TERM FORMULA*

No. Points from Foot of Section	Un-smoothed Observations $u(x)+\epsilon$ (1)	Divided by $7=u$ (2)	Sum Col. 2 in 3's (3)	(2)+(3) (4)	$u_{-3}+u_3$ (5)	(4)−(5) (6)	Sum Col. 6 in 7's (7)	Divided by 5 (8)	Sum Col. 8 in 5's (9)	Smoothed Values u'_x Sum in 5's Sum in 5's Cut Down as Far as Necessary (10)
276	1.4	0.2	...	....	...	......	......	...	....	....
277	1.4	0.2	0.6	0.8	...	......	......	...	....	....
278	1.4	0.2	0.6	0.8	0.4	0.4	......	...	....	....
279	1.4	0.2	2.6	2.8	1.4	1.4	......	...	....	....
280	15.4	2.2	2.6	4.8	0.4	4.4	......	...	....	....
281	1.4	0.2	3.6	3.8	3.4	0.4	17.8	3.6	....	....
282	8.4	1.2	1.6	2.8	3.4	− 0.6	25.8	5.2	....	....
283	1.4	0.2	4.6	4.8	2.4	2.4	27.2	5.4	20.0	....
284	22.4	3.2	6.6	9.8	0.8	9.4	17.2	3.4	19.7	....
285	22.4	3.2	6.6	9.8	1.4	8.4	11.8	2.4	19.3	9.4
286	1.4	0.2	3.6	3.8	1.0	2.8	16.6	3.3	18.1	9.1
287	1.4	0.2	0.6	0.8	6.4	− 5.6	24.2	4.8	17.3	9.0
288	1.4	0.2	1.2	1.4	6.4	− 5.0	21.2	4.2	17.0	9.1
289	5.6	0.8	4.2	5.0	0.4	4.6	13.0	2.6	18.0	9.6
290	22.4	3.2	7.2	10.4	0.4	10.0	10.6	2.1	20.4	10.4
291	22.4	3.2	6.6	9.8	3.4	6.4	21.6	4.3	23.2	11.2
292	1.4	0.2	3.6	3.8	4.0	− 0.2	36.0	7.2	25.0	11.8
293	1.4	0.2	3.6	3.8	3.4	0.4	34.8	7.0	25.5	11.7
294	22.4	3.2	6.6	9.8	3.4	5.4	22.2	4.4	24.0	10.8
295	22.4	3.2	6.6	9.8	0.4	9.4	13.2	2.6	19.6	9.4
296	1.4	0.2	3.6	3.8	0.4	3.4	13.8	2.8	14.4	....
297	1.4	0.2	0.6	0.8	3.4	− 2.6	13.8	2.8	10.0	....
298	1.4	0.2	0.6	0.8	3.4	− 2.6	8.8	1.8	....	....
299	1.4	0.2	0.6	0.8	0.4	0.4	− 0.2	0.0	....	....
300	1.4	0.2	0.6	0.8	0.4	0.4	......	...	....	....
301	1.4	0.2	0.6	0.8	0.4	0.4	......	...	....	....
302	1.4	0.2	0.6	0.8	0.4	0.4	......	...	....	....
303	1.4	0.2	0.6	0.8	...	......	......	...	....	....
304	1.4	0.2	...	....	...	......	......	...	....	....

* See equation (2), p. 706. After Whitteker and Robinson (1926).

u'_x from raw data in column 1. Thus column 2 shows that values of numbers in column 1 must be divided by 7. Figure 0.6 at the beginning of column 3 is obtained from adding the first three figures in column 2. The comparison of the titles of the columns with figures in these columns shows the way to computate these values. Column 10 contains smoothed values (u'_x) which substitute for the raw values u'_x in column 1. It is obviously very inconvenient for smoothing discrete values in column 1, to substitute in column 10 the rather regular succession of figures u'_x. We replaced stepped function in

vals of smoothing with different origins, we must obtain each time the value of u', which lies in an interval

$$u'_x - \epsilon < u'_x < u'_x + \epsilon,$$

where ϵ is small. At present we cannot precisely determine what is "small" and must take it by intuition. It is also obvious that for different problems, ϵ can be different.

At present we do not have any general theory for solution of this question, but we have tried to resolve it experimentally as far as it is possible. For this purpose we have made some experiments with the sections of sedimentary deposits taken from different types of deposits. The first series of sections were taken from the Tcheleken and Apsheron peninsulas and were composed of deposits in a brackish lagoon. Second series dealt with sections from the Taurian suite in Crimea—Flysch-like turbidities of open-sea origin. The third type of section was obtained from slaty series of Jurassic deposits of Eastern Caucasus. These deposits contain traces of coal measures.

Experiment 1.—The range numbers of the Tcheleken red-bed section (general thickness is 24.7 m., average [$\bar{x}$] thickness of bed is 62.1 cm., standard deviation [s] is equal to 110.7 cm., and number of beds [n] is equal to 40), described in table 3 are smoothed five times under different conditions. The curves obtained are plotted on figure 2.

The first smoothing was produced with 25-cm. intervals of smoothing without averaging range numbers within intervals; the beginning of smoothing is in point $x = 0$ (curve I). Second smoothing performed with 25-cm. intervals of smoothing, averaging range numbers within intervals (curve II). The third smoothing (curve III) produced as the first one but the beginning of smoothing taken at point $x = 12.5$. The first and fifth smoothings were carried out with 50-cm. intervals of smoothing (without averaging within intervals) with the beginning of smoothing at $x = 0$ (curve IV) and at $x = 25$ cm. (fifth smoothing, curve V).

Experiment 2.—Range numbers of the description of the section of Middle Pliocene composed of 400 beds in Bastanar-Shore saline in Azerbaidjan were twice smoothed. The interval between points is 1 m., the average thickness of beds 22.6 cm., their standard 70.5 cm. The first smoothing was made after averaging the range number within intervals of smoothing. The second smoothing was obtained directly from the range numbers observed (without their averaging within intervals). Results of experiments are plotted in figure 3. Similarity between curves is preserved.

Experiment 3.—The range numbers of red-bed section located on Tcheleken Peninsula composed of 220 beds (average bed thickness is 77.0 cm. [its standard equals 125.7 cm.]) were twice smoothed. The first smoothing is obtained with 25-cm. intervals between the points of observation. The second smoothing is performed with 250-cm. intervals between the points of

observation. No averaging within the intervals was made. The experimental data are plotted in figure 4. There seems to be some similarity between the curve configurations.

Experiment 4.—Near the village of Lutchistoe in the vicinity of Alushta (Southern Crimea) the deposits of the Taurian suite belonging to the Lower Jurassic–Upper Triassic age occur. These deposits consist of thin beds intercalated with sands, silts, and argillites. The investigated section has a 15-meter thickness and is composed of 72 beds, with an average bed thickness of 22.4 cm. and with a standard of 27.2 cm. It is characterized by a relatively wide occurrence

TABLE 3

SHORT DESCRIPTION OF BED SUCCESSION OF
TCHELEKEN RED-BEDS SMOOTHED
FIVE TIMES*

No. of Beds from Foot of Section	Bed Composition	Range Number	Thickness (Cm.)
270....	Clay	1.4	40
271....	Arenaceous clay	5.6	6
272....	Argillaceous silt	8.4	12
273....	Arenaceous clay	5.6	139
274....	Fine sand	22.4	52
275....	Clay	1.4	103
275....	Fine sand	22.4	351
277....	Arenaceous clay	5.6	88
278....	Sand	29.4	222
279....	Sand	29.4	170
290....	Clay	1.4	37
281....	Sand	29.4	38
282....	Argillaceous silt	5.6	7
283....	Arenaceous clay	5.6	19
284....	Sand	29.4	41
285....	Arenaceous clay	5.6	44
286....	Argillaceous silt	8.4	13
287....	Arenaceous clay	5.6	64
288....	Silt	15.4	2
289....	Clay	1.4	76
290....	Silt	15.4	27
291....	Arenaceous clay	5.6	65
292....	Argillaceous silt	8.4	7
293....	Arenaceous clay	5.6	18
294....	Argillaceous silt	8.4	16
295....	Fine sand	29.4	37
296....	Clay	1.4	23
297....	Argillaceous silt	8.4	4
298....	Clay	1.4	50
299....	Silt	15.4	20
300....	Sand	29.4	87
301....	Clay	1.4	1
302....	Silt	15.4	10
303....	Arenaceous clay	5.6	91
304....	Sand	29.4	173
305....	Argillaceous clay	8.4	6
306....	Clay	1.4	45
307....	Silt	15.4	15
308....	Sand	29.4	49
309....	Clay	1.4	206

* See Experiment 1.

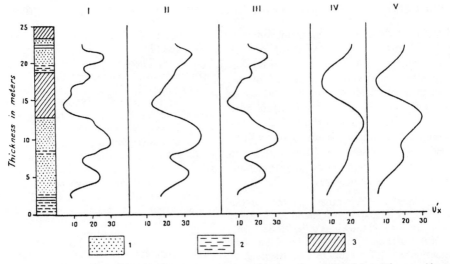

FIG. 2.—Five smoothings of the range numbers of the red-bed section of Tcheleken Pliocene. Along the horizontal axes there are range numbers. *1* = sands; *2* = silts; *3* = intercalation of beds with thickness of a single bed less than 0.5. Details in text.

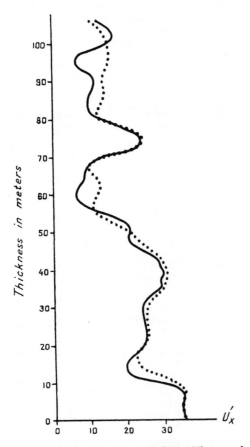

FIG. 3.—u'_x curves of the Middle Pliocene of Apsheron Peninsula. Dots are u'_x curve obtained by averaging of range numbers within intervals of smoothing, Δ; the thick line is the same curve obtained without averaging.

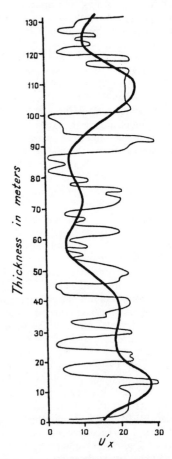

FIG. 4.—Two smoothings of range numbers of the same section. The thick line is the curve obtained with $\Delta = 2.00$ m.; thin line is the curve obtained with $\Delta = 0.25$ m.

76

of argillites and thin layers of sandstones and siltstones. We could observe no time trend in the composition of beds.

The range numbers of the section were smoothed six times, and the curves obtained were plotted in figure 5. The first smoothing was carried out with 10-cm. intervals of grading and at the beginning of the smoothing $x = 0$; the second smoothing was performed with the same interval of smoothing as in the first case but with the averaging of bed composition within the intervals; the third and fourth

ard 8.0 cm. The results of smoothing are plotted in figure 6. There is a stable rise in the middle part of every curve. The smoothed curve is stable and has a typical rise for every grading.

Experiment 6.—We have studied the Lower Jurassic deposits to the south of the village Gunib in Daghestan (Caucasus), where we found some shales, sandstones, and silts with thin lenses of coal. The thickness of the section investigated was 52.2 m. It was composed of 526 beds with an average thickness of the bed equalling 10.6 cm., and the standard 12.6 cm.

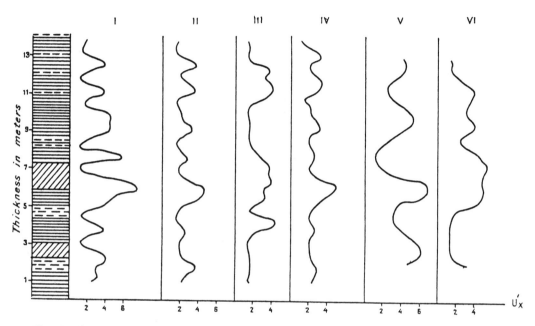

FIG. 5.—Six smoothings of the range numbers of a section of Taurian suite near village Lutchistoe. Parallel lines are clays. Other designations as in fig. 2.

smoothings, were made with an interval length equal to 10 cm. but with the beginning of an adjustment from $x = 5$ cm. and with an averaging within intervals for the fourth smoothing. In the next computations the interval of smoothings was taken equal to 20 cm., with the beginning of smoothing at $x = 0$ (the fifth computation) and at $x = 10$ cm. (the sixth computation).

Experiment 5.—The section of Taurian suite was described by G. I. Sokratov (1955) and accepted as typical for the Taurian suite by some geologists. We smoothed the sequence of beds reported by G. I. Sokratov, and, having obtained the thickness equal to 9.32 m., the average thickness of the bed is 6.9 cm. and the standard

The section investigated occupied a part of a thick Jurassic formation in the lower part (where the section was located), of which no trend of sediment composition was observed. The arithmetized sequence of beds was smoothed nine times with different intervals and different beginnings of smoothing. The results are plotted in figure 7. The investigation of the curves listed indicates that there is some rise of u'_x rather stable under different conditions of smoothing.

The sequential smoothing of different sections under different conditions shows:

a) The general features of curves obtained are stable.

b) Details of configurations of obtained

curves are connected with the length of the smoothing intervals.

c) The boundaries of thick beds have some different phases on the curves obtained under different conditions.

d) The effect of averaging within smoothing intervals is rather small. Averaging is not necessary in routine work.

Experiment 7.—This experiment was done to check mistakes of different operators determining the composition of one and the same rocks being used for construction of trend function. The red-beds of Tcheleken have badly marked boundaries between beds. The bed composition has many subtle variations which it was very difficult to determine unequivocally. One section of red-beds was described three times by different operators (for the first time in 1947; the second time in 1953; and the third in 1955) who previously had worked on the investigation of different deposits (metamorphosed Paleozoic of Kasakhstan; Tertiary deposits of the Caucasus; and conglomerates of Middle Asia). The beginning of the smoothing

of the section under investigation was also a little different. The smoothing was performed with 25-cm. intervals between the points of observations. The results are plotted on figure 8, which shows that curves u_z', have preserved many similar features, although they have some operator variations.

Experiment 8.—Stability of a trend function along the strike of a packet of beds was checked. For this purpose the packet of clay and sand beds between two thick sand beds was used. This packet was exposed by seven ditches, and afterward it was photographed from the air. Exactness of correlation of underlying and covering beds, which is very important, can be checked by the reader on plate 1. The average thicknesses of beds, standards, and number of beds for every section described along the ditches are listed in table 4.

The lower, middle, and upper parts of the outcropped sections have beds of sands which are traced throughout all the sections. The parts between the sand layers were composed of unstable thin-stratified lamellae of brick-red clays,

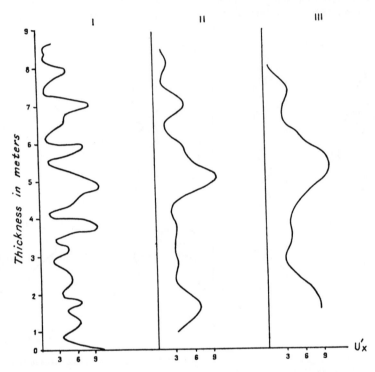

FIG. 6.—Three smoothings of the range numbers of a typical section of Taurian suite in the Crimea. $I - \Delta = 5$ cm.; $II - \Delta = 10$ cm.; $III - \Delta = 15$ cm.

silts, and sands. The composition of beds in every section was arithmetized and graded. The results are plotted on figure 9. On each curve there is a certain stable rise of u'_x which can be sufficiently identified and compared with the others. So the composition of unstable lamellae have some trends which can be used for the correlation of sections, e.g., $x \ldots x. \ldots$

The successions of range numbers obtained from the sections of sedimentary deposits were repeatedly smoothed (graded) by means of a special process. The repeated smoothing was done with sections of different deposits investigated in different regions.

The smoothing was produced with different starting points of smoothing and with different interval length between the points.

We also smoothed some sections exposing one unstable state along the strike.

The results of the smoothing of the repeated descriptions of one and the same section by different operators were obtained.

The stability of the trend function is somewhat different for different sections. The red-bed and Middle Pliocene sections with the general trend of sedimentation have more stable trend functions. The sections of Jurassic deposits with a very slight trend of sedimentation give trend functions of a smaller stability.

It seems that the interval of smoothing should be close to an average thickness of the bed of the section treated.

In general the results of investigation indicate that the smoothing function u'_x was rather stable for each section or group of sections. In other words we shall obtain curves rather alike after smoothing one and the same bed sequence under different conditions of smoothing. So the function u'_x obtained by equation (2) can be used for the generalization of the sedimentary section description and as a generalized function of the sections for the correlation of strata among different exposures or bore holes. The last conclusion will be checked in one of the following sections.

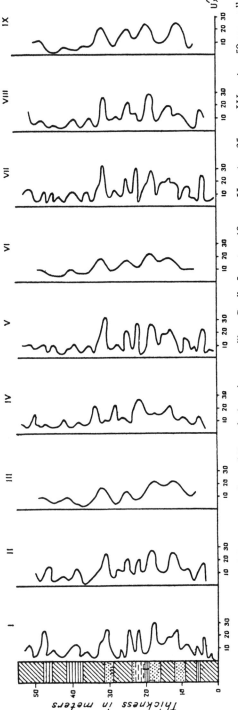

Fig. 7.—Nine smoothings of the range numbers of Jurassic deposits near village Gunib. $I - \Delta = 10$ cm.; $II - \Delta = 25$ cm.; $III - \Delta = 50$ cm.—all without averaging within intervals. $IV - \Delta = 25$ cm.; $V - \Delta = 10$ cm.; $VI - \Delta = 50$ cm.—all with averaging within intervals; $VII - \Delta = 10$ cm., beginning of smoothing at 5 cm. from the foot of the section; $VIII - \Delta = 25$ cm. beginning of smoothing at 12.5 cm. from the foot of the section; $IX - \Delta = 50$ cm., beginning of smoothing at 25 cm. from the foot of the section.

ORGANIZATION OF FIELD AND
COMPUTATIONAL WORK

The success of trend-method applications is partly connected with the possibility of making speedy computations of trend-function u'_x. These computations can be performed with the greatest speed if the field work as well as the computations are organized according to a special scheme. We suggest some organizational advice derived from the application of the trend method during a few years work.

It has already been shown that this meth-

od requires the detailed description of bed successions. Thus the field work would be successful if we could describe a sufficient number of beds every day. These descriptions may be short, but they should contain the most precise determination of thickness of bed and its rock composition. In such a case it is very convenient to make the description of bed sequences on a special form. The form is given in table 5.

The left side of the form is filled out in the field. The right side of it is kept blank and is filled out in the course of the treatment of the

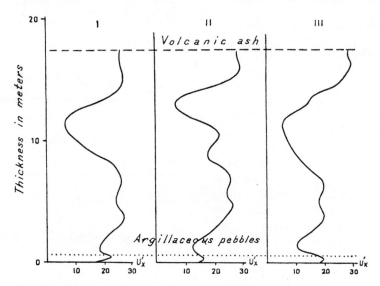

FIG. 8.—Results of smoothing of the same section by three different operators

TABLE 4

AVERAGES ($\bar{x}$), STANDARDS (s), AND NUMBERS OF BEDS (n) IN
SEVEN PARALLEL SMOOTHED SECTIONS

$\bar{x}$......	17.6	16.1	16.5	17.8	17.3	18.9	18.8
s......	39.2	37.3	39.4	40.7	40.2	39.1	39.9
n......	82	87	89	86	92	85	87

PLATE 1

A, Relative position of the sections being investigated. Straight lines are the ditches along which the sections were described.

B, Aerophoto outcrops of beds being compared. *Dots* = sands: *circles* = clay conglomerates; *straight lines* = ditches along which the descriptions of sections were done.

C, General view of exposures of bed sections II and *K*II being correlated. Broken lines mark the main faults.

D, Aerophoto exposures of sections *P3* and *P*. Only the main faults are deciphered.

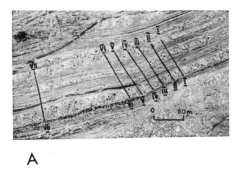

A

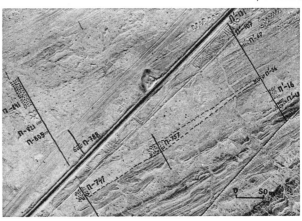

B

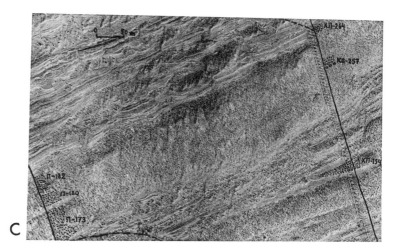

C

D

Aerophotos of beds and section locations

data. The column "bed composition" is filled out by means of special signs for rock composition. We used the following signs:

Clay γ
Silt a
Sand π
Crossbedding   ///
Induration   —
Fine grains   —
Small grains   <
Middle grains   ×
Coarse grains   >
Interbedding   ∫
Lensels   ⊖

If these signs do not cover all the types of rocks investigated, their number may be easily enlarged. These signs are simple in writing and can be easily kept in mind.

The complex description of bed is given in the following order (a) the noun, (b) the adjectives. Thus the sentence "Fine-grained argillaceous sandstone with cross lamination" can be designated as $\overline{\pi}\gamma$///.

The application of symbolic description saves much time (raising the productivity of labor). We have described nearly 900 beds during one working day in well-exposed regions with the aid of such a symbolic system.

All the descriptions of beds are ranged and smoothed out.

This work requires standardized calculations. The computation schemes (as given in table 2), the auxiliary tables and stencils with silts for

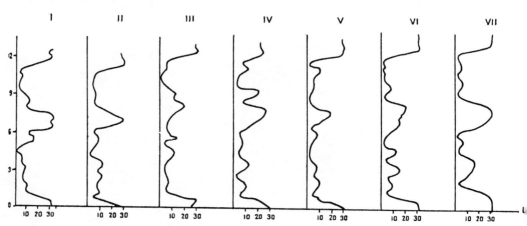

FIG. 9.—Seven u'_x curves taken along the strike of the same sequence of beds

TABLE 5

FORM OF STANDARDIZED DESCRIPTION OF THE SECTION*

| | | | | This Side Filled in Process of Treatment | | |
No. of Beds from Origin	This Side Filled in the Field	Thickness Measured (Cm.)	True Thickness (Cm.)	Cumulated Thickness from the Beginning of Section to Boundary of a Bed (Cm.)	No. of Observations (x) from Beginning of Section	Range Number of Bed
	Description of Beds†					
281	Brid-red clay (γkk)	492	103	7372–7475	276–279	1.4
282	Silt (a)	70	15	7475–7490	276–280	15.4
283	Brick-red clay (γkk)	58	12	7490–7502	276–281	1.4
284	Argillaceous silt (aγ)	185	39	7502–7541	276–282	8.4
285	Brick-red clay (γkk)	159	33	7541–7574	276–283	1.4
286	Fine-grained sand (π̄)	214	45	7574–7619	284–285	22.4
287	Brick-red clay (γkk)	274	58	7619–7677	285–287	1.4
288	Arenaceous clay (γa)	126	26	7677–7703	285–289	5.6
289	Fine-grained sand (π̄)	320	67	7703–7770	290–291	22.4
290	Brick-red clay (γkk)	210	44	7770–7814	292–293	1.4
291	Fine-grained sand (π̄)	243	51	7814–7865	294–295	22.4
292	Brick-red clay (γkk)	1467	308	7865–8173	296–307	1.4

* It occupies the right page of diary; left pages of diary are used for comments (angle of dip, peculiarities of the texture of beds, etc.).

† Symbolic designation of rock composition shown in parentheses.

separation of the summarizing figures (in columns 3, 5, 7, 9, and 10 of table 2) must be widely used. If we deal with thin beds and with intervals between the points being smoothed (the case of averaging) more than the thicknesses of the beds (Example 2, case 2, p. 709), the tables of the products of the thicknesses multiplied by the range numbers are very useful. Such tables should be computed at the beginning of the work.

The method described was used during several years under expeditional conditions. The work was organized so that one group described the beds and another group calculated the trend functions and correlated the bed. We used the

zon under very complicated conditions of uniform intercalation of thin laminae. The correlation of Apsheron Pliocene deposits demonstrates the application of trend functions for the sake of comparison of deposits not distorted by faults, well exposed over great distances. The Tcheleken example reveals a complete process of restoration and correlation of sections of uniform deposits over short distances under conditions of exceedingly wide occurrence of faults. The Permian deposits were investigated by means of trend functions after being bored under usual conditions.

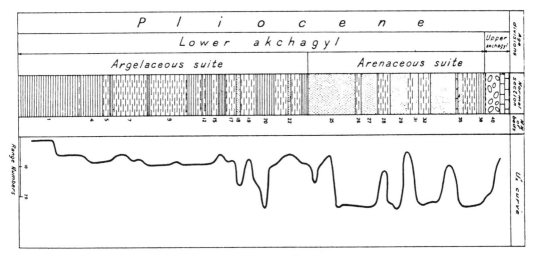

FIG. 10.—Normal section and its u'_x curve (Sulmen region)

simplest computation equipment and worked with auxiliary tables, stencils, and so on. Under such conditions one operator smoothed nearly 600 points per day. Thus computations did not detain the investigations.

EXAMPLES OF USING THE TREND METHOD

The proposed method applications are given below. The following examples are given for different conditions. Thus the generalization of the description of Sulmen Pliocene section illustrates a very simple case of using the trend function with a big bios of the curve rise relative to the bed boundaries. The generalization of the Potaskuevo section shows the separation of hori-

It is necessary to point out that all our examples contain trend curves and final conclusions about their comparison only. We omit all discussion of reasons of correlation among different parts of the curves being compared. We understand that such a position is most unfavorable for our correlations. So if attentive readers find our correlation right under such unfavorable conditions, it will show that our method will prove better than is claimed in our paper. At the same time, we tried to make clear all steps of the correlation and documented as carefully as possible every transition from one bed to another for sections being restored. Working with such complicated deposits as Tcheleken red-beds, we used aerophoto for precise documentation of such transitions. It made our transitions more clear for geologists accustomed to aerophotos but unaccustomed to curves and functions.

TERRIGENOUS SEDIMENTS OF SULMEN COUNTY (KRASNOVODSK PLATEAU, NORTHWESTERN TURKMENIA)

The simplest application of the trend function is its use for the generalization of the description of thin stratified deposits. In such a case the trend functions can be used for definition of the boundaries between horizons or marking suites.

The section of thin, stratified terrigenous sediments is exposed on a precipice located to the south of Sulmen village. In 1956 it was necessary to investigate the possibility of separating suites in the sequence of beds exposed in the precipice. The solution was found in the following way: A description of section was made, the bed compositions were ranged and smoothed out with 10-cm. intervals of smoothing. In figure 10 a geological column of the section described and the u'_x curve are given. The u'_x curve gives a graphic demonstration of the possibility of separating two suites—the lower argillaceous suite and the upper arenaceous one. The boundary between suites can be marked at the foot of the sand bed N. 25. There is a big bios of the phase of the lower boundary of sand bed to the u'_x curve.

PERMIAN DEPOSITS OF POTASKUEVO VILLAGE

The Lower Permian deposits of the North Ural Mountains are composed of different thin-layered terrigenous deposits: conglomerates, sands, silts, and clays. In many cases it would be useful to separate some series of beds for the separation of suite boundaries or marking surfaces. It is a difficult problem for thin-bedded strata.

The section of Lower Permian deposits was described near Potaskuevo village, located in the middle stream of the Vishera River of the North Urals. The general view of the section does not reveal any specific feature for separating some marking series of beds (fig. 11). The arithmetizing of rock determinations in Potaskuevo sequence of

beds and the smoothing out of the sequence of range numbers indicate that there is one series of beds with small variation of smoothing-function figures. This rather stable segment of smoothing curve can be checked as a marking series (fig. 11).

MIDDLE PLIOCENE OF THE APSHERON PENINSULA

The Apsheron Peninsula is located in the southwestern part of the Caspian Sea. The sedimentary deposits of the peninsula are composed of Tertiary beds with a series of great thickness formed by intercalation of

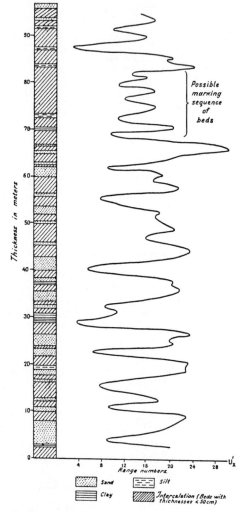

FIG. 11.—Normal section and its u'_x curve (Potaskuevo region).

sand and clay layers with very rare conglomerate beds. This series rests upon the Pontian Stage and is covered with the Akchagylian Stage of Middle Pliocene age.

The Middle Pliocene series is subdivided into the lower and the upper division separated by a nonconformity. The upper division of the Middle Pliocene, as a result of many years' labor of numerous geologists, is divided into Balachansky, Sabunchinsky, and Surachansky suites.

The beds of the upper division are well exposed; the deposit thickness is great; they are composed of a terrigenous thin-layered sequence with a marked suite boundary, the above being good material for the checking of trend correlation.

In 1946–1948 the author described three sections of the upper division; the first one

geologists. Figure 12 shows an interrelation between the suite boundaries and the peculiarities of the trend curve u_x'. Thus the trend curve u_x' represents a useful means for section correlation in the particular case of the Apsheron Middle Pliocene.

RED-BEDS OF THE TCHELEKEN PENINSULA

The Tcheleken Peninsula is located opposite the Apsheron Peninsula on the southeastern shore of the Caspian Sea. The central part of the peninsula is composed of Tertiary red-beds, which had been investigated since 1901 (Ivanov, 1903) without any positive results until 1953 when the author in collaboration with M. A. Romanova applied the trend-correlation method in the study of this series. The systematic application of the trend-correlation method per-

TABLE 6

SHORT CHARACTERISTIC OF MIDDLE PLIOCENE SECTIONS OF APSHERON PENINSULA

NO. BEDS AND THICKNESS

SECTION	General Thickness in Meters	Sands	Silts	Clays	Conglom- erates	General Quantity
Yasmalsky Valley	1142	2319	2517	935	4	5771
Bastanar-shor	436	1045	643	446	..	2137
Zhiloy Island	463	1329	993	463	..	2725

is located in the southwestern part of the peninsula in the Yasamalsky Valley; the second section is situated in the central part of the peninsula in the Bastanar-shor saline. The latter section was investigated in the Zhiloy Island. The description was given in great detail, namely, we described separately every lamellae with a thickness equal to 1 cm. and more. The composition of rocks was determined in the field by comparing them with typical rocks which had previously been analyzed.

Some characteristics of the sections under investigation are listed in table 6 taken from our previous paper (Vistelius, 1957).

The bed succession of every section was smoothed out after ranging by means of equation (2) with 1-meter interval length between the points of smoothing. The obtained values of u_x' were plotted on the graph with boundaries between suites taken from normal sections previously studied by the Apsheron

mitted us to reconstruct the natural succession of beds in the deposits under investigation and to map the red-bed outcrops (1953–1955).

In this case we tried to reconstruct a normal section and to correlate three similar sections, which may be of general interest.

The Tcheleken red-beds are a uniform succession of greenish and grayish fine-grained graywackes—with sand and silt particle dimensions and brick-red, hydromica clays colored by hydrogoethite and hydrohematite. In the main, beds dip at $5°–25°$ to the northwest. For more than half a century no marking beds, cyclothemes, typical fauna remains, or mineral associations were found in these deposits. Numerous faults exist in the red-bed area. The total number of faults seems to be very great as even those which have already been investigated exceed 512.

The trend-correlation method was ap-

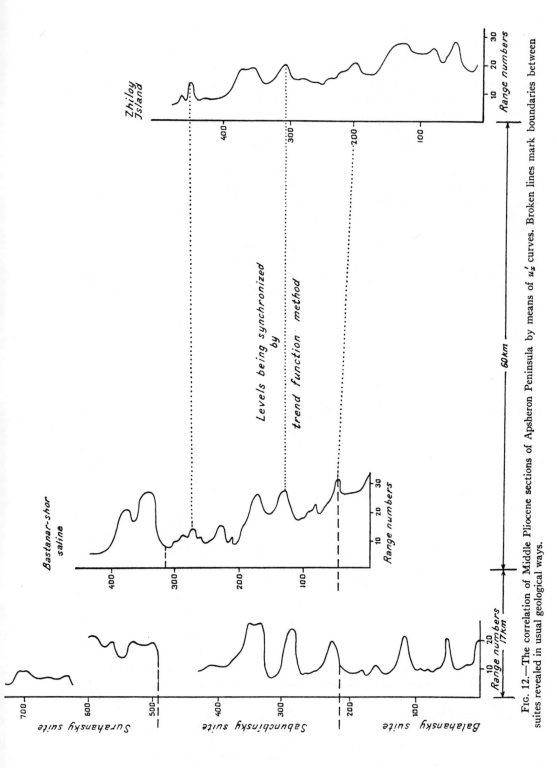

FIG. 12.—The correlation of Middle Pliocene sections of Apsheron Peninsula by means of u'_z curves. Broken lines mark boundaries between suites revealed in usual geological ways.

plied as follows. In order to find out the largest area of red-beds without faults we used the aerophoto analysis. The section in the areas without faults was correlated with the help of the trend-correlation method, the aerophoto analysis, and computation of correlation coefficients between the thicknesses of the correlated beds. Thus we obtained restored sections that show the natural bed succession. The trend curves u'_x were drawn for each restored section separately

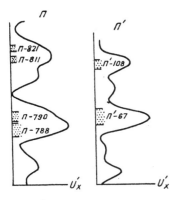

FIG. 13.—u'_x curves being compared; the boundaries of typical sand beds are marked by traits.

and were intercorrelated. The typical process of section restoring can be shown in the Section II in the following way.

Example of the composite section restoration.—The restoration of the composite section (NII) under investigation started from the foot of the Aktchagilian Stage. At first, the section was described by a continuous succession of beds—across their strike—up to bed II-746, behind which there is a road. The further extension of the section was continued with the aid of a ditch II′ located 0.4 km; to the east (along the strike of beds). As can be seen in plate 1, B, the peculiarities of the aerophoto of the site near the exposure of bed II-788–II-811 and of that with beds II′-67–II′-108 are similar. We check this similarity by means of the trend curve u_x which is plotted in figure 13. The comparison of aerophotos and curves u'_x shows that the beds II-811 and II′-108 are similar on the aerophoto and have a similar u_x curve con-

figuration. Thus beds II-811 and II′-108 seem to be different outcrops of the same bed. The continuous bed succession reaches bed II′-36. Bed II′-36 is traced over a distance of 0.2 km. on the surface to the west and is designated as bed II-741, in a ditch which is the extension of our first ditch II. We traced the continuous sequence of beds within the ditch II up to bed II-173, which is terminated by a great fault. The further extension of the section was done by means of the ditch KII. Plate 1, C, shows that a part of ditch KII with beds KII-257, KII-285 is located along the strike line of beds II-180–II-224; moreover, the aerophoto (pl. 1, C) shows that the parts of the red-beds being compared belong to the same sandy bed series. The trend

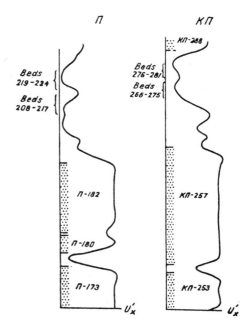

FIG. 14.—u'_x curves compared. The brackets mark the position of beds the thicknesses of which were correlated.

curves u'_x in figure 14 have many features which show that beds II-182 and KII-257 represent the same bed. The check of the trend correlation by means of u'_x was made by the computation of correlation coefficients between bed thicknesses. The correlated beds are shown in table 7.

The computations made (by means of scheme, Vistelius, p. 32, 1958) give the values shown in the accompanying tabulation. Here r is the coefficient of correlation

No. Correlated Beds	Π-208–KΠ-266 Π-216–KΠ-275	Π-219–KΠ-276 Π-224–KΠ-281
r	0.903	0.999
t	3.9	6

and t is the ratio of r by σr after Fisher's transformation.

correlation coefficient between bed thicknesses in section P with those of section $P3$. The bed successions in the above sections with the greatest correlation between bed thicknesses are listed in table 8.

The calculations give $r = +0.760$, $t = 4.3$; if we had a succession of independent values, there would be a quite clear result. But we do not know which correlation exists between the thicknesses of beds in each of our bed successions. On the whole, the esti-

TABLE 7

COMPOSITION AND THICKNESSES OF BEDS BEING CORRELATED

	SECTION Π			SECTION KΠ	
No. Beds	Composition of Beds	Thicknesses	No. Beds	Composition of Beds	Thicknesses
		FIRST BED PACKET			
208	Arenaceous clay	16	266	Arenaceous clay	20
209	Clay	5	267	Clay	15
210	Arenaceous clay	14	268	Sand	23
211	Silt	9	269	Arenaceous clay	26
212	Arenaceous clay	5	270	Sand	6
213	Clay	68	271	Clay	54
214	Arenaceous clay	15	272	Argillaceous sand	33
215	Clay	131	273	Clay	92
216	Sand	55	274	Argillaceous sand	17
217	Argillaceous silt	9	275	Clay	9
		SECOND BED PACKET			
219	Silt	25	276	Silt	18
220	Clay	2	277	Clay	11
221	Argillaceous silt	7	278	Clay	13
222	Clay	11	279	Argillaceous silt	14
223	Argillaceous sand	22	280	Silt	17
224	Clay	204	281	Clay	106

The correlated bed series, as shown in figure 14, are located along the curves u'_x in the intercepts of curves with identical configuration. Thus we associated the bed succession of ditch Π with that of ditch KΠ.

The further extension of the section was carried out in a similar way and is of no great interest. Only the last part of the section was extended in another way. The end of ditch $P3$ was connected with the short bed succession in ditch P.

Since we had the end of the bed succession in $P3$ and a short bed succession in P, it was impossible to apply the trend method. Numerous faults separated the bed successions being compared and made it impossible to use aerophoto pattern as can be seen on plate 1, D. In this case we calculated the

mation of parameter values ρ by means of statistics r is a very complicated problem.

The matrixes of transitional frequencies in table 9 reveal that succession of the thickness in the P succession has no autocorrelation. The bed thicknesses in the $P3$ sequence have a non-linear autocorrelation. Since the geological situation is such that most probably P sequence and $P3$ sequence are the same and since the value of r is not very close to the unit, the value of t is very high (4.3), and the sequence P is non-autocorrelated, we take a liberty with some risk in considering the bed sequences P and $P3$ as the same bed sequence.

Thus we restored the section of red-beds with its general thickness equal to 520 m.

The fragments of red-bed deposits separated by faults were restored in composite sections as is shown in the example, and the curve u'_x was plotted for every section (fig. 15). A careful examination of the u'_x curves in figure 15 permits us to mark the boundaries of horizons. These boundaries were obtained by comparing the parts of u'_x curves with a maximum similarity of their configurations when intercepts of curves compared were located near the bottom of thick sand series. The general correlation of three composite sections is shown in figure 15. We

nary correlation coefficients. The difference is in the nature of our observations because every bed in the succession of beds of normal section occupied one and only one quite definite place in this intercalation. Thus, if we have two sections, we have two successions of bed thicknesses:

$$y_1, y_2, \ldots, y_n$$

for one section and

$$w_1, w_2, \ldots, w_n$$

TABLE 8

COMPOSITIONS AND THICKNESSES OF THE BEDS BEING CORRELATED

SECTION P3			SECTION P		
No. Beds from Foot	Composition of Beds	Thicknesses	No. Beds from Foot	Composition of Beds	Thicknesses
61......	Argillaceous clay	8	1015.....	Clay	8
62......	Clay	109	1014.....	Sand	34
63......	Sand	15	1013.....	Clay	18
64......	Arenaceous sand	4	1012.....	Sand	13
65......	Silt fine stratified	10	1011.....	Gray clay	24
66......	Clay	7	1010.....	Brick-red clay	20
67......	Argillaceous Clay	2	1009.....	Sand	14
68......	Clay	97	1008.....	Clay	29
69......	Argillaceous clay	6	1007.....	Sand	7
70......	Clay	15	1006.....	Clay	23
71......	Sand	6	1005.....	Sand	3
72......	Clay	34	1004.....	Clay	15
73......	Sand	2	1003.....	Sand	18
74......	Arenaceous clay	8	1002.....	Argillaceous sand	11
75......	Clay	33	1001.....	Clay	20
76......	Arenaceous clay	7	1000.....	Clay fine stratified	9
77......	Sand	29	999.....	Argillaceous silt	18
78......	Clay	7	998.....	Clay	8
79......	Argillaceous sand	2	997.....	Sand	2
80......	Clay	78	996.....	Clay	29
81......	Sand	6	995.....	Argillaceous sand	11
82......	Clay	5	994.....	Clay	2

divided all the sections under investigation into six cyclothems in a similar way. The comparison of separate sections of red-beds terminated by faults with the composite sections produced with the aid of u'_x curves permits us to determine the relative position of these sections on the general lithostratigraphic scale of Tcheleken red-beds and to build up a geological map of the red-bed area in this way.

Some explanations about significance of correlation coefficients.—An attentive reader must find that correlation coefficients used in the course of investigation of Tcheleken red-beds are different from ordi-

for another section. If we compute correlation coefficients between y and w for different steps of displaced ν, one series of observation relative to another series, we receive value $R_{yw}(\nu)$. This value will be the value of cross-correlation function (Davenport and Root, 1958). If we want to estimate the significance of $R_{yw}(\nu)$, we must compare it with $\sigma R_{yw}(\nu)$. The common way of estimation

$$t = \frac{|R_{yw}(\nu)|}{\sigma_{R_{yw}}(\nu)},$$

accepting $R_{yw}(\nu)$ as "significant for $t \geq 3$," is invalid here if there is an autocorrelation

TABLE 9

MATRICES OF TRANSITIONAL FREQUENCIES FROM Kth BEDS TO $K+1$st AND $K+2$d BEDS

Section "P3"

(K, $K+1$, $K+2$ are thicknesses of beds in cm.)

K to $K+1$

K	0–11	11–22	22–33	33–44	44–55	55–66	66–77	77–88	88–99	99
0–11	6	1	1							
11–22	2									
22–33	1									
33–44	2			2						
44–55										
55–66										
66–77	1						1			
77–88	1						1			
88–99		1								
99									1	1

K to $K+2$

K	0–11	11–22	22–33	33–44	44–55	55–66	66–77	77–88	88–99	99
0–11	8	1								
11–22	1			1				1		
22–33	1			1						
33–44	1									
44–55			1							
55–66										
66–77									1	
77–88	1									1
88–99	1	1								
99										

Section "p"

(K, $K+1$, $K+2$ are thicknesses of beds in cm.)

K to $K+1$

K	0–3.5	3.5–7.0	7.0–10.5	10.5–14.0	14.0–17.5	17.5–21.0	21.0–24.5	24.5–28.0	28.0–31.5	31.5
0–3.5										
3.5–7.0	1			1						
7.0–10.5	1		2			1				
10.5–14.0			2	2		1				
14.0–17.5					1	1				
17.5–21.0	1			1		1				
21.0–24.5									1	
24.5–28.0			1	1					1	
28.0–31.5						1				1
31.5										

K to $K+2$

K	0–3.5	3.5–7.0	7.0–10.5	10.5–14.0	14.0–17.5	17.5–21.0	21.0–24.5	24.5–28.0	28.0–31.5	31.5
0–3.5										
3.5–7.0				1						
7.0–10.5		1	1		1				1	
10.5–14.0		1	1		1					
14.0–17.5				1					1	
17.5–21.0	1				2				1	
21.0–24.5						1				
24.5–28.0	1						1			
28.0–31.5				1			1			
31.5										

90

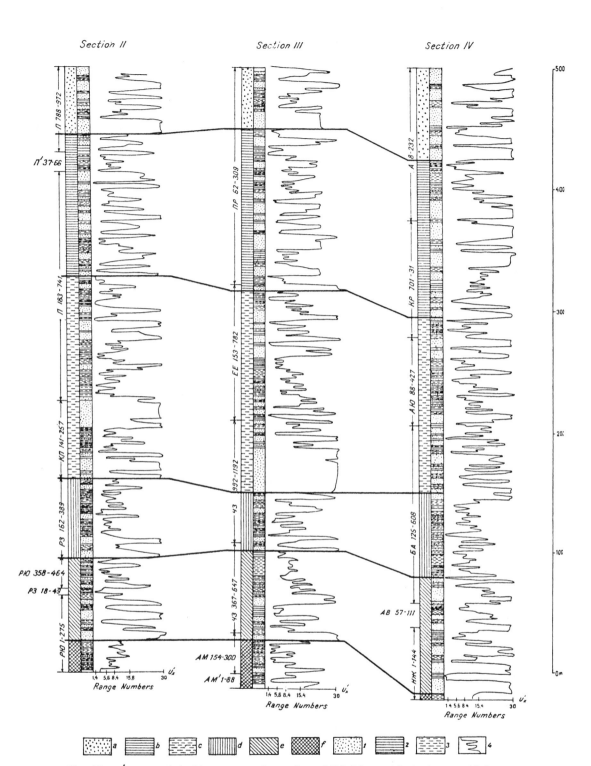

FIG. 15.—u_x' curves of the biggest composite sections of Tcheleken red-beds. Arrows with letters and figures indicate normal sections included in composite ones. Thick lines are boundaries between horizons revealed by trend method. Horizons are as follows: a = kurtepinsky; b = Karakhelsky; c = kutuburunsky; d = koshajushinsky; e = Kishmishlinsky; and f = kushkuslansky. Parts of section with average composition approximately 1 = sands; 2 = clays; 3 = silts; 4 = u_x' curve.

91

between some values of y_i and y_j and between w_e and w_k in series w values. If the autocorrelation is absent, we can use the common way for the estimation of the significance of $R_{yw}(\nu)$. So we must investigate the existence of autocorrelation in intercalations of bed thicknesses being compared. For short series of observations it is impossible, and we check significance of correlation from the geological point of view. For rather long series of observations we can build up a matrix of transitional frequencies. If the investigated series has values of observations autocorrelated, points of our observations would occupy a field of elliptical or curvilinear configuration on the field of matrix. If our observations have no autocorrelation, we will obtain a rather isometrical field of observations on our matrix.

Thus a matrix (as shown in the accompanying tabulation), where $\nu = 1$ and $x^{(j)}$

x_k	$x^{(1)}$	$x^{(2)}$	x_{k+1} ...	$x^{(n)}$
$x^{(1)}$	P_{11}	P_{12}	...	P_{1n}
$x^{(2)}$	P_{21}	P_{22}	...	P_{2n}
...	...	...	...	...
$x^{(n)}$	P_{n1}	P_{n2}	...	P_{nn}

are concrete values of observed variable, gives a check of an autocorrelation. In other words, we check the possibility of applying usual methods of estimation of a significance of correlation coefficients. A matrix of transitional frequencies in table 2 has such a sense as was explained. These matrices are named "Markov's matrices."

About the reading of such matrices: matrix for the section "P3" (table 9) has figure 6 in the point of intersection row titled 0–11 and column 11. The geological mean is as follows: ' We have six cases where a bed with thickness less than 11 cm. is followed by a bed with a thickness again less than 11 cm."

Markov's matrices are very useful tools in some geological investigations, but these special problems are not within our subject.

TARTARIAN BEDS OF ORENBURG DISTRICT

In many cases there is a necessity to correlate geological sections obtained by boring. Below we give an example of the trend-

method application to the treatment of the geological sections carried out by boring sedimentary deposits.

The western slope of the Ural Mountains and the eastern margin of the Russian platform consist of Upper Paleozoic deposits with Tartarian Stage of the Upper Permian–Lower Triassic age at the top.

The Tartarian deposits contain an intercalation of clay, marl, silts, and sandstones in red-beds. These beds are thin, and they vary much in composition along the strike. It is very difficult to correlate the sections of these deposits between different bore holes or exposures.

The beds under investigation build up a gentle fold near hamlet Yablonya in the northwest of the Orenburg district. This fold was bored in several places. Bore-hole cores were taken at 3-meter intervals. Yielded core thickness relative to the thickness of boring intervals was rather low. The description of sections was produced with aid of cores. These cores as mentioned above did not fill up the whole depth within intervals and could not be fixed at a definite depth. These circumstances lead to many indefinite errors in the description of geological sections obtained by boring. Thus we can reveal only the general character of beds with an overstated estimation of the role of hard rocks in the restored sections.

The above characterized description of sections obtained by means of bore cores were smoothed, having been ranged by the method described above. The u'_x curves for investigated sections (holes) are listed on figure 16.

SUMMARY AND CONCLUSIONS

There are some regularities in the evolution of the products of sedimentation in the course of time. These regularities create a trend of sedimentation throughout the time.

The trend in sedimentation products makes up individual features for short sequences of sedimentary products in sections of sedimentary deposits. The trend of sedimentation is disguised by aleatory (accidental) components of sedimentation.

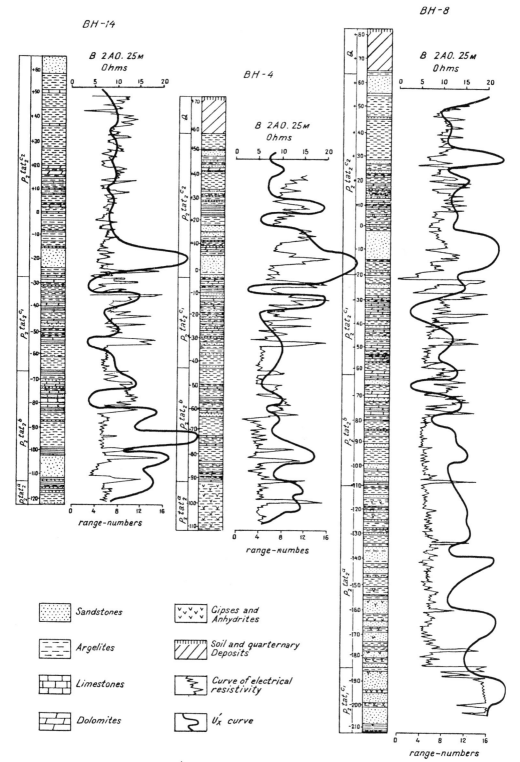

Fig. 16.—Comparison of three u'_x curves obtained from the bore-hole sections in Yablonya region. Figures indicate the distance to sea level in meters. (Sections and resistivity curves from the data of Oil Prospecting Organization.)

Use of a special process weakens the aleatory components of sedimentation and accentuates its trend component. The revealed trend-component function can be used for the correlation of sedimentary sections.

The stability of trend function under different geological conditions was checked. The investigation of different sections of Jurassic deposits in Crimea and the Caucasus and the Pliocene deposits in Apsheron and Tcheleken peninsulas indicates that the trend component is a stable one and is rather free from computational conditions.

The application of trend functions for correlation of strata of Pliocene deposits of Apsheron and Tcheleken peninsulas and Upper Paleozoic red-beds of the western slope of the Ural Mountains in natural sequences of beds and in bore-hole materials indicates that the method proposed can be used in different cases as a means of correlating strata.

Finally, it is necessary to point out some contradiction between the trend-function method and the usual method for correlating reference horizons. So in table 2 some figures are striking by their high values. As key horizons these beds are of special interest; from the point of view of trend function they are of some interest if they make a stable feature of a function. Thus trend function must be used for correlation by a whole set of its points. The difference between the trend-function method of correlation and the method of key horizons is similar to the difference in correlation by cyclothem method and key-horizon method. They use different sides of one and the same object.

There is a possibility of using more subtle properties of sequences of beds, such as autocorrelation of their thicknesses and separation of secular components of bed composition for restoring conditions of sedimentation. This requires special consideration.

REFERENCES CITED

BROONS, E. P., 1935, K litologii pestchano-glinistoy tolstchi Borovitchskogo rayona: Izvest. Leningradskogo geologo-gidro-geodezitcheskogo tresta. Nos. 2–3, p. 52–63.

CHAMBERLIN, T. C., 1898, The ulterior basis of time divisions and the classification of geological history: Jour. Geology, v. 6, p. 449–462.

——— 1909, Diastrophism as the ultimate basis of correlation: ibid., v. 17, p. 685–693.

DAVENPORT, W. B., JR., and ROOT, W. L., 1958, An introduction to the theory of random signals and noise: New York, McGraw-Hill Book Co., Inc.

DAWSON, J. W., 1855, Acadian geology, an account of the geological structure and mineral resources of Nova Scotia and portion of the neighbouring provinces of British America: Edinburgh.

FORSH, N. N., 1935, Novye dannye o stroenii okskoy tolstchi Borovitchskogo i Tichvinskogo rayonov i znatchenie etich voprosov dlja problemy C₁: Izvest. Leningradskogo geologo-gidro-geodezitcheskogo tresta. Nos. 2–3, p. 18–32.

——— 1940, Khazanskiy iarus v okrestnostjach g. Kuybysheva: Materialy po geologii permskoy sistemy Evropeyskoy tchasti SSSR, p. 59–75: Leningrad.

GILBERT, G. K. 1895, Sedimentary measurement of Cretaceous time: Jour. Geology., v. 3, p. 121–128.

HULL, E., 1863, On Izo-diametric lines, as means of representing the distribution of sedinentary clay and sandy strata, as distinguished from calcareous strata, with special reference to the carboniferous rocks of Britain: Quart. Jour. Geol. Soc. London, v. 18, p. 127–137.

IVANOV, A. P., 1903, Tchelekenskoe mestorozhdenie: Neftjanoe delo, v. 6. p. 328–341.

JEWETT, J. M., 1933, Evidence of cyclic sedimentation in Kansas during the Permian period: Trans. Kansas Acad. Sci., v. 36, p. 137–140.

KORN, H., 1939, Schichtung und absolute Zeit, u.s.v.: Neues Jahrbuch f. Min. Geol. Pal., v. 74, p. 51–166.

MICHAILOV, N. N., 1940, Utchet iskazhaiustchich anomalii pri robotach s gravitatsionnym variometrom: Problemy Arktiki, no. 1, p. 88–93.

MILLER, R. L., 1956, Trend surfaces: their application to analysis and description of environments of sedimentation. The relation of sediment-size parameters to current-wave system and physiography: Jour. Geology, v. 64, p. 425–446.

MILNE, W. E., 1949, Numerical calculus, approximation, interpolation, finite differences, numerical integration and curve fitting: Princeton, N.J., Princeton University Press.

NEWBERRY, J. S., 1874, Circles of deposition in American sedimentary rocks: Am. Assoc. Adv. Sci. Proc., v. 22, pt. 2, p. 185–196.

NOINSKY, M. E., 1924, Nekotorye dannye otnositelno stroenija i fatzialnogo charaktera kazhanskogo jarusa v prikazhanskom raione: Izvest. Geologitcheskogo Komiteta, v. 19, no. 6, p. 565–623.

ROMANOVA, M. A., 1957, Geologia verchnei tchasti krasnotsvetnych otlozhenii poluostrova Tcheleken: Trudy Obstchestva Estestvoispytatelei pri Leningradskom Universiteti, v. 49, no. 2, p. 116–126.

SOKRATOV, G. I., 1955, Nekotorye osobennosti litologii i ckladtchatoi struktury Tavritcheskoi tolstchi Kryma: Zapiski Leningradskogo Gornogo Instituta, v. 30, no. 2, p. 3–23.

UDDEN, J. A., 1912, Geology and mineral resources of the Peoria quadrangle, Illinois: U.S. Geol. Survey Bull. 506, p. 1–103.

ULRICH, E. O., 1911, Revision of the Paleozoic systems: Geol. Soc. America, Bull., v. 22, p. 281–680.

VISTELIUS, A. B., 1944, Zametki po analititcheskoi geologii: Doklady Academii Nauk S.S.S.R., v. 64, no. 1, p. 27–31.

——— 1944a, Fizitcheskie svoistva neftenosnych otlozhenii permskoi sistemy, Nautchno-issledovatelskie raboty neftijnikov, vypusk 1: Geologia, v. 24, p. 81–82.

——— 1945, O vyrazhenii resultatov fossilizatsii kolebatelnich dvizhenii zemnoi kory s pomostchiu rijda: Doklady Akademii Nauk S.S.S.R. v. 69, no. 7, p. 531–535.

——— 1946, Ritmy poristosti i javlenie fazovoi differentsiatsii osadotchnich tolstch: ibid., v. 54, no. 6, p. 519–521.

——— 1947, O korrelatsii mesoritmov v nizhnepermskich otlozhenijach Zakamskoi Tatarii i ich stratigraficheskom znatchenii: ibid., v. 55, no. 3, p. 241–244.

——— 1948, K geologii nizhnekazanskich otlozhenii Buguruslanskogo neftenosnogo raiona (issledovanie v oblasti analititcheskoi geologii): Symposium, Sovetskaja Geologia, no. 28, p. 48–63.

——— 1948a, O nekotorych analitchicheskich metodach issledovanija ritmitchnosti: ibid., p. 174–182.

——— 1949, K voprosu o mechanizme sloeobrazovania. Doklady Akademii Nauk S.S.S.R., v. 65, no. 2, p. 191–194.

——— 1949a, K voprosu o mechanizme svijzi pri sloeobrazovanii: ibid., no. 4, p. 535–538.

——— 1950, K voprosu o paleogeografitcheskom znatchenii svijzi mezhdu mostchnostijmi sloiev: Litologitcheskii sbornik Neftijnogo Instituta, no. 3, p. 61–73.

——— 1951, Ritmy poristosti v nizhnekazanskich otlozhenijach iuzhnoi Tatarii: Trudy Obtchestva Estestvoispytatelei pri Leningradskom Universiteti, v. 68, no. 2, p. 150–167.

——— 1952, Kirmakinskaija svita vostotchnogo Azerbaijana: Doklady Akademii Nauk Azerbaijana, v. 8, no. 1, p. 17–23.

——— 1957, Regionalnaija litostratigrafija i uslovija formirovanija produktivnoi tolstchi ijugovostochnogo Kavkaza: Trudy Obstchestva Estestvoispytatelei pri Leningradskom Universiteti, v. 69, no. 2, p. 126–150.

WANLESS, H. R., 1939, Pennsylvanian correlation in the eastern interior and Appalachian coal fields: Geol. Soc. America Special Paper 17, p. 1–130.

——— 1946 Pennsylvanian geology of a part of the southern Appalachian coal field: Geol. Soc. America Mem. 13, p. 1–162.

——— and WELLER, J. M., 1932, Correlation and extent of Pennsylvanian cyclothems: Geol. Soc. America Bull., v. 43, p. 1003–1016.

WELLER, J. M., 1930, Cyclical sedimentation of the Pennsylvanian period and its significance: Jour. Geology, v. 38, p. 97–135.

——— 1931, Conception of cyclic sedimentation during the Pennsylvanian period. Bull. Illinois State Geol. Survey, v. 60, p. 163–193.

——— 1931, Sedimentary cycles in the Pennsylvanian Strata: A Reply: Am. Jour. Sci., 5th ser., v. 21, no. 124, p. 311–329.

——— 1956, Diastrophic control of Late Paleozoic cyclothems. Am. Assoc. Petroleum Geologists Bull., v. 40, p. 17–51.

———, WANLESS, H. R., CLINE, L. M., and STOOKEY, D. G., Interbasin Pennsylvanian correlation, Illinois and Iowa. Am. Assoc. Petroleum Geologists Bull., v. 26, p. 1585–1593.

WHITTEKER, E. T., and ROBINSON, G., 1926, The calculus of observations: A treatise of numerical mathematics, 2d ed.: London, Blackie & Son, Ltd.

10

Reprinted from *Am. Assoc. Petrol. Geol. Mem.* 1:253–272 (1962)

CLASSIFICATION OF MODERN BAHAMIAN CARBONATE SEDIMENTS[1]

JOHN IMBRIE[2] AND EDWARD G. PURDY[3]

New York, New York, and Houston, Texas

ABSTRACT

An optimum empirical classification is defined as one in which there is exactly one category for each group of samples separated from other groups by discontinuities in the ranges of their observed properties. A statistical scheme for identifying discrete sample groupings (if they exist) in a set of data is developed which gives equal weight to any number of properties, and treats nonhomogeneous properties simultaneously. Essential features of the scheme are (1) representation of each sample as a vector in an n-coordinate system, where n is the number of attributes considered, (2) use of the angle of separation between sample vectors as an inverse measure of similarity, (3) application of factor analysis to determine the minimum number of coordinates necessary to express main features of the data efficiently, and (4) inspection of the resulting vector array for discrete vector clusters.

Applied to 200 samples of modern Bahamian carbonate sediments studied by Purdy, this scheme identifies five discrete sample groups (oölite, oölitic, grapestone, coralgal, and lime mud facies) whose discrete character is not apparent if only a small number of attributes are considered simultaneously.

The same 200 samples are treated according to Folk's limestone classification, and the results compared with the optimum classification scheme. Application of Folk's criteria yields nine named categories belonging to Type I (sparry allochemical limestone) and Type II (microcrystalline allochemical limestone), although it is difficult to make the I–II distinction in many samples. In general, correspondence between the two classifications is good in Type I, and poor in Type II.

INTRODUCTION

PURPOSE

As illustrated by other papers in this symposium, limestone classifications may be formulated in various ways. The purpose of this paper is to indicate one procedure by which modern carbonate sediment data may be used to evaluate classifications designed to reflect characteristics of the depositional environment. The criterion we suggest is that of *simplicity*, and in applying it we ask—Is the classification being evaluated as simple as possible? By a simple scheme we mean a classification in which there is exactly one category for each group of samples found to be separated from other groups by discontinuities in the ranges of their observable properties.

OUTLINE OF PROCEDURE

The classification evaluation procedure followed in this paper involves four steps.

1. *Designation of study area.*—A considerable variety of carbonate sediment types are now forming over a wide area in shallow waters of the Bahama Banks (Fig. 1). A number of investigators have published studies of this region and at least the general characteristics of the sediments and the depositional environments are fairly well known (*see* Black, 1933; Illing, 1954; Newell and others, 1959; Newell, Purdy, and Imbrie, 1960; Newell and Rigby, 1957; and Smith, 1940). A paper by Purdy (Ms.) will make available for the first time detailed, quantitative, petrographic observations on 200 bottom samples taken on the Andros lobe of the Great Bahama Bank. These data form the factual basis of the present study.

2. *Development of an optimum classification of samples from study area.*—An optimum classification is here defined as a classification providing one, and only one, category for each group of samples having discrete properties—in short, a

[1] Part of a symposium arranged by the Research Committee, and presented at Denver, Colorado, April 17, 1961, under joint auspices of the Association and the Society of Economic Paleontologists and Mineralogists. Manuscript received, May 6, 1961.

Dr. Norman D. Newell introduced the authors to the study of Bahamian sediments and organism communities, and has stimulated in many ways the growth of ideas expressed here. All the computations contained in this paper were run at the Watson Scientific Computing Laboratory of Columbia University. Programs used in these calculations were written and executed by Mr. Jonathan E. Robbin, whose experience with factor analysis greatly assisted us in developing special applications of these techniques to geological problems. Dr. George R. Orme, Mr. Hugh Buchanan, and Mrs. Marjorie Darland gave valuable assistance in various phases of data processing. The statistical part of this paper was supported by grant PRF-687-A from the Petroleum Research Fund of the American Chemical Society.

[2] Columbia University.

[3] Rice University.

classification in which the designated categories correspond to natural rather than artificial units. This definition is illustrated by hypothetical data of Figure 2-*I*. For simplicity of presentation, the assumption is made that the system to be classified is completely described by three observational parameters visualized as end members of a triangular diagram. The plotted points reveal two distinct clusters representing two groups of samples with discrete properties. Each cluster represents a category in what is here defined as an optimum classification scheme for these data. Note that the boundary between the clusters is a natural one and (assuming adequate sampling) may be taken to reflect a significant discontinuity in the dynamic system responsible for the observed properties of the sediment. The facies map corresponding to the classification locates the geographic position of this discontinuity.

In the simple case being illustrated, no *a priori* decisions are required and the data, in a sense, classify themselves. In dealing with real sediments and rocks, a very large number of properties may be observed, so that the characteristics of a classification may be predetermined by the selection of a limited number of observational parameters. In order to arrive empirically at an optimum classification, therefore, it is necessary to overcome or minimize this arbitrary element. One technique for accomplishing this objective is described below.

3. *Application of the classification to be evaluated to samples from study area.*—This is possible only with limestone classifications which are based, at least in part, on properties observable in unconsolidated sediments. Figure 2-*II* illustrates the application of a classification scheme based on three 1:1 ratio lines to the same data treated in Figure 2-*I*.

4. *Comparison of results.*—The facies patterns and sample groupings resulting from application of the two classification schemes now may be di-

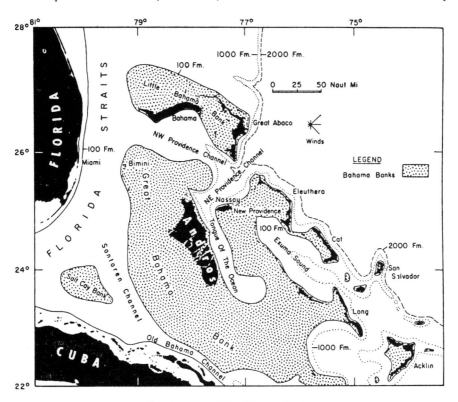

Fig. 1.—Map of the Bahama Banks.

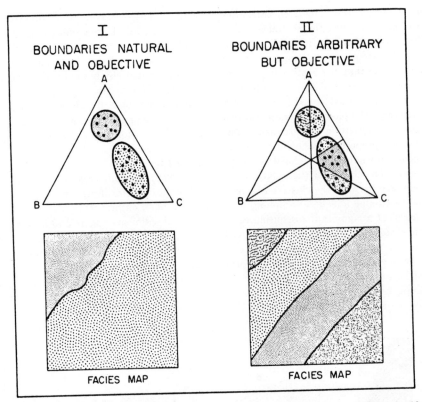

FIG. 2.—Diagram illustrating an optimum classification (I) and an objective but arbitrary classification (II).

rectly compared, and the degree of correspondence between them taken as one measure of the validity of the system being evaluated.

Figure 2 illustrates such a comparison. Although boundaries between the five categories of system II are objectively determined, the categories themselves are in a sense artificial—subdivisions are indicated where no natural discontinuity exists, and the real discontinuity is not reflected by any boundary between designated categories. None of the boundaries on facies map II corresponds with a natural discontinuity, although the mapped units do reflect a real gradient. For some purposes it might be useful to portray such a gradient by arbitrary subdivisions of natural facies; but in general it seems to us unwise to use a classification as a basis for detecting environmental or petrographic gradients. In the present example it would be more efficient to map the ratio A/C.

OPTIMUM CLASSIFICATION OF
BAHAMIAN SEDIMENTS

OBJECTIVES

In the example illustrated in Figure 2, it is possible to identify discrete clusters by simple visual inspection of plotted points. As the system is here assumed to be completely described by three components, an optimum classification is automatically achieved. If we had selected only two properties, however, our results would have been satisfactory or unsatisfactory depending on which pair of properties had been selected. A–C will yield essentially the optimum classification; but A–B or B–C would give the false impression that only one cluster is present. In a real situation, with an endless number of rock properties available, the possible effect of adding information on other components must be considered. By plotting points in a tetrahedron, simultaneous consideration can of course be given to four end members.

The limitations of Euclidean space prevent us from giving simultaneous consideration to more than four coordinates, by means of standard triangle or tetrahedron plots. We therefore seek a descriptive technique which will enable us to give equal and simultaneous consideration to any number of rock properties. In addition, it will be useful to construct a system capable of treating nonhomogeneous rock properties, that is, combinations of parameters which do not meaningfully add to one hundred per cent. It may be desirable, for example, to give simultaneous consideration to data on fossil content, sorting, perme-

ability, grain size, and trace elements, as well as allochem proportions. Although the immediate objective of the proposed system is to identify discrete clusters in an n-coordinate system, other useful features will appear as a by-product of this primary objective.

METHOD

Vector representation.—The position of point X in the plane triangle of Figure 3 represents the three ratios X_A/X_B, X_A/X_C, and X_B/X_C. The same information can be displayed by representing sample X as a unit vector OX in a three-coordinate system with orthogonal reference vectors OA, OB, and OC. Here, the coordinates X_A, X_B,

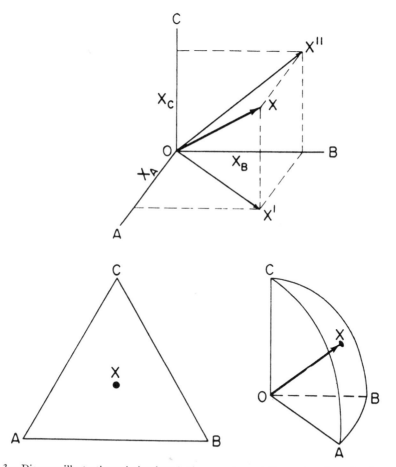

FIG. 3.—Diagram illustrating relationships between vector notation and a triangular diagram.

and X_C fix the direction of the vector OX. Note that any set of coordinates with the same proportions will determine the same vector. Ratio X_A/X_B is represented by the position of OX', the projection of OX on plane OAB. As OX' varies from OA to OB, this ratio will vary from infinity to zero. Thus, the 90° angle between vectors OA and OB symbolizes the maximum possible difference in the ratio X_A/X_B. Any two collinear vectors (angular separation of zero) represent two samples with identical values of the ratio. As noted by Krumbein (1957), angular transformations of ratio and percentage data have definite advantages over the untransformed measures.

As indicated in Figure 3, any sample vector OX in a three-coordinate system can be represented as a point X on the surface of a right spherical triangle with apical points A, B, and C. Here X is construed as the intersection of vector OX with the surface of a sphere of unit radius. The right spherical triangle is thus analogous to a plane triangle of conventional notation, although the plane figure distorts distance relationships to some extent.

In general, the degree of difference between any two samples X and Y (as regards the proportions of their constituent properties) is represented in vector notation by the size of the angle between their respective vectors, θ_{xy}; θ is thus the general measure of dissimilarity in the system under discussion. It may range from 0°, corresponding to identical composition, to 90°, corresponding to complete dissimilarity, and take any value between these extremes; 45° corresponds to a neutral or random relationship in which the two samples tend to be neither similar nor dissimilar.

One of the chief advantages of vector notation is that the algebraic formulation of key concepts can be easily extended into a system of any desired number of coordinates, even though our geometric intuition does not extend beyond three. Given a set of observations of n rock properties on N samples, and the generalized observation x_{ij}, of the i'th property on the j'th sample ($i = 1, 2, 3, \ldots n; j = 1, 2, 3, \ldots p, q, \ldots N$) the formula for computing $\angle\theta$ between any pair of samples, p and q, is

$$\cos \theta_{pq} = \frac{\sum_{i=1}^{n} x_{ip} x_{iq}}{\sqrt{\sum_{i=1}^{n} x_{ip}^2 \sum_{i=1}^{n} x_{iq}^2}}$$

The use of an angular measure of similarity strikes most geologists as bizarre, and it is convenient for some purposes to transform θ into a dimensionless parameter θ' by the following formula

$$\theta' = \frac{45° - \theta}{45°}$$

Note that θ' ranges from $+1.000$ through 0.000 to -1.000, with the three stated values corresponding respectively to θ values of 0°, 45°, and 90°. Negative values reflect dissimilarity, positive values similarity.

Figure 4 illustrates the use of θ' as a measure of similarity between six samples based on observations of five parameters. In each case θ' has been computed between sample a and one of the six samples.

Factor analysis of Bahamian samples.—Having defined a measure of similarity between any given pair of samples, we may now proceed to identify clusters of sample vectors in an n-coordinate system. The first step in the procedure used here is to perform an algebraic operation known as factor analysis. As the theory and practice of these techniques are fully described elsewhere (*see* especially Cattell, 1952, for a good nontechnical summary; and Harman, 1960, for

		VARIABLES					θ'_{ax}
		A	B	C	D	E	
SAMPLES	a	10	20	30	40	0	+1.00
	b	5	10	15	20	0	+1.00
	c	5	10	15	50	0	+0.49
	d	10	30	20	4	0	0.00
	e	40	30	20	10	0	−0.07
	f	0	0	0	0	80	−1.00

FIG. 4.—Hypothetical data matrix and calculated values of θ' between sample a and other samples.

detailed algebraic instructions), only an outline of objectives and procedures will be included here. A three-step procedure is involved.

1. The first step is to construct a data matrix in which every observation is recorded for each sample. Table I is a data matrix representing twelve petrographic variables observed at each of 40 Bahamian localities chosen to represent a wide range of bottom sites. A detailed description of the grain types recognized, location of samples, and genetic interpretations of grain types are to be published by Purdy (ms.). For present purposes it is necessary only to note that each bottom sample was sieved to remove the size fraction $< \frac{1}{8}$ mm; the weight per cent of that fraction was recorded; and by point-count analysis of thin sections made from plastic-impregnated samples of the coarse fraction, the volume per cent of 11 grain types was determined.

In order to give each parameter equal weight in determining the location of any sample vector, it is necessary to transform each row of the raw data matrix so that observations are recorded in per cent of the maximum value of that variable observed over the 200 Bahamian samples.

Variable	Maximum Value in 200 Bahamian Samples (wgt. or vol. %)
Molluscs	33.0
Peneroplids	28.6
Other foraminifera	13.2
Halimeda	39.8
Coralline algae	7.4
Corals	23.8
Pellets	76.8
Mud aggregates	30.2
Grapestone	58.6
Cryptocrystalline grains	57.0
Oölite	98.8
<1/8 mm	92.3

If this had not been done, variables constituting a large fraction of the volume would tend to exert a disproportionate influence on the position of any sample vector. Although this result might be desirable in some applications, the objective here is to give each variable equal consideration.

2. The second step is to calculate $\cos \theta$ for each pair of sample vectors, and arrange the results in a square matrix. As $N = 40$, the number of different calculations involved is $(N-1)\,N/2 = 780$. Although this would be an incredibly tedious task

TABLE I. CONSTITUENT PARTICLE COMPOSITION OF FORTY BAHAMIAN SEDIMENT SAMPLES
(Observations expressed as 0/00 of maximum range)

Sample No.	1	2	3	4	5	6	7	8	9	10	11	12	13	14	15	16	17	18	19	20
Molluscs	067	012	054	115	297	430	315	145	091	067	127	152	164	315	358	370	085	242	048	036
Peneroplidae	349	021	014	028	168	202	147	084	119	098	216	147	349	530	447	063	007	049	007	028
Other Foraminifera	666	015	106	212	227	091	136	136	091	091	151	197	409	318	545	484	273	136	045	015
Halimeda	085	015	025	040	131	281	407	005	020	005	050	075	085	136	256	643	346	316	010	025
Coralline Algae	270	000	027	000	000	000	000	000	000	000	000	000	000	000	000	000	000	459	000	000
Corals	134	000	000	017	017	000	000	000	000	000	000	000	008	000	000	000	000	319	000	000
Pellets	013	047	096	044	793	619	611	894	998	803	876	868	559	416	291	335	286	055	242	341
Mud Aggregates	126	099	046	225	205	205	299	278	139	291	245	649	563	556	245	305	278	119	132	
Grapestone	248	218	439	751	000	007	007	000	000	000	000	000	000	003	000	000	003	082	595	180
Crypt. Grains	434	312	284	217	010	052	032	052	007	056	007	024	021	042	035	014	164	196	354	164
Oölite	004	596	392	222	026	006	030	042	028	202	042	032	016	004	012	053	331	081	162	450
< 1/8 mm.	249	076	070	027	338	433	561	635	380	343	561	688	980	942	926	963	054	024	082	060

Sample No.	21	22	23	24	25	26	27	28	29	30	31	32	33	34	35	36	37	38	39	40
Molluscs	036	048	085	273	364	158	121	042	054	315	315	206	155	048	009	000	000	000	003	321
Peneroplidae	000	000	000	021	035	098	007	077	014	056	070	251	192	028	000	000	000	000	000	021
Other Foraminifera	000	045	030	212	136	409	015	197	136	091	242	166	053	038	000	000	000	000	000	068
Halimeda	020	070	025	090	663	397	040	025	020	236	999	356	131	023	000	000	000	000	000	562
Coralline Algae	000	027	000	000	486	270	378	000	000	513	027	378	000	000	000	000	000	000	000	203
Corals	000	000	000	000	538	260	025	000	000	286	042	580	034	000	000	000	000	000	000	567
Pellets	260	151	333	057	003	023	083	042	039	029	010	000	008	006	000	000	000	001	000	027
Mud Aggregates	060	086	079	060	344	212	046	113	040	060	086	126	040	053	000	000	000	007	000	000
Grapestone	082	054	218	245	048	031	653	833	826	054	054	095	624	194	005	000	000	005	008	241
Crypt. Grains	108	658	262	798	136	616	388	560	536	301	346	476	276	208	037	024	040	051	023	205
Oölite	644	368	372	137	000	014	182	022	079	044	000	000	251	695	978	992	986	974	988	132
< 1/8 mm.	046	090	059	048	026	204	027	049	050	022	044	027	031	064	015	014	025	016	014	006

by hand, the entire calculation may be accomplished in a matter of minutes on modern digital computers. Table II displays the cos θ matrix for the 40 Bahamian sediment samples. Decimal points have been omitted and should be understood to precede each three digit sequence. This matrix displays in compact form the similarities and dissimilarities between every pair of samples considered. By scanning a column or a row, one may select the sample most similar or least similar to any given sample. Sample 26, for example, is most similar to sample 1 (cos θ = .827). If samples are arranged according to position in field traverses, the values in one column will constitute a gradient of affinity between the first and each successive sample in the traverse.

3. Step three is to calculate a centroid factor matrix. This calculation is the key step for factor analytic techniques in general, for it provides an objective determination of the degree of complexity in the system being analyzed. Each dimension of complexity is spoken of as a factor.

Composite end members are a geologically meaningful way of expressing the results of a factor analysis. In the plagioclase feldspar suite, for example, it is standard practice to express the composition of any given plagioclase crystal as the sum of two end members, albite and anorthite. Mineralogically, these end members are single crystal species; chemically, they might equally well be considered composite end members with compositions $NaAlSi_3O_8$ and $CaAl_2Si_2O_8$. If a factor analysis were performed on chemical data from a varied set of plagioclase crystals, the results would indicate that the data could be expressed in terms of two factors, that is, two composite end members. If pure specimens of albite and anorthite had been included, the technique would specify exactly the composition of the end members. If such pure specimens had not been included, the results would show that all samples could be expressed as mixtures of end members having compositions of the two most distant vectors. Factor analyses of more complicated, artificially constructed test examples have shown the ability of this technique to recover simple relationships in scrambled data.

In dealing with actual examples, we can hardly expect to express perfectly all of the information

with only three or four factors. Sampling error, identification error, as well as the natural perversity of geological data, all contribute to disorder. Conveniently, centroid factor analysis is an iterative procedure, so that factors (composite end members) are extracted one at a time, and their individual influence assessed. If the investigator wishes to explain every nuance exactly, then N end members will ordinarily be required (in this case forty). If he is willing to accept a more modest achievement, say explaining 90 per cent of the observed relationships with a simple hypothesis, a small number of factors may suffice. In the present example, 89 per cent of the information can be explained by four factors, a surprisingly simple and satisfactory result.

Like the computation of the cos θ matrix discussed above, and the quartimax rotation described below, centroid analysis is far too tedious for hand computation. Fortunately, digital computers of the IBM 650 class or higher perform the calculations rapidly and accurately. There is, however, an upper limit to the number of samples which can be handled; on the IBM 650, this limit is 40. On more powerful computers, the number is 200.

4. The fourth step is a rotation procedure, wherein the orthogonal reference vectors (in this case four) are placed in convenient locations with respect to the 40 sample vectors situated in a four-coordinate system. The locations used in this paper are given by the quartimax rotation procedure (Harman, 1960, p. 294–301) in which the reference axes are located so as to maximize the sums of the fourth powers of the projections of the sample vectors on the reference vectors. The final result of the factor analysis is thus a rotated quartimax factor matrix shown in Table III. Here, for each of the 40 samples, four coordinates are given for the end of the sample vector, corresponding to the projections of the end of the sample vector on the four reference vectors (Factor axes I–IV). If the position of the end of a sample vector can be located precisely in the four coordinate system, the squares of the four projections will sum to unity as indicated by the Pythagorean theorem. This sum of squares is known as the communality, symbolized as h^2, and entered in the last column of the factor matrix. This figure

TABLE II. COSINE θ MATRIX FOR FORTY BAHAMIAN SEDIMENT SAMPLES

Sample No.	1	2	3	4	5	6	7	8	9	10	11	12	13	14	15	16	17	18	19	20
1	1000																			
2	338	1000																		
3	515	893	1000																	
4	540	616	895	1000																
5	375	163	256	184	1000															
6	364	171	252	185	947	1000														
7	373	174	252	177	907	968	1000													
8	311	198	258	140	943	883	875	1000												
9	229	158	222	114	952	851	819	968	1000											
10	259	331	351	176	930	830	817	956	974	1000										
11	339	184	245	139	960	899	884	990	975	958	1000									
12	361	181	249	147	948	903	909	993	951	942	991	1000								
13	533	193	241	172	805	809	808	862	760	736	874	889	1000							
14	555	187	232	176	772	827	812	799	687	666	824	837	976	1000						
15	636	182	244	213	731	791	795	728	602	585	751	779	955	976	1000					
16	514	173	244	204	707	809	888	700	565	571	696	759	833	836	895	1000				
17	486	594	553	361	644	626	634	542	550	604	566	556	566	522	583	637	1000			
18	590	299	365	318	353	431	397	235	217	211	254	252	340	369	421	440	579	1000		
19	492	665	900	907	370	354	328	384	375	424	368	363	305	282	256	236	441	331	1000	
20	281	862	817	560	567	501	476	570	595	718	570	541	410	364	316	308	732	320	707	1000
21	135	892	730	409	377	327	327	385	397	566	381	361	246	212	182	212	657	226	488	937
22	486	813	731	441	281	305	296	307	265	390	277	285	251	245	239	251	610	402	635	713
23	366	853	869	644	570	522	493	564	577	696	550	533	381	343	304	321	704	355	806	970
24	654	608	702	603	253	313	274	203	145	215	175	197	216	248	287	284	492	456	704	483
25	454	131	193	198	298	420	412	154	138	107	180	186	277	320	391	481	531	948	175	157
26	827	368	440	363	364	438	458	291	214	237	294	326	449	469	554	579	639	798	420	297
27	538	605	835	852	160	193	170	150	129	186	134	137	116	130	126	130	305	556	870	544
28	619	525	815	918	151	164	145	142	111	137	134	140	186	193	210	161	292	322	927	443
29	557	564	844	932	117	137	126	118	084	127	100	112	127	132	146	132	266	293	938	470
30	584	268	342	288	268	368	336	153	121	140	156	169	193	251	294	342	402	936	306	238
31	431	190	260	259	300	468	526	128	104	104	159	186	230	293	400	608	645	614	237	174
32	679	278	352	320	243	331	313	133	107	119	160	161	242	305	345	339	428	880	347	221
33	533	686	896	941	185	257	235	121	096	171	142	136	176	237	247	222	392	395	884	595
34	270	977	840	564	119	130	141	132	091	280	125	124	140	144	154	162	566	258	548	828
35	026	846	599	276	035	019	043	048	031	228	046	038	023	017	022	052	471	116	234	722
36	018	839	589	267	031	014	039	045	030	226	045	036	021	013	019	049	467	109	224	717
37	028	847	597	271	035	020	045	052	034	231	050	043	029	022	026	057	472	113	232	722
38	033	853	604	279	035	019	042	050	034	230	049	040	027	019	024	052	477	119	240	727
39	020	841	594	274	032	015	040	046	030	226	045	036	022	014	019	049	467	110	230	719
40	437	321	415	423	265	391	401	122	106	124	139	152	175	227	291	430	523	849	383	309

gives (in parts per thousand) the fraction of information accounted for by four factors. For sample 38, for example, the communality is 99.5 per cent. The mean communality is 89 per cent.

If a factor analysis reduces to two or three end members, a Euclidean vector model can be constructed directly from the factor matrix, and visual inspection suffices to locate vector clusters. In the present example this cannot be done, and it is necessary to view the four dimensional vector display two dimensions at a time. Four of the possible six views of the vector model are plotted in Figure 5. Each point on the figure is a graphic plot of the coordinates contained in one row of the factor matrix, and corresponds to the end of a sample vector. If different symbols are used for each point it is clear that the 40 vectors are arranged in 5 discrete clusters. Each cluster has been named by reference to the chief constituents of samples composing the cluster. Although some clusters seem to overlap in one view (for example, the grapestone and oölitic clusters on the II–IV plot), this apparent overlap is a consequence of

viewing the scheme two dimensions at a time. By examining other views it is clear that all clusters are in fact discrete groups separated by discontinuities. Two subgroups are indicated in the lime mud cluster. Although the names applied to the clusters are subjective, once the data matrix has been compiled, their definition is entirely objective.

Compositions of the four samples lying closest to the reference axes may now be taken as the compositions of four theoretical composite end members. By inspection of Table III, these samples are identified as 38, 12, 29, and 25. Within a tolerance limit indicated by the communalities, each of the 40 samples can be represented as a simple mixture of the 4 end members.

The procedure of factor analysis outlined above is known as a Q-type factor analysis, in which samples are compared with each other and factored. Alternatively, using different techniques, variables may be compared and clustered by a technique known as R-type factor analysis. The reader is cautioned that in the published literature

TABLE II (*continued*)

Sample No.	21	22	23	24	25	26	27	28	29	30	31	32	33	34	35	36	37	38	39	40
1																				
2																				
3																				
4																				
5																				
6																				
7																				
8																				
9																				
10																				
11																				
12																				
13																				
14																				
15																				
16																				
17																				
18																				
19																				
20																				
21	1000																			
22	649	1000																		
23	868	801	1000																	
24	344	875	650	1000																
25	085	235	183	320	1000															
26	171	663	403	762	738	1000														
27	387	570	651	666	375	494	1000													
28	223	558	580	756	188	501	858	1000												
29	271	571	609	750	160	457	879	992	1000											
30	168	452	326	545	859	786	603	327	318	1000										
31	102	380	247	500	728	726	268	290	269	562	1000									
32	116	490	305	588	852	865	418	385	892	613	1000									
33	460	542	675	665	293	433	854	890	902	386	383	451	1000							
34	917	703	789	484	113	272	516	401	448	226	161	208	631	1000						
35	910	501	620	186	008	043	228	047	104	075	015	021	340	929	1000					
36	908	489	612	170	003	032	217	035	092	066	008	012	329	923	1000	1000				
37	910	504	619	185	006	045	224	044	102	073	013	020	335	928	1000	1000	1000			
38	913	514	626	196	009	052	234	055	112	078	017	027	344	932	1000	1000	1000	1000		
39	909	489	615	172	004	032	223	041	099	067	009	013	336	925	1000	1000	1000	1000	1000	
40	240	363	365	455	913	708	499	378	379	795	737	863	527	301	150	143	147	151	146	1000

most of the work to date has been done with R-type analyses, and that the present paper is the first application of cos θ to Q-type problems.

Representation of hierarchical relationships.—As four dimensions are required to display relationships among the five clusters, it is impossible to view the entire system simultaneously. By a slight modification of a system of hierarchical representation proposed by Sokal and Michener (1958) it is possible, however, to extract the main features of the inter-cluster relationships and display them in a two dimensional figure, as follows.

1. Represent each cluster by a single vector of unit length passing through the centroid of the cluster. This is equivalent to representing the cluster by a unit vector extended in the direction of a vector sum of all members of the cluster. In a sense, this unit centroid vector represents the average qualities of the cluster.

2. Compute the angle between all pairs of unit centroid cluster vectors. This can be done conveniently by using a formula derived from an expression given by Spearman (1913)

$$\cos \theta qQ = \frac{\Box qQ}{\sqrt{q + 2\triangle q}\sqrt{Q + 2\triangle Q}}$$

where $\Box qQ$ is the sum of all cos θ's between members of one group and the other group, $\triangle q$ is the sum of cos θ's between all possible pairs formed from members of the first group, $\triangle Q$ is a similar sum between members of the second group, q is the number of vectors in group one, and Q the number of vectors in group two. The necessary values of cos θ are contained in the original cos θ matrix.

3. The results of these calculations are then displayed in a 5×5 matrix, representing cosines of angles between the unit centroids (*see* Table IV). The location of the highest number in the matrix (in this case 0.810) identifies the pair of clusters which is most closely related (oölite and oölitic clusters). The value of cos θ is then converted into θ' ($\theta' = 0.20$), and on a hierarchy diagram (Fig. 6) a horizontal line drawn at the appropriate level linking the two facies. A measure of the homogeneity within each cluster can be achieved by taking $\triangle q$ and dividing by the number of sum-

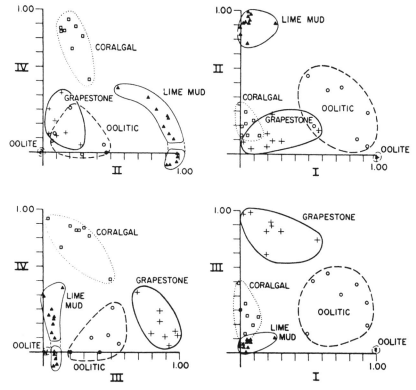

Fig. 5.—Two-dimensional plots of the ends of 40 Bahamian sample vectors as located by the rotated quartimax factor matrix in Table III. Descriptive terms refer to facies names.

mated pairs (q) $(q-1)/2$. The resulting $\cos \theta$ is then converted to θ' and the lower line of the plotted rectangle drawn to symbolize the value. For the oölite facies, this value is 1.00; for the oölitic facies, 0.27.

4. The highest number in the matrix must represent a pair of samples having mutually strongest resemblance. Other such pairs may exist and can be discovered by scanning each row

5. Combined facies (in this example only the oölite facies plus the oölitic facies) are now each considered as a unit and represented in the next matrix by an appropriate number of collinear unit vectors (one for each vector in the new combined cluster) extended along the direction of the combined centroid. The reduced (4×4) matrix of cosines is conveniently computed from the previous matrix, using the formula

$$\cos \theta_{AB} = \frac{Na(Nb\theta ab + Nb'\theta ab') + Na'(Nb\theta a'b + Nb'\theta a'b')}{\sqrt{Na^2 + Na'^2 + 2NaNa'\theta aa'}\sqrt{Nb^2 + Nb'^2 + 2NbNb'\theta bb'}}$$

and each column for the highest value outside the principal diagonal. Any location that is the highest value in its row and column will constitute a pair having highest mutual resemblance and should be grouped together on the hierarchy diagram.

where A and B stand for two groups of vectors each composed of one or two subgroups, a and a' and b and b' respectively; and N_a, $N_{a'}$, N_b, and $N_{b'}$, stand respectively for the number of vectors in each subgroup.

6. Steps three and five are repeated until a

TABLE III. ROTATED QUARTIMAX FACTOR MATRIX OF FORTY BAHAMIAN SAMPLES
(Values expressed in 0/00)

Problem Code	Field Sample Code	Facies	Projections on Factor Axes				
			I	II	III	IV	h^2
38	413	Oolite	997	-008	030	012	995
37	414		995	-005	020	008	991
35	418		995	-009	023	009	991
39	438		994	-009	019	002	988
36	417		994	-010	012	003	988
21	334	Oolitic	927	293	197	-005	984
34	411		924	059	359	101	996
02	31		851	103	482	120	982
20	330		738	474	413	007	940
23	312		638	452	559	062	928
22	310		529	197	505	311	670
17	374		491	549	138	449	763
12	369	Pellet Mud	052	989	071	-007	986
11	368		060	986	072	-028	982
08	360		068	973	088	-048	961
10	367		248	905	104	-087	899
09	366		062	930	074	-086	882
06	174	Skeletal Mud	035	918	075	238	906
13	370		017	915	058	127	857
07	380		059	909	037	257	897
05	171		047	956	082	097	932
14	371		-001	876	067	191	808
15	372		-001	826	057	296	773
16	373		042	774	006	394	756
29	317	Grapestone	079	039	982	122	987
28	514		022	081	974	149	978
19	328		227	280	916	047	971
04	59		231	095	889	105	864
33	409		299	093	854	219	875
27	512		195	051	850	293	849
03	43		569	166	789	126	990
24	502		195	141	688	415	703
25	504	Coralgal	001	192	043	927	898
18	207		095	236	196	882	881
32	7-47		000	127	304	870	865
30	537		054	130	256	847	803
40	250		138	127	250	846	813
26	279		040	294	344	814	869
31	37		038	213	132	733	602
01	1		-012	344	487	509	615

2×2 matrix is left. The results of the above procedure are displayed in Figure 6.

Sequential grouping system.—From the factor analysis of 40 Bahamian sediments, 5 distinct groups (and 2 subgroups) have been identified.

One hundred and sixty Bahamian samples remain to be examined and classified. In principle, this objective can be accomplished by factoring the entire 200×200 matrix. Such an approach strains the storage capacity of computers now available,

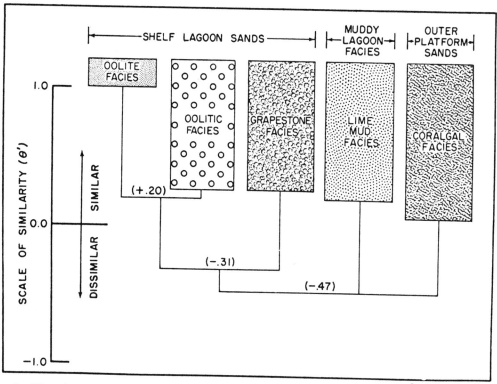

FIG. 6.—Hierarchy diagram depicting similarities among five major Bahamian facies. For explanation, see text.

however; and supplementary grouping procedures must therefore be sought. One such procedure (described below) views the clusters provided by factor analysis as the skeletons of denser clusters to be filled in by subsequent comparisons.

1. Compute a cos θ matrix for all samples. In this case, a 200×200 matrix was computed on an IBM 7090 in approximately 12 minutes. For obvious reasons, it is not reproduced in this paper.

2. Identify the 40 elements of the skeletal clusters with suitable colored notations on the large matrix.

3. Choose one facies, and examine, in turn, each member. Designate a sample under scrutiny as a "host sample." For each host determine the unclassified sample most similar to it. Designate this sample a "trial" sample. If the trial sample is related to any member of another facies more closely than it is to the host sample, do not include the trial sample in the host sample facies. If the trial sample is not related to any member of another facies more closely than it is to the host sample, admit the trial sample as a new member of the host facies.

TABLE IV. VALUES OF COSINE θ AMONG UNIT CENTROID FACIES VECTORS

	Oölite	Oölitic	Lime Mud	Grapestone	Coralgal
Oölite	1.000	.810	.091	.268	.062
Oölitic	.810	1.000	.411	.674	.338
Lime Mud	.091	.411	1.000	.256	.399
Grapestone	.268	.674	.256	1.000	.492
Coralgal	.062	.338	.399	.492	1.000

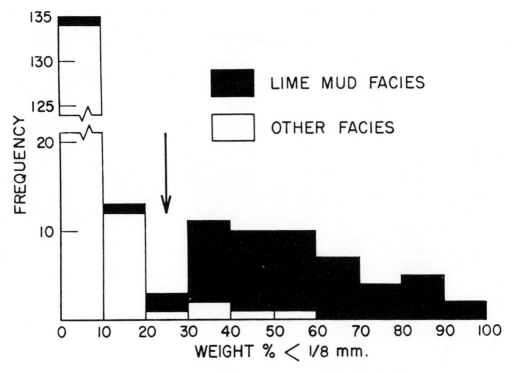

Fɪɢ. 7.—Frequency distribution of weight per cent less than ⅛ mm in 200 samples of Bahamian sediments. Arrow indicates optimum dividing line between lime mud and other facies.

4. When each member of the first facies has been examined in this way, proceed to the next facies and repeat step 3.

In effect, this procedure allows each skeleton cluster to grow or be filled in independently of other such clusters, and natural discontinuities in the vector configuration are discovered, if they exist. In the present case, each of the initially unclassified samples was shown to belong clearly with one of the skeletal clusters. Definite discontinuities exist among the six clusters.

RESULTS

The principal results of applying the methods outlined above to the 200 samples of Bahamian bottom sediments can be summarized as follows

1. Based on an equal and simultaneous consideration of the 12 sediment properties examined by Purdy (Ms.), the samples group themselves into five discrete groups, with definite discontinuities among them. These boundaries are considered to reflect significant discontinuities in the dynamic system responsible for the observed properties. Note that they are not reflected in the distributions shown on Figures 7, 8, 9, and 11.

2. Eighty-nine per cent of the observational data gathered by Purdy can be accounted for by considering each sample as a mixture of four complex end members (samples 38, 12, 29, and 25). Considering that sampling and identification errors are inherent in the data, and that twelve parameters were measured, the simplicity of this result is notable. Its meaning is clear—the many processes responsible for the distribution and abundance of grain types are highly interdependent, integral parts of a tightly organized marine ecosystem. *"Natura simplex est."*

3. Each of the recognized facies is characterized by distinctive properties. Purdy (Ms.) has described and interpreted these in detail. For present purposes it is necessary only to review the salient characteristics of each facies.

Oölite facies.—Nearly pure oölite sand.

Oölitic facies.—Cleanly washed sand with a high portion of oölite grains and considerable quantities of other grain types.

Grapestone facies.—Cleanly washed sand characterized chiefly by grapestone lumps and recrystallized grains, but containing significant proportions of oölite and skeletal grains.

Lime mud facies.—Poorly sorted sediments with a sufficient quantity of grains smaller than $\frac{1}{8}$ mm to keep the modal class of sand grains from being in contact. Two subfacies are recognized, skeletal mud and pellet mud, the former with relatively more fine material and skeletal grains, the latter with relatively more pellets and less fine material.

Coralgal facies.—Cleanly washed sands characterized by relatively abundant corals and algae, including *Halimeda* and coralline algae.

4. As indicated in the hierarchy diagram (Fig.

6) three facies families are recognizable—shelf lagoon sands, muddy lagoon facies, and outer platform sands. The families are equally distinct from each other ($\theta' = 0.47$) and represent sharply contrasting bottom types. The grapestone facies, although distinct from the combined oölite facies, has its closest relationship with that facies. Note that the petrographic relationships symbolized in the hierarchy diagram are borne out by the distribution of the corresponding map patterns (Fig. 10). The field distribution of the three facies families corresponds with three distinct hydrographic environments.

5. Figures 7, 8, and 9 illustrate the most efficient means of classifying the 200 samples into the 5 major facies by observing four characteristics—per cent of fines (wgt. per cent less than $\frac{1}{8}$ mm); and volume per cent of oölite, grapestone, and corals plus algae.

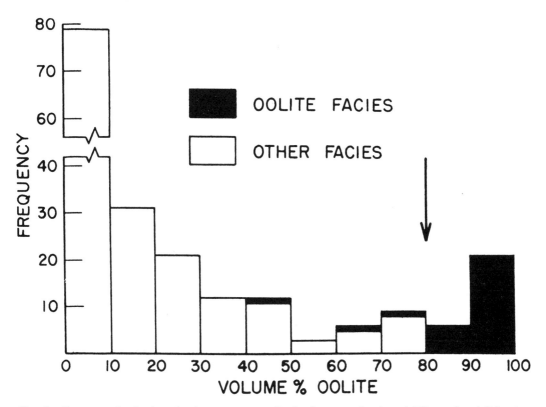

Fig. 8.—Frequency distribution of volume per cent oölite in the coarse fraction of 200 samples of Bahamian sediments. Arrow indicates optimum dividing line between oölite and other facies.

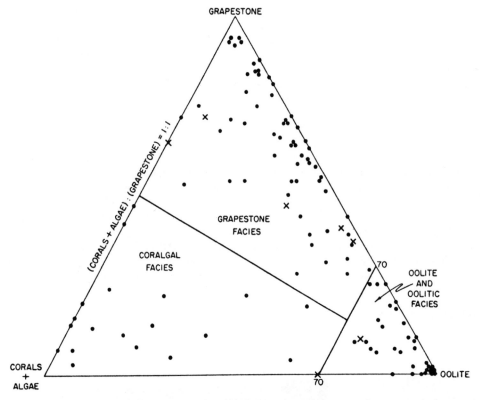

FIG. 9.—Triangular plot of Bahamian samples classified by an optimum procedure as coralgal, grapestone, oölite, and oölitic facies. Using the suggested ratio boundaries, samples plotted as points are correctly classified. Samples plotted as crosses are incorrectly classified.

(1) Classify all samples having more than 25 per cent fines as lime mud. A total of six identification errors are committed.

(2) Classify samples having more than 80 per cent oölite as oölite facies. Three errors are committed.

(3) Classify the remaining samples according to the ratio boundaries in Figure 9. Nine identification errors are committed.

With this scheme, identification errors total eighteen, or just under 10 per cent.

6. Figure 10 represents the facies map corresponding to the sample groupings identified by the optimum classification procedure described in this paper. Mapped facies boundaries have been drawn between control points to conform with hydrographic and other information available to the writers. As discussed by Purdy (Ms.), the facies

patterns bear systematic relationships to hydrographic parameters.

APPLICATION OF FOLK'S CLASSIFICATION TO BAHAMIAN SEDIMENTS

CRITERIA

Many of the limestone varieties recognized in the classification proposed by Folk (1959) are undoubtedly represented in unconsolidated sediments forming today on the Bahama Banks. Several of these are not considered in detail here, as they are forming in areas not covered by the 200 sample localities. Algal biolithite or dismicrite is clearly accumulating in intertidal mud flats, for example, and coral-algal biolithite is being formed in some areas of reef growth and on rocky shoals. In this paper, attention is directed principally to

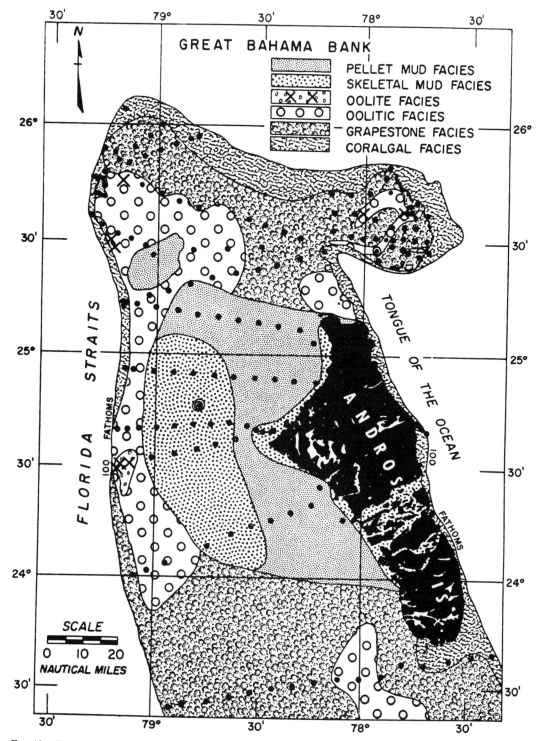

Fɪɢ. 10.—Facies map of the Andros lobe of the Great Bahama Bank based on optimum classification procedure. Heavy dots show location of 200 sample stations.

111

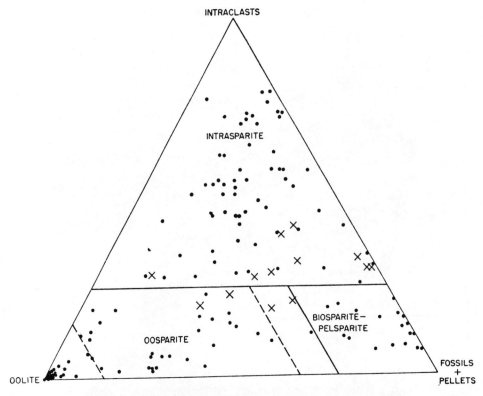

Fig. 11.—Triangular plot of Bahamian samples according to the classification proposed by Folk. Only samples judged to represent Type I limestones are considered. Heavy lines are Folk's boundaries; dashed lines are revised boundaries suggested in this paper. Samples plotted as points are correctly classified according to the optimum classification procedures used in present paper. Samples plotted as crosses are incorrectly classified. Compare with Fig. 12.

infratidal areas covered by the 200 localities indicated on Figure 10. These clearly fall into Folk's limestones of Type I (sparry allochemical limestone) or Type II (microcrystalline allochemical limestone). Possibly a few of the samples should be classified as Type III (micrite), indicating that the lithified and compacted equivalent would contain less than 10 per cent allochems.

In judging whether these modern samples should be classified as Type I or Type II, decision must be based on a judgment as to the relative volumes of sparry calcite cement and microcrystalline calcite matrix in their unknown lithified and compacted equivalents. For samples consisting of cleanly washed sand this decision can be made without difficulty, and the samples classified as Type I. For samples containing significant

quantities of fine material, however, this decision is unfortunately very difficult to make, at least from data and samples available to the authors. For this reason, we have somewhat arbitrarily chosen to classify every sample of the lime-mud facies as Type II, and the remaining samples as Type I. Two exceptions were made for samples designated by the optimum classification system as belonging to the lime-mud facies, but which contained so little fine material relative to intergranular voids that the resulting rock can be expected to be a Type I limestone.

By making this arbitrary division between Type I and Type II, we have maximized the degree of correspondence between Folk's classification and the optimum system. From an inspection of wet and dried samples of the lime-mud facies, however, and from an inspection of three thin sections prepared from unsieved, undisturbed, plastic-impregnated samples, we are reasonably certain that a considerable number of samples from this facies would be classified by Folk's scheme as Type I, and thus be considered as cleanly washed limestones. If this suspicion is borne out by future work, it would suggest a flaw in the basis for distinguishing Types I and II, for it would class together some samples coming from environments distinct in terms of circulation and bottom agitation. Illustrations of this are particularly prevalent in the pellet-mud facies, where many samples are relatively cleanly washed below a grain size within the silt grade.

After distinguishing Type I from Type II samples, each group was subdivided according to the criteria proposed by Folk. Figure 11 shows the properties of Type I samples. In plotting these points, mud aggregates and grapestone were lumped as intraclasts, and the total skeletal grain content was lumped with pellets. Using the definition of oölite employed by Newell, Purdy, and Imbrie (1960), any grain is classified as an oölite if a recognizable oölitic coating of any size is present. Solid ratio lines on this figure correspond to those recommended by Folk. The nine samples plotted as crosses (6 per cent of 151 samples) correspond to Type I samples misclassified according

to the category equivalence indicated in Figure 12. Disregarding the lack of subdivision of the two oölitic facies in Folk's scheme, and disregarding the subdivision of the coralgal facies into biosparite and biopelsparite, this figure for identification error can be compared with a corresponding identification error of seven achieved by application of the four-parameter optimum classification scheme described above. This result is a striking confirmation of the correctness of the boundary locations proposed by Folk for subdividing Type I limestones. A slight improvement can be achieved by moving the oösparite-biosparite boundary to 35 per cent oölite as suggested by the dashed line in Figure 11. Also, if an oölite per cent boundary of 85 per cent is used to partition off a high-oölite subdivision of oösparite, the distinction in Bahamanian sediments between oölite and oölitic facies can be recognized with only three errors.

The distinction between biosparite and pelsparite does not correspond with a natural dividing line among Bahamian sediment samples treated in this paper.

As indicated in Figure 12, Folk's classification subdivides Type II samples into five categories—oömicrite, intramicrite, biomicrite, biopelmicrite, and pelmicrite. These do not correspond to the skeletal mud and pellet mud subfacies, and sug-

PRESENT PAPER			FOLK (1959)	
SHELF LAGOON SANDS	OOLITE FACIES		OOSPARITE	TYPE I
	OOLITIC FACIES			
	GRAPESTONE FACIES		INTRASPARITE	
OUTER PLATFORM SANDS	CORALGAL FACIES		BIOSPARITE	
			BIOPELSPARITE	
MUDDY SANDS OF THE SHELF LAGOON	SKELETAL MUD SUBFACIES	LIME MUD FACIES	OOMICRITE	TYPE II
			INTRAMICRITE	
			BIOMICRITE	
	PELLET MUD SUBFACIES		BIOPELMICRITE	
			PELMICRITE	

FIG. 12.—Comparison between categories of the optimum classification scheme of the present paper and those of Folk's classification.

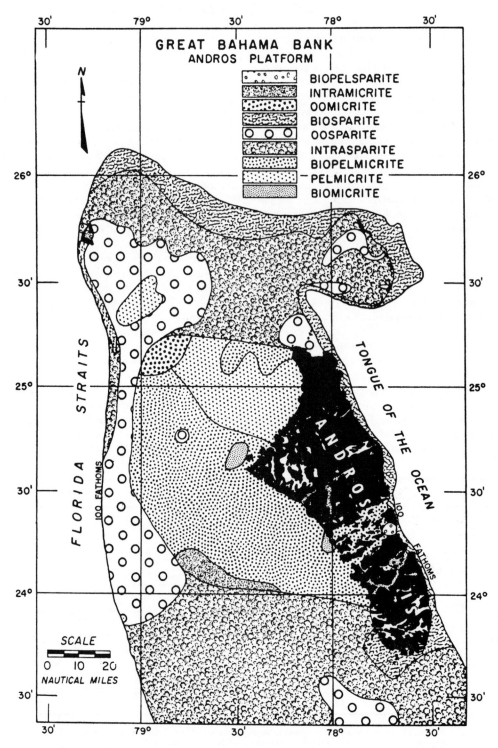

FIG. 13.—Facies map of the Andros lobe of the Great Bahama Bank based on Folk's classification. For discussion of criteria, see text.

114

gest (for Bahamian data at least) that this group is oversplit. The situation here resembles that indicated in Figure 2, where objective gradients have been subdivided along *a priori* boundaries. If, instead of the two fossil-pellet ratio boundaries proposed by Folk, a single boundary is placed at a skeletal:pellet ratio of 1:2, the Bahamian samples can be split into groups corresponding to skeletal mud and pellet mud with an identification error of only 10 per cent. It is therefore suggested that empirical tests be made in limestones using a corresponding fossil:pellet ratio.

COMPARISON OF RESULTS OF THE TWO CLASSIFICATIONS

If Folk's classification is applied to the Bahamian sediment samples considered here, according to the criteria discussed above, the samples are subdivided into nine named categories. As indicated on Figure 12, this result compares with six named categories of the optimum scheme. In general, the correspondence is excellent within Type I, and poor within Type II.

Perhaps a more meaningful way of comparing the results of the two systems is to examine the corresponding facies maps. Figure 13 represents the map distribution of the categories yielded by Folk's scheme, with facies boundaries between control points drawn to correspond as closely as possible to those of Figure 10. For the areas mapped as Type I (sparites) the correspondence between the two systems is excellent, and strongly supports Folk's classification. Three differences, however, are worth noting.

1. Folk's classification does not reflect the high-oölite shoals. As suggested above, this discrepancy could be removed by adding an appropriate category.

2. In the Berry Islands area (north of Andros Island) a portion of an oölite shoal is ignored by Folk's classification.

3. Some sediment samples along the western edge of the Banks on the outer platform are misclassified by Folk's scheme.

For areas mapped as Type II (micrites), a considerable discrepancy exists between the results of the two classifications. The facies boundary between skeletal and pellet mud is missed by Folk's scheme, although it should be remembered that his subdivisions do correspond to objective, systematic differences in allochem proportions and reflect actual compositional gradients.

REFERENCES

Black, Maurice, 1933, The precipitation of calcium carbonate on the Great Bahama Bank: Geol. Magazine, v. 70, no. 10, p. 455–466.

Cattell, R. B., 1952, Factor analysis: New York, Harper and Bros.

Folk, R. L., 1959, Practical petrographic classification of limestones: Am. Assoc. Petroleum Geologists Bull., v. 43, no. 1, p. 1–38.

Harman, H. H., 1960, Modern factor analysis: Chicago, Ill., The Univ. Chicago Press.

Illing, Leslie, 1954, Bahaman calcareous sands: Am. Assoc. Petroleum Geologists Bull., v. 38, no. 1, p. 1–95.

Krumbein, W. C., 1957, Comparisons of percentage and ratio data in facies mapping: Jour. Sed. Petrology, v. 27, no. 3, p. 293–297.

Newell, N. D., Imbrie, J., and others, 1959, Organism communities and bottom facies, Great Bahama Bank: Am. Mus. Nat. History Bull., v. 117, art. 4, p. 181–288.

———— Purdy, E. G., and Imbrie, J., 1960, Bahaman oölitic sand: Jour. Geology, v. 68, no. 5, p. 481–497.

———— and Rigby, J. K., 1957, Geological studies on the Great Bahama Bank, *in* Regional aspects of carbonate deposition: Soc. Econ. Paleontologists and Mineralogists Spec. Pub. No. 5, p. 15–72.

———— Whiteman, A. J., and Bradley, J. S., 1951, Shoal water geology and environments, eastern Andros Island, Bahamas: Am. Mus. Nat. History Bull., v. 97, p. 1–29.

Purdy, E. G., Ms., Recent calcium carbonate facies of the Great Bahama Bank. (Scheduled for publication in 1963.)

Smith, C. L., 1940, The Great Bahama Bank: Jour. Marine Research, v. 3, p. 147–189.

Sokal, R. R., and Michener, C. D., 1958, A statistical method for evaluating systematic relationships: Univ. Kansas Sci. Bull., v. 38, pt. 11, no. 22, p. 1409–1438.

Spearman, C., 1913, Correlations of sums and differences: British Jour. of Psychology, v. 5, p. 417–426.

11

Reprinted from *J. Anim. Ecol.* **32**:535–547 (1963)

MULTIVARIATE ANALYTICAL TREATMENT OF QUANTITATIVE SPECIES ASSOCIATIONS: AN EXAMPLE FROM PALAEOECOLOGY*

By R. A. REYMENT

Department of Geology, University of Stockholm†

INTRODUCTION

A common observation made by micropalaeontologists in analysing series of borehole samples is that the proportions of species may fluctuate widely from sample to sample. It would not seem unreasonable to assume that the oscillations in numbers of individuals were caused by ecological factors. Although it seems hardly likely that the palaeoecologist will ever be in a position to identify such factors specifically, the statistical methods of principal component analysis and factor analysis, applied to frequency data, would seem to be suitable for supplying a picture of the order of complexity of the environment in which these species lived, and their reactions to it. Both methods attempt to divide up the total variation into entities, which one tries to provide with a meaningful interpretation. Both were originally devised for application to psychometric problems; factor analysis has been used for some 30 years in the analysis of biological variation, while principal component analysis has only more recently been applied to such problems.

Numerous examples of the taxonomic use of principal component analysis with respect to the ostracod carapace are given in Reyment (1963). A discussion of factor analysis as applied to an ecological problem in botany, comparable to that treated in the present paper, is to be found in Greig-Smith (1957, pp. 147–9, 159, 161–3). A general text on multivariate statistical analysis, which contains a detailed account of the statistical theory involved in the following, is that of Kendall (1957).

Ostracods are useful indicators in palaeoecology as they are amongst the most common animal fossils in borehole samples, they are benthonic (in, on, or close to the substratum) and, as opposed to their greatest potential rivals, the Foraminifera, they occur in a wide range of aquatic environments.

A note on the interpretation of the mathematical results might be in place. The correlation coefficients between pairs of variables (species), ordered in a correlation matrix, are subjected to the mathematical procedure known as a transformation (equation (1)). The new matrix **D** in equation (1) has all off-diagonal elements equal to zero and its diagonal elements add up to the same total as the diagonal elements of the original correlation matrix, in our example 17. This diagonalization process is of wide application in applied mathematics. It was used by Jacobi over a hundred years ago in his study of the orbits of the planets and his solution is the one used in most electronic computer programmes today. Many statistical procedures make use of the process, including the statistical treatment of certain taxonomical problems and a problem in population dynamics.

The element d_1 of matrix **D** is larger than any of the succeeding elements. We shall here refer to it as an *eigenvalue*; other terms in use for it are 'latent root' and 'characteristic root'. Each eigenvalue is associated with two *eigenvectors* (for an asymmetric

---

* Publication No. 6 of the Department of Geology, University of Ibadan, Nigeria.
† Present address: Department of Geology, University of Ibadan, Ibadan, Nigeria.

116

matrix there are two different eigenvectors, but in this paper we shall only be concerned with symmetric matrices, where the eigenvectors do not differ); other terms are 'latent vector' and 'characteristic vector'. The relationship is indicated in equation (2).

It is easy to see that the elements of **D** represent another distribution of the total variance of the original matrix (the sum of its diagonal elements—this is called the *spur* or the *trace* of a matrix; thus, spur **A** = spur **D** in equation (2)).

In the taxonomical analysis of, for example, ostracods, the first eigenvalue is usually many times greater than the second eigenvalue and we say that most of the variation is concentrated to the *first principal component*. This is given by the elements of the first eigenvector, which are employed as the coefficients in an equation. Thus, if the elements of the eigenvector are (b_1, b_2, b_3), and the dimensions measured, x_1, x_2, x_3, the first principal component may be written out as

$$U_1 = b_1x_1 + b_2x_2 + b_3x_3.$$

The variance of U_1 is given by d_1, the first element of **D**.

In our palaeoecological problem the x_i are species and the b_i denote the importance of species x_i in a particular principal component. We should therefore be able to recognize associations of species, yielded by the principal components, the relative significance of each species being indicated by the corresponding coefficient of the pertinent eigenvector (in statistical parlance this coefficient is sometimes called a 'weight' or 'weighting'; the corresponding psychometric term of factor analysis is 'loading').

It will usually be found that the first few eigenvalues account for almost all of the variance. The question is, then, whether any importance is attached to the remaining eigenvalues. This may be tested by seeing if the remaining eigenvalues are indistinguishable from each other (isotropic residue), which also implies that they are unimportant with respect to magnitude. Obviously, all eigenvalues must, in a sense, be significant (*cf.* Kendall 1957), unless we have a matrix, the rank of which is less than its order, which is a remote possibility in the kinds of biological problems we consider here. What the aforementioned test helps us to do is to decide the number of important eigenvalues; that is, in the ecological problem under review, the ones that have been most important in determining the frequencies in the various species associations, indicated by the elements of the corresponding eigenvalues. One would like to be able to test these coefficients for significance, but, as far as the author is aware, no such tests appear to be available. Consequently, there must be an element of subjectivity attached to decisions regarding the importance of any vector element.

Up to now the discussion has been in terms of principal component analysis. The ecological interpretation of the results of factor analysis is made in much the same way. However, the structure of factor analysis differs somewhat from principal component analysis, as indicated by equation (3), for it includes a term for random variation. The 'communalities' of factor analysis, discussed further on, are derived from the random matrix, as is indicated by equation (4). The method of factor analysis employed in the present paper would appear to be an improvement over previous methods, as it provides a more realistic assumption concerning the matrix of residual variances (equation (5)).

STATEMENT OF THE PROBLEM

We have, say, k samples from boreholes in which p different species occur (not necessarily all p species occur in each of the k samples). For each sample a certain arbitrary total number of individuals, N, are picked (randomly), this N being the same for all samples.

A matrix $\mathbf{T}$ of 'scores' (relative frequencies) results, each row of which consists of the score for each of the p species in the j-th sample ($j = 1, \ldots, k$).

$$\mathbf{T} = \begin{bmatrix} \nu_{11}s_1 & . & \nu_{12}s_2 & . & . & . & . & \nu_{1p}s_{1p} \\ . & . & . & . & . & . & . & . \\ \nu_{p1}s_1 & . & . & . & . & . & \nu_{pp}s_p \end{bmatrix}$$

here ν_{ik} represents the score for species s_k ($\nu_{ik} \geqslant 0$).

It is desired to use these observations on the frequencies of the species to ascertain the extent of relationship in their occurrence. That is, one would like to see if species that react in the same way to all factors in their environment can be marshalled into one group, and those that are adversely affected by some environmental facet can be made to show up in another group or groups.

METHOD 1

It would seem that a possible solution of the problem might be offered by the well-known linear algebraic procedure, whereby a matrix may be reduced by means of transformations to a simple form (*canonical form*), the rank of which is equal to the rank of the original matrix. The new matrix is a diagonal matrix. This procedure forms the basis of '*principal component analysis*', introduced by Hotelling (1933) for the study of psychometrical data.

For any symmetric ($k \times k$) matrix $\mathbf{A}$, there exists an orthogonal matrix $\mathbf{B}$ such that

$$\mathbf{B'AB} = \mathbf{D} = \begin{bmatrix} d_1 & 0 & . & . & . & . & 0 \\ . & . & . & . & . & . & . \\ 0 & . & . & . & . & . & d_k \end{bmatrix} (d_1 > d_2 \ldots > d_k) \quad (1)$$

If the matrix $\mathbf{A}$ is positive definite, then the $d_i > 0$. (In other words a positive definite symmetric matrix has only positive eigenvalues.)

Corresponding to each eigenvalue there will be an eigenvector, $\mathbf{c}$, such that if d is any eigenvalue of $\mathbf{A}$, $\mathbf{c}$ is a non-zero vector satisfying the equation

$$(\mathbf{A} - d\mathbf{I}) = \mathbf{0} \quad (2)$$

These eigenvalues and eigenvectors provide the basis for the ecological analysis. Primarily we are interested in the components of the eigenvectors, which we may regard as *ecospectra*; i.e. the observational vectors are broken up into functionally connected spectra of ecological reaction. The same reasoning is applicable to the matrix of factor loadings of *factor analysis*, treated as method 2.

Computational procedure for method 1

1. Note the frequencies of each of the p species in k samples for a total of N individuals.

2. Arrange the data for each sample in rows the one above the other; the order of the rows with respect to each other should be random (important for palaeoecological data in order to eliminate the possibility of serial correlation).

	s_1	.	.	.	.	.	.	.	s_p
Sample 1	v_{11}	.	.	.	.	.	.	.	v_{1p}
	.	.	.	..	.	.	.	.	.
Sample k	v_{p1}	.	.	.	.	.	.	.	v_{pp}

The v_{ij} represents the score for the j-th species in the i-th sample. Because of the constant $\Sigma v_{ij} = N$, these scores are proportions which may be directly compared between samples. Logically, the greater one takes the value of N, the more likely are the components of the observational vectors to approach stability.

3. Compute all combinations of correlation coefficients between frequencies of species. It may be more suitable to express the data as decimal proportions at this stage.

4. The correlation coefficients are arranged into a matrix of correlations and this correlation matrix is reduced to its canonical form. This is best done on an electronic computer as the arithmetical labour involved for more than three or four variables is prohibitive.

Interpretation

Each eigenvalue gives a quantitative indication of the relative importance of the species combination represented by the components of the associated eigenvector. Particularly in large matrices, it may be found that only the first few eigenvalues will be meaningful; the remaining eigenvalues can be tested to see whether they differ significantly from an isotropic residue (Lawley 1956) The eigenvector may be written out in equation form as

$$z = a_1b_1 + a_2b_2 + \ldots \ldots + a_pb_p$$

The usual computer solution of an eigenvalue problem provides normalized vector components so that an a_i may receive any value between 0 and 1, depending on the importance of the contribution of the variable. Hence, the variables (= species) of the above equation that have large coefficients are said to be causing most of the variation represented in the eigenvalue. The reasoning is further elucidated in the worked example.

Example using method 1

The example considered here is palaeontological. It will be apparent, however, that the same reasoning may also be applied to data concerning relative abundances of living associations (*cf.* Greig-Smith 1957, p. 147).

The first 600 randomly selected individuals of twenty-eight borehole samples from western Nigeria, containing seventeen species of Paleocene (Cenozoic) ostracods were used to construct the correlation matrix given in Table 1. (Significant correlations printed in bold face type; these are 'total' correlations not taking the influences of the other $k-2$ variables into account.)

The species involved, in order of representation in the matrix, are *Cytherella sylvesterbradleyi* Reyment (x_1), *Bairdia ilaroensis* Reyment and Reyment (x_2), *Ovocytheridea pulchra* Reyment (x_3), *Iorubaella ologuni* Reyment (x_4), *Brachycythere ogboni* Reyment (x_5), *Dahomeya alata* Apostolescu (x_6), *Leguminocythereis bopaensis* (Apostolescu) (x_7), *L. lagaghiroboensis* Apostolescu (x_8), *Trachyleberis teiskotensis* (Apostolescu) (x_9), *Veenia warriensis* Reyment (x_{10}), *Buntonia (Buntonia) fortunata* Apostolescu (x_{11}),

Table 1. *Correlation matrix based on collections of seventeen species of fossil ostracods*

1	2	3	4	5	6	7	8	9	10	11	12	13	14	15	16	17
1·0000																
−0·3176	1·0000															
0·1363	−0·1902	1·0000														
0·1804	−0·1775	−0·3495	1·0000													
−0·2547	0·1606	−0·1285	−0·1838	1·0000												
−0·0267	−0·2299	−0·1599	0·2067	−0·0948	1·0000											
−0·0038	−0·1499	−0·2642	0·0481	−0·1279	**0·4711**	1·0000										
−0·1540	−0·1756	−0·2558	−0·1547	0·0639	0·0332	0·0808	1·0000									
−0·3659	−0·1807	−0·1862	−0·2026	−0·0435	0·0945	0·2368	−0·0499	1·0000								
−0·0509	−0·0028	0·2996	−0·0920	−0·2624	0·3722	−0·1809	−0·0083	0·0675	1·0000							
−0·1097	−0·2288	−0·2509	**0·5888**	−0·3699	0·0400	−0·2040	−0·1603	−0·1889	−0·2021	1·0000						
−0·2104	−0·1734	−0·1110	0·0320	−0·2349	−0·1206	−0·0903	−0·2735	−0·1878	0·3229	**0·7303**	1·0000					
−0·3085	−0·3166	0·3419	0·0981	−0·1927	−0·1685	−0·1421	−0·0871	−0·1408	−0·3229	−0·2679	0·0075	1·0000				
−0·0213	−0·2386	−0·2331	−0·0183	−0·1567	−0·0722	−0·0079	−0·3098	−0·0515	−0·1285	−0·1212	−0·1286	0·2957	1·0000			
−0·0130	−0·2420	−0·1388	−0·0855	−0·3088	0·0792	−0·3362	−0·3636	−0·2613	0·0320	−0·0136	−0·2747	0·0277	−0·0811	1·0000		
−0·1513	−0·1549	**−0·5050**	**−0·4269**	0·2614	−0·3008	0·2442	0·0315	**0·4612**	0·0118	0·0118	−0·2447	−0·1761	−0·0099	**0·4261**	1·0000	
−0·1389	−0·1714	**−0·4500**	0·1173	0·2841	0·2005	0·3086	−0·1857	**0·4612**	0·0315	**0·4265**	−0·3249	−0·3127	−0·1052	−0·0165	**0·4563**	1·0000

B. (Buntonia) beninensis Reyment (x_{12}), *B. (Buntonia) bopaensis* Apostolescu (x_{13}), *B. (Quasibuntonia) livida* Apostolescu (x_{14}), *Ruggieria tattami* Reyment (x_{15}), *Schizocythere* spp. (x_{16}), *Xestoleberis kekere* Reyment (x_{17}). Data from Reyment (1960, 1963). The results of the principal component analysis are shown in Table 2.

We note that all eigenvalues are positive, which is a result of the correlation matrix being symmetric and positive definite. A very interesting aspect of the analysis is that the first component only accounts for 18% of the total variation. Compared with applications of principal component analysis in taxonomic work this is low indeed (*cf.* Reyment 1963). The first three eigenvalues add up to 46·57% of the total variation and the first four to 57%. The slight differences between successive eigenvalues indicate the great computational difficulties that would have been experienced had the work been done manually. The differences between successive eigenvalues are smaller between the first and last few eigenvalues than between the intermediate ones. We note a further feature of the results, notably, that the elements of the first eigenvector are not all positive, which is a situation differing from that usually found in principal component analysis of biological materials (*cf.* Reyment 1963).

One interpretation that suggests itself is that the several eigenvalues of roughly the same magnitude may indicate the existence of more than one ecological factor and that all are of roughly the same importance.

Using Lawley's (1956) test of significance it was found that the first five eigenvalues are significantly different from each other, while the remaining twelve may be regarded as being isotropic.

Approximate 95% confidence intervals for the eigenvalues were found as follows:

The procedure is to choose l and u so that

$$\Pr[nl<\chi_n^2] = \sqrt{(1-\epsilon)}; \ \Pr[\chi_n^2<nu] = \sqrt{(1-\epsilon)}.$$

$$\text{Then } \Pr\left[\frac{\lambda_m(\mathbf{S})}{u} <\text{all } \lambda\ (\boldsymbol{\Sigma})< \frac{\lambda_M(\mathbf{S})}{l}\right]\geq 1-\epsilon.$$

(λ = the eigenvalues and $n = N-1$; λ_m = smallest eigenvalue and λ_M = largest eigenvalue)

These intervals are $0·07<\text{all } \lambda_i<35·08$, with a probability of 0·96.

A complication encountered when interpreting the meaning of the eigenvectors is that the standard deviations of the variables are frequently·very different, which is an outcome of the nature of the data. In the present problem it might have been advisable to perform some sort of scaling operation on the input material. The vector of standard deviations is:

$$\mathbf{s}' = (0·1384, \ 0·1593, \ 0·0609, \ 0·0475, \ 0·0247, \ 0·0123, \ 0·0425, \ 0·0714, \ 0·0829, \ 0·0213,$$
$$0·0514, \ 0·0850, \ 0·0733, \ 0·0808, \ 0·0431, \ 0·0421, \ 0·0101)$$

Dividing each component of the eigenvectors by the appropriate standard deviation, as is often done in principal component analysis based on a correlation matrix, did not lead to intelligible results.

A rough interpretation of the eigenvectors corresponding to the first five eigenvalues is that the components of the first vector are mainly suggestive of some environmental factor which influences species 16 and 17 in one direction and species 3 in another. The second eigenvector is mainly concerned with species 5, 11 and 12 and suggests the exist-

Table 2. *The principal component analysis*

Eigenvalue	Percentage of total variation and diff. $\lambda_i - \lambda_{i+1}$		Eigenvectors
3·1052	18·27		(0·0832, −0·0973, −0·4252, 0·2438, 0·0065, 0·2300, 0·2360, −0·2753, 0·1080, −0·1759, 0·1954, 0·0852, −0·1990, −0·1003, 0·1690, 0·4427, 0·4435)
2·8655	16·86	1·41	(−0·1007, 0·1494, −0·0737, −0·1465, 0·4621, −0·2244, −0·1195, −0·1376, −0·2157, 0·0651, 0·3962, 0·4695, −0·2127, −0·1497, −0·2506, −0·2386, 0·1672)
1·9441	11·44	5·42	(−0·0323, 0·1795, −0·0531, −0·2826, 0·0705, 0·1092, 0·4530, 0·2376, 0·0294, 0·5075, −0·0079, −0·2717, −0·3912, −0·3246, 0·1172, −0·0308, −0·0144)
1·7764	10·45	0·99	(0·4703, 0·3122, −0·0128, 0·2283, −0·0938, −0·1424, −0·0681, −0·2084, −0·4765, −0·0761, −0·2868, −0·2290, −0·3481, −0·0224, −0·2487, 0·0270, −0·0126)
1·6070	9·45	1·00	(−0·4411, 0·5877, −0·3302, 0·0077, 0·0149, −0·1952, −0·2028, −0·0827, 0·2877, −0·2192, −0·1959, −0·1202, 0·0128, −0·2083, 0·0313, 0·1065, −0·1587)
1·1766	6·92	2·53	(−0·3485, 0·1197, −0·1118, 0·3392, 0·1588, 0·4946, 0·3246, 0·1046, −0·2307, 0·0051, −0·2213, 0·0641, 0·1974, 0·1327, −0·3858, −0·1939, 0·0337)
1·0167	5·98	0·94	(−0·1522, 0·2167, 0·1220, −0·4273, 0·0597, 0·0898, 0·0825, −0·4874, −0·0272, 0·1011, −0·0061, −0·1876, −0·0137, 0·6258, 0·0580, 0·0531, 0·1796)
0·8610	5·06	0·92	(0·0465, 0·0717, 0·0209, −0·1252, 0·3317, 0·4378, −0·2595, 0·0706, −0·3452, −0·2520, −0·1079, 0·0190, −0·0469, −0·0974, 0·6129, −0·0976, −0·1053)
0·6940	4·08	0·98	(0·0110, −0·0613, 0·2005, 0·3263, 0·3224, −0·1049, 0·0111 −0·5049, 0·0432, 0·4600, −0·2023, 0·0037, 0·3202, −0·2689, 0·1878, 0·0864, −0·0869)
0·6050	3·56	0·52	(−0·1580, −0·1108, −0·3080, 0·3368, 0·1484, −0·2197, −0·0507, 0·1772, −0·1719, 0·2609, 0·1767, −0·0202, −0·2246, 0·4767, 0·1991, 0·1872, −0·4107)
0·4302	2·53	1·03	(0·1081, −0·1690, 0·0023, −0·0274, 0·5267, 0·0105, −0·3173, 0·2839, 0·3200, 0·0632, −0·3887, −0·1422, −0·1655, 0·1482, −0·2202, 0·1739, 0·3072)
0·3077	1·81	0·72	(−0·2783, 0·2141, 0·5282, 0·1785, −0·2111, 0·2332, −0·3230, 0·1314, −0·1002, 0·1839, 0·2030, 0·0830, −0·1806, −0·0689, −0·0704, 0·4424, 0·1234)
0·2286	1·34	0·47	(0·2314, 0·0514, 0·0591, 0·1078, 0·0234, 0·4292, −0·0955, −0·2610, 0·4973, −0·0047, 0·1418, 0·1084, −0·3362, 0·0570, −0·1597, −0·1986, −0·4542)
0·2025	1·19	0·15	(0·2141, 0·1178, −0·4328, −0·0664, −0·1403, 0·2547, −0·4470, 0·0370, −0·0612, 0·4007, 0·2759, −0·2355, 0·3607, −0·0298, −0·1124, −0·0899, 0·1232)
0·1194	0·70	0·49	(−0·1768, −0·0193, 0·0062, 0·3237, −0·2712, −0·0804, −0·1662, −0·0321, 0·1293, 0·1536, −0·1420, 0·0268, −0·2809, 0·1001, 0·3031, −0·5810, 0·4188)
0·0403	0·24	0·46	(−0·1869, −0·1720, 0·1738, 0·1889, 0·2332, −0·0474, −0·0011, −0·0830, −0·0181, −0·2729, 0·4286, −0·7038, −0·0342, −0·1215, −0·0848, −0·1680, 0·0301)
0·0198	0·11	0·13	(0·3735, 0·5383, 0·1971, 0·2544, 0·1939, −0·1027, 0·2350, 0·2889, 0·2147, −0·0601, 0·2427, 0·0343, 0·2514, 0·2109, 0·1967, −0·0496, 0·1472)

ence of some environmental factor that influences all of these in the same way. The third eigenvector indicates the existence of an environmental factor that influences species 7 and 10 in one way and species 13 and 14 oppositely. The fourth vector suggests a factor that influenced species 1 and 2 in one direction and species 9 and 13 in another. The fifth component would appear to indicate an influence on species 1 and 5 in one direction and on species 2 in another.

Now, in principal component analysis it is assumed that the variation observed is mainly due to factors having characteristic effects on each of the variates. It is therefore open to question whether the foregoing analytic model is really the best way of attacking the problem. The related technique of factor analysis takes the possibility of extraneous variation into account. The $r_{ii} = 1$ terms of the correlation matrix are replaced by so-called 'communalities' that represent the proportion of the variation of each variate which is due to factors operating on some or all of the other variables, the rest being assumed due to chance. It should be pointed out that the communality concept of factor analysis is not accepted by all quantitative biologists as being valid. Also, in principal component analysis one is primarily interested in studying the *shape* of the scatter of the observations.

METHOD 2

The method of *factor analysis* here employed is that recently proposed by Jöreskog (1962). As it involves certain modifications it is given detailed discussion.

The fundamental postulate of factor analysis is

$$\mathbf{S} = \alpha\beta + \epsilon \tag{3}$$

where $\alpha = (a_{it})$, $\beta = (\beta_{tv})$ and $\epsilon = (\epsilon_{iv})$. Here a_{it} and β_{tv} are non-random quantities, while matrix ϵ is a set of random variables for which $E(\epsilon) = 0$ and $E(\epsilon\epsilon') = n\Delta$, where Δ is a positive diagonal matrix, the diagonal elements of which are the residual variances. We note that the a_{it} are termed factor loadings (= coefficients) of k common factors and the β_{tv} are the factor values of the individuals.

Factor analysis assumes the covariance matrix to be made up of two matrices:

$$\Sigma = \alpha\alpha' + \Delta \tag{4}$$

The diagonal elements of $\Sigma - \Delta$ are called *communalities*.

Under the common assumption of factor analysis that $\Delta = \sigma^2 I$, where σ^2 is a positive constant and I is the unit matrix, a test for significant eigenvalues may give a large number of common factors, since the $p - k$ eigenvalues of the population correlation matrix (which is what is used in most analyses in place of the covariance matrix) are not likely to be nearly isotropic unless k is large. In order to find the least number of common factors Jöreskog (1962) has proposed another assumption for Δ, namely, that

$$\Delta = \theta(\text{diag } \Sigma^{-1})^{-1} \tag{5}$$

where θ is a positive scalar and Σ is non-singular.

Computational procedure for method 2

We shall write $\mathbf{C}$ for the correlation matrix. Beginning with the sample correlation matrix the following steps are required:

1. Compute $\mathbf{D} = \mathrm{diag}\ \mathbf{C}^{-1}$.
2. $\mathbf{C}^* = \mathbf{D}^{\frac{1}{2}}\mathbf{C}\mathbf{D}^{\frac{1}{2}}$.
3. $|\mathbf{C}^* - l^*\mathbf{I}|$, where the l^*_i are the eigenvalues of $\mathbf{C}^*$. At the same time the corresponding eigenvectors are found, $\mathbf{z}^*_i = (z^*_{1i}, z^*_{2i}, \ldots, z_{pi})'$.
4. The diagonal elements of $\mathbf{L}^*$ are tested for significance. $\mathbf{L}^*_k$ is the diagonal matrix of the first k eigenvalues of $\mathbf{L}^*$.
5. Then one computes

$$\mathbf{A}^* = \mathbf{Z}^*_k(\mathbf{L}^*_k - t_k\mathbf{I})^{\frac{1}{2}}$$

where t is defined as

$$t_k = \frac{1}{p-k}\sum_{i=k+1}^{p}l^*_i$$

and $\mathbf{A}$, the estimate of α, is $\mathbf{A} = \mathbf{D}^{-\frac{1}{2}}\mathbf{A}^*$.
6. The diagonal matrix of estimated *residual* variances $\mathbf{R}$ is

$$\mathbf{R} = t_k\mathbf{D}^{-1}$$

7. In order to find a lower confidence limit for k one first tests the hypothesis $k = 1$ against $k > 1$; if this hypothesis is rejected one tests $k = 2$ against $k > 2$, and so on. The number of factors for which the test does not show significance is determined. If the level of significance is 5% in each test, this method leads to a lower confidence limit for the number of common factors. The test criterion is (after Lawley 1956 and Whittle 1952)

$$c_k = -(n-1)[\log (l^*_{k+1} \ldots l^*_p) - (p-k)\log t]$$

where t is defined as in a foregoing equation and the l^*_i are the eigenvalues of $\mathbf{C}^*$. If the hypothesis is true and if n is large, c_k is approximately distributed as χ^2 with degrees of freedom

$$d_k = \tfrac{1}{2}(p-k+2)(p-k-1)$$

Example using method 2

We use the same correlation matrix as in the previous example, that is, the data of Table 1. We observe that the number of observational vectors, twenty-eight, is relatively small for an analysis in which five factors appear to be of some consequence. Furthermore, the observations deviate widely from normality, which is not a desirable situation in a factor analysis. Component analysis does not necessarily require normally distributed variables but the tests of significance for it as well as for factor analysis do. The results of the computations appear in order below.

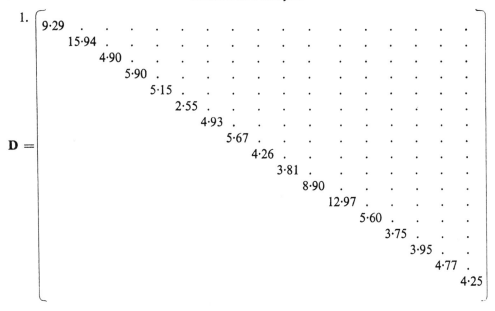

1.
$$\mathbf{D} = \begin{bmatrix} 9\cdot29 \\ & 15\cdot94 \\ & & 4\cdot90 \\ & & & 5\cdot90 \\ & & & & 5\cdot15 \\ & & & & & 2\cdot55 \\ & & & & & & 4\cdot93 \\ & & & & & & & 5\cdot67 \\ & & & & & & & & 4\cdot26 \\ & & & & & & & & & 3\cdot81 \\ & & & & & & & & & & 8\cdot90 \\ & & & & & & & & & & & 12\cdot97 \\ & & & & & & & & & & & & 5\cdot60 \\ & & & & & & & & & & & & & 3\cdot75 \\ & & & & & & & & & & & & & & 3\cdot95 \\ & & & & & & & & & & & & & & & 4\cdot77 \\ & & & & & & & & & & & & & & & & 4\cdot25 \end{bmatrix}$$

2. Computing $\mathbf{C}^{*} = \mathbf{D}^{\frac{1}{2}}\mathbf{C}\mathbf{D}^{\frac{1}{2}}$ and extracting the eigenvalues gives the matrix

$$\mathbf{L}^{*} = \begin{bmatrix} 23\cdot76 \\ & 20\cdot35 \\ & & 14\cdot62 \\ & & & 11\cdot17 \\ & & & & 9\cdot66 \\ & & & & & 5\cdot79 \\ & & & & & & 4\cdot80 \\ & & & & & & & 3\cdot60 \\ & & & & & & & & 3\cdot17 \\ & & & & & & & & & 2\cdot73 \\ & & & & & & & & & & 2\cdot32 \\ & & & & & & & & & & & 1\cdot57 \\ & & & & & & & & & & & & 1\cdot12 \\ & & & & & & & & & & & & & 0\cdot90 \\ & & & & & & & & & & & & & & 0\cdot57 \\ & & & & & & & & & & & & & & & 0\cdot32 \\ & & & & & & & & & & & & & & & & 0\cdot13 \end{bmatrix}$$

Up to this point factor analysis does not differ essentially from principal component analysis. Compared with the results of the principal component analysis it is interesting to note that the eigenvalues are percentually of the same order of magnitude. More essential, however, is the fact that the $p-k$ smallest eigenvalues of $\mathbf{C}^{*}$ tend to be more like each other than do the $p-k$ smallest eigenvalues of $\mathbf{C}$.

The first eigenvalue represents 22·29% of the total variation, the second 19·09% and the third 13·72%. The first three eigenvalues add up to 55·1% of the total variation, and the first four to 65·6%; these figures indicate a somewhat greater concentration of variation to the first eigenvalues than was found for the correlation matrix alone.

3. The results of the tests of significance for the number of common factors are given in the following table. (If $d_k > 100$, c_k is a λ-value, otherwise c_k is a χ^2-value.)

k	t	d_k	c_k
1	5·18	135	5·35***
2	4·16	119	4·17***
3	3·42	104	3·27***
4	2·82	90	126·47**
5	2·25	77	95·67
6	1·93	65	79·26
7	1·64	54	64·69
8	1·43	44	54·30
9	1·21	35	44·04
10	0·99	27	33·84

The approximate 95 % confidence interval for the eigenvalues is, using the same formula as before, $0.08 < \text{all } \lambda_i < 45.73$, with a probability of 0·96.

Hence the smallest number of factors that may be considered to fit the model is $k = 5$, which agrees with the principal component analysis. However, since the sample size is small and the data not truly multivariate normal, it might be wise to consider a further five factors.

4. To find the unrotated factor loadings one computes

$$A = D^{\frac{1}{2}}A^*$$

where $A^* = Z_k^*(L_k^* - t_k I)^{\frac{1}{2}}$. It is common in psychometry to 'rotate' these vectors by some graphical device or other in order to present the data in 'standard' terms; rotation is not considered useful in most biological situations by the present writer. For $k = 10$, the matrix of factor loadings is

$$A = \begin{bmatrix}
-0.15 & -0.47 & 0.30 & -0.72 & -0.03 & 0.09 & 0.09 & 0.01 & -0.04 & -0.06 \\
-0.25 & 0.90 & 0.21 & -0.02 & -0.06 & 0.03 & 0.00 & -0.03 & 0.00 & -0.01 \\
-0.21 & -0.12 & -0.69 & -0.36 & 0.02 & 0.13 & -0.18 & 0.03 & -0.13 & -0.05 \\
-0.02 & -0.30 & 0.38 & 0.07 & -0.54 & -0.48 & 0.09 & 0.00 & -0.09 & 0.20 \\
0.59 & 0.40 & -0.06 & -0.10 & 0.10 & -0.14 & -0.11 & 0.44 & 0.02 & 0.05 \\
-0.10 & -0.31 & 0.21 & 0.28 & 0.12 & -0.29 & -0.25 & 0.20 & 0.32 & -0.24 \\
-0.09 & -0.21 & 0.35 & 0.19 & 0.56 & -0.37 & -0.31 & -0.11 & -0.06 & -0.08 \\
-0.25 & -0.09 & -0.52 & 0.06 & 0.36 & -0.32 & 0.41 & -0.10 & 0.22 & -0.03 \\
-0.09 & -0.18 & -0.01 & 0.69 & 0.12 & 0.26 & 0.17 & -0.07 & -0.27 & -0.08 \\
-0.09 & 0.08 & -0.21 & -0.14 & 0.66 & -0.16 & -0.21 & 0.07 & -0.31 & 0.27 \\
0.85 & -0.02 & 0.08 & 0.02 & 0.27 & 0.18 & 0.05 & -0.17 & 0.02 & 0.10 \\
0.93 & 0.09 & -0.13 & -0.02 & -0.17 & -0.07 & 0.04 & 0.00 & -0.01 & -0.04 \\
-0.11 & -0.26 & -0.57 & 0.38 & -0.44 & 0.01 & -0.14 & 0.05 & -0.14 & -0.10 \\
-0.11 & -0.24 & -0.23 & 0.01 & -0.26 & 0.20 & -0.49 & -0.20 & 0.37 & 0.25 \\
-0.16 & -0.32 & 0.16 & 0.33 & 0.22 & 0.35 & 0.17 & 0.46 & 0.15 & 0.09 \\
-0.10 & -0.36 & 0.69 & 0.34 & -0.06 & 0.16 & 0.05 & 0.00 & -0.02 & 0.12 \\
0.47 & -0.18 & 0.58 & 0.05 & 0.07 & 0.03 & -0.23 & -0.02 & 0.01 & -0.29
\end{bmatrix}$$

$$\begin{array}{cccccccccc}
2.47 & 1.93 & 2.54 & 1.70 & 1.66 & 0.95 & 0.84 & 0.56 & 0.54 & 0.40
\end{array}$$

The last row gives the contribution of each factor to the total test variance.

The results show clearly that these ten factors reproduce both the correlations and variances of the original matrix very well and it is quite possible that even a lesser number of factors would do this too. The communalities are found from finding diag $\mathbf{C}-\mathbf{R}$, where $\mathbf{R} = t\mathbf{D}^{-1}$. That is, (1—estimated residual variance). There are in order:

$$0 \cdot 90, \ 0 \cdot 94, \ 0 \cdot 80, \ 0 \cdot 84, \ 0 \cdot 81, \ 0 \cdot 62, \ 0 \cdot 80, \ 0 \cdot 83, \ 0 \cdot 77, \ 0 \cdot 74, \ 0 \cdot 89,$$
$$0 \cdot 93, \ 0 \cdot 83, \ 0 \cdot 74, \ 0 \cdot 75, \ 0 \cdot 80, \ 0 \cdot 25.$$

The first factor indicates that species 5, 11, 12 and possibly 17 are affected in the same way by some environmental factor; the other elements of the vector do not differ significanlty from zero. The second factor suggests an environmental control that mainly influences species 2 but also to a lesser degree, in the same direction, species 5. Species 1, 15 and 16 are influenced in the opposite direction. The third factor seems to represent an environmental control that affects species 3, 8 and 13 in one direction and species 16 and 17 in another direction. The fourth vector would appear to indicate the influence of an environmental factor which rather strongly affects species 1 in one direction and about equally as strongly species 9 in the reverse direction; other affected species are 3, 13, 15 and 16. The fifth vector suggests relatively strong influences on species 4, 7 and 10 in opposite directions, and lesser influences on species 8 and 13. These five factors seem to be the most informative. The remaining five factors included in the matrix of factor loadings are remarkable in that none of them suggests a strong reaction of any environmental agent on any of the species.

Hence, it would seem that most of the variation in frequencies of the seventeen species may have been controlled by five environmental stimuli of some kind or other (for example, temperature, light, salinity, variation in chemical proportions of sea water, pH, redox).

If as a criterion of non-reactivity to environment we take small factor loadings it may be suggested that species 6, 14 and 15 are euryoic and this agrees extremely well with what is to be observed qualitatively in the material. Judging from occurrences in the borehole samples one gains the impression that species 1 and 9 are also euryoic. Both of these are strongly affected only by the fourth factor, which may indicate that this factor is an unimportant one with respect to actual distribution. None of the species appears to give the impression of being stenoöic.

It will be seen that the results of the factor analysis tend to differ somewhat in detail from the component analysis, although both give a strong impression of the operation of several approximately equally important environmental factors.

Inasmuch as the factor analysis model would seem to be more suitable for the kind of ecological data here treated (remembering we are not studying the shape of the distribution) the results obtained with its aid could be more descriptive of the actual situation.

ACKNOWLEDGMENTS

The computations for method 1 were performed on the computer FACIT of the Swedish Board for Computing Machinery, Stockholm, and those for method 2 were made on the IBM computer 1620 of the University of Uppsala. Thanks are due to these institutions for providing free computing time. The investigation was supported financially by the

Swedish Natural Science Research Council. Dr K. G. Jöreskog, Uppsala, and Fil. mag. K. Arle, Stockholm, gave invaluable assistance in processing the data.

SUMMARY

1. Variations in the relative frequencies of different species in samples may be interpreted in terms of the major environmental factors to which the organisms react. The number of these factors, if not their nature, may be deduced from a multivariate statistical analysis of the relative frequencies. These frequencies are set in a matrix of covariances or correlations and the number of factors whose interplay could have accounted for the observed matrix is computed by factor analysis or principal component analysis.

2. In the present investigation the results indicate the influence of five major environmental factors on seventeen species of fossil ostracods. Although the study is based on palaeoecological material, the reasoning can equally well be applied to other ecological situations on which insufficient chemical and physical data are available, or where it may not be exactly known which environmental factors are influencing the relative abundances of the species.

3. The procedures discussed, although unable to disclose whether a species is, for example, stenohaline or euryhaline, stenothermal or eurythermal, can indicate whether the species is stenoöic or euryoic. Although both principal component analysis and factor analysis are made use of here, it would seem that the underlying model of the latter method might be more suited to such problems.

REFERENCES

Greig-Smith, P. (1957). *Quantitative Plant Ecology.* London.

Hotelling, H. (1933). Analysis of a complex of variables into principal components. *J. Educ. Psychol.* **24**, 417–41, 498–520.

Jöreskog, K. G. (1962). On the statistical treatment of residuals in factor analysis. *Psychometrika*, **27**, 335–54.

Kendall, M. G. (1957). *A Course in Multivariate Analysis.* Griffin's Statistical Monographs, No. 2.

Lawley, D. N. (1956). Tests of significance for the latent roots of covariance and correlation matrices. *Biometrika*, **43**, 128–36.

Pearce, S. C. & Holland, D. A. (1961). Analyse des composantes, outil en recherche biométrique. *Biométrie-Praximétrie*, **2**, 159–77.

Rao, C. R. (1955). Estimation and tests of significance in factor analysis. *Psychometrika*, **20**, 93–111.

Reyment, R. A. (1960). Studies on Nigerian Upper Cretaceous and Lower Tertiary Ostracoda. Part I. Senonian and Maestrichtian Ostracoda. *Stockh. Contr. Geol.*, vol. 7.

Reyment, R. A. (1963). Studies on Nigerian Upper Cretaceous and Lower Tertiary Ostracoda. Part II. Danian, Paleocene and Eocene Ostracoda. *Stockh. Contr. Geol.*, vol. 10.

Whittle, P. (1952). On principal components and least squares methods of factor analysis. *Skand. Akt.-Tidskr.* **36**, 223–29.

12

Reprinted from *J. Geol.* 74:703–715 (1966)

CLUSTER ANALYSIS APPLIED TO MULTIVARIATE GEOLOGIC PROBLEMS

JAMES M. PARKS

Union Oil Company of California, Research Center, Brea, California

ABSTRACT

In some geologic studies it is desirable to group together similar samples on which many measurements have been made, and to measure the degree of similarity between the groups. Using either the product-moment correlation coefficient, the matching coefficient, cosine θ, or the distance function, the resulting matrix is usually too large for direct interpretation.

Cluster analysis, a technique developed by psychologists, is a method of searching for relationships in a large symmetrical matrix. A logical pair-by-pair comparison of samples results in a two-dimensional hierarchical diagram on which the natural breaks between groups are obvious. The observer can also pick off groups at any desired level of similarity. Non-overlapping clusters are used.

A computer program has been written that will handle up to 200 measurements on as many as 1,000 samples. The hierarchical cluster diagram is printed out by an off-line printer. Several highly correlated variables can bias the results, so the variables are clustered first by correlation coefficient. Specified variables or groups of variables can then be used in clustering the samples by distance function. Important variables may be weighted by using them more than once. An alternative procedure is to perform an R-type factor analysis on the variables and cluster the samples by distance function on factor measurements estimated by regression methods. An example problem demonstrates the options and output of the program.

The cluster analysis solution of a published twelve-variable, forty-sample problem based on constituent particle composition of Bahamian sediment samples shows clear-cut groups similar to the facies described by Imbrie and Purdy derived from factor analysis. Multivariate facies maps can easily be constructed from the cluster analysis results.

INTRODUCTION

It is becoming a common practice in many geologic studies to make many different kinds of measurements of each of a large number of samples for the purpose of classifying the samples into related groups. The variables may be percentages or amounts of various constituents such as oölites, heavy minerals, foraminifera or pollen species, or trace elements; or they may be measurements such as grain size, porosity, or API gravity. The samples may be thin sections of carbonate or clastic rocks, or bottom samples of recent sediments, or depth intervals of drill cuttings, or zones of sample logs. When as many as twenty or thirty or more measurements are made on one hundred or more samples, the resulting table of data is so large that interpretation "by eye" becomes difficult (see table 7 below).

Classification is a process of putting similar objects into an unknown number of distinct categories, with the objects in each category being more similar to each other than to the objects in all the other categories. Several measures of similarity have been used: the product-moment correlation coefficient; the matching coefficient for data coded as to presence or absence (Sokal and Michener, 1958, p. 1417); cosine θ (Imbrie and Purdy, 1962, p. 257); and the simple distance function (Sokal, 1961). Imbrie (1963, p. 22) pointed out some of the disadvantages of the product-moment correlation coefficient when applied to samples rather than variables, and proposed the cosine θ function, which is computed the same way as the correlation coefficient but without the correction terms. This results in a function that ranges from $+1.0$ (exact similarity) to 0.0 (complete dissimilarity), which is then transformed to θ' to give a function ranging from $+1.0$ to -1.0, as does the correlation coefficient. Both θ' and the correlation coefficient are measures of the angle between two samples in a multidimensional space; when the data are standardized (transformed to have a mean of 0.0 and a standard deviation of 1.0), the two functions are identical.

Several workers (Mahalanobis, 1936; Rao, 1952; Sokal, 1961; Jizba, 1964; Harbaugh, 1964; Edwards and Cavalli-Sforza, 1965) have used a distance function as a measure of similarity. As used in this paper, the formula for the distance (D) between two objects in a multidimensional hyperspace of M dimensions (M = number of variables) is

$$D_{1,2} = \sqrt{\left[\sum_{i=1}^{M} (X_{i1} - X_{i2})^2 / M \right]}. \tag{1}$$

where X equals the normalized (transformed to range from 0.0 to 1.0) values of the variables. The actual distance between samples would be a function of the number of variables used, so the sum of the squared differences is divided by M, the number of variables used. This gives a mean squared difference, which is really the variance of the differences between the measurements on the two samples. The simple distance function is the square root of this variance, and is thus the standard deviation of the difference.

This Euclidean distance of normalized variables assumes that all input variables are orthogonal. As this is not generally the case, certain steps to be described later are taken to insure orthogonality or near-orthogonality of the variables used. The simple distance function computed in this manner will range from 0.0 (shortest distance equals closest similarity) to +1.0 (longest distance equals greatest dissimilarity). This function is not directly amenable to the techniques of factor analysis without some form of transformation, but it is eminently suitable for cluster analysis.

The matrix resulting from computing any of the similarity measures for a large number of samples is usually too large for direct interpretation. When each sample is compared with all other samples in a two hundred sample problem, the matrix contains 40,000 terms; however, the matrix is symmetrical and the diagonal terms are meaningless, so only 19,900 terms are used. This is not a large problem in present-day studies.

Several other methods of classification by computer techniques have been described (Sokal and Michener, 1958; Imbrie and Purdy, 1962; Imbrie, 1963; Ward, 1963; Bonner, 1964; Casetti, 1964; Howd, 1964; Jizba, 1964; Edwards and Cavalli-Sforza, 1965; Bonham-Carter, 1965). Most of these methods are limited to fairly small numbers of samples or to fairly small numbers of variables, and none have an easily readable diagrammatic output of the relationships.

CLUSTER ANALYSIS

Cluster analysis is a simple form of correlation analysis, a method of searching for relationships in a large symmetrical matrix. Like factor analysis, cluster analysis was first developed by psychologists; unlike factor analysis, it does not involve artificial or abstract "factors." Cluster analysis is a straightforward, logical, pair-by-pair comparison between samples, objects, or variables. The results of a cluster analysis can be presented in an easily understood two-dimensional hierarchical diagram on which the "natural breaks" between groups will be obvious. The observer can also pick off groups at any desired level of similarity or dissimilarity.

In an R-type analysis where variables are compared, factor and cluster analysis give similar results. However, in cluster analysis each variable as a unit is placed in a cluster, whereas in factor analysis, different portions of the variance of a variable may be assigned to different factors.

In Q-type analysis where samples are compared, factor and cluster analysis also give similar results if the variables are uncorrelated or orthogonal. One way to achieve this is to perform an R-type analysis on the variables first, and select only the uncorrelated or least correlated variables for use in a Q-type cluster analysis. Several highly correlated variables can bias the results of a Q-type (sample) cluster analysis. Intentional bias or "weighting" of important variables can be achieved by using those variables more than once in the Q-type analysis.

An alternative procedure is to perform an R-type principal factor analysis on the variables, compute or "estimate" factor measurements on each sample by linear regression of each factor on the observed variables by the square-root method of solving systems of linear equations (Harman, 1960, p. 338–348), and cluster the samples by distance function on these factor measurements. No rotation of the R-type factor solution is necessary. The factors and factor measurements are orthogonal.

A computer program package for cluster analysis with all of the above options has been written that will handle up to 200 measurements on as many as 1,000 samples. The programs, originally written in MAD language (Michigan Algorithm Decoder) to take advantage of a SYMM function to utilize only one-half of a symmetric matrix, are now being rewritten in FORTRAN to make them more widely usable. The program prints out the two-dimensional hierarchical cluster diagram on an off-line printer, making it unnecessary to use an X-Y plotter. Non-overlapping clusters are used; that is, a sample can appear in only one cluster.

TABLE 1

RAW-DATA MATRIX

SAMPLE	VARIABLE							
	A	B	C	D	E	F	G	H
1.......	100	0	0.8	0.0014	69	9.3	0	2
2.......	90	670	1.4	.0012	25	7.7	440	13
3.......	50	20	0.5	.0006	30	5.5	40	2
4.......	95	740	0.3	.0026	23	6.8	340	8
5.......	23	880	5.0	.0073	18	0.6	550	42
6.......	98	10	1.0	.0009	55	8.1	40	6
7.......	47	140	0.3	.0024	7	4.3	170	9
8.......	38	420	6.1	.0066	72	1.4	510	38
9.......	18	260	3.2	.0082	46	3.3	780	66
10.......	21	470	5.5	0.0025	5	1.0	190	40

EXAMPLE PROBLEM

A simple example problem will demonstrate the main features and options of this method of cluster analysis. Table 1 is the data matrix: eight variables (A–H) are assumed to be measured or counted on each of ten samples (1–10). It is obvious that samples 1 and 6 are quite similar in almost all their measurements. At the other extreme, samples 5 and 6 are very different: almost every measurement that is high on sample 5 is low on sample 6, and nearly every measurement that is low on sample 5 is high on sample 6. However, these "eyeball" comparisons are non-quantitative and subjective, and are very difficult to make as the data matrix becomes larger.

CLUSTERING THE SAMPLES

The raw-data matrix is normalized column by column in order to give equal weight to each of the variables, which may be measured in quite different sized units as in this example (table 2). Using the product-moment correlation coefficient, a similarity matrix is calculated between all possible pairs of samples. This produces a square matrix, for this problem of ten samples, a 10 × 10 matrix. However, this matrix is symmetrical about its principal diagonal, so that only one-half of the matrix need be used. The diagonal itself, composed of the comparison of a sample with itself, is all 1 and may be disregarded. The initial similarity matrix for this example is shown in table 3.

This matrix is systematically searched for the highest correlation coefficient; in this case, 0.99 between samples 1 and 6. The normalized data for these two samples are then combined and averaged, using a weighting factor of 1 for each sample, forming a new sample 1. Sample 6 is deleted from the matrix, and the correlation coefficients of the new sample 1 with each of the other samples are calculated and inserted in the similarity matrix. This revised matrix, after the first pairing, is shown in table 4.

The highest correlation coefficient now is 0.96, between samples 1 and 3. The normalized

TABLE 2

NORMALIZED-DATA MATRIX

SAMPLE	VARIABLE							
	A	B	C	D	E	F	G	H
1.......	1.000	0.000	0.086	0.105	0.955	1.000	0.000	0.000
2.......	0.878	0.761	0.190	0.079	0.298	0.816	0.564	0.172
3.......	0.390	0.023	0.034	0.000	0.373	0.563	0.051	0.000
4.......	0.939	0.841	0.000	0.263	0.269	0.713	0.436	0.094
5.......	0.061	1.000	0.810	0.882	0.194	0.000	0.705	0.625
6.......	0.976	0.011	0.121	0.039	0.746	0.862	0.051	0.062
7.......	0.354	0.159	0.000	0.237	0.030	0.425	0.218	0.109
8.......	0.244	0.477	1.000	0.789	1.000	0.092	0.654	0.562
9.......	0.000	0.295	0.500	1.000	0.612	0.310	1.000	1.000
10.......	0.037	0.534	0.897	0.250	0.000	0.046	0.244	0.594

TABLE 3

INITIAL SIMILARITY MATRIX (SAMPLES)

	1	2	3	4	5	6	7	8	9	10
1.....	....									
2.....	0.47	....								
3.....	.96	0.57	....							
4.....	.45	.93	0.49	....						
5.....	− .94	− .46	− .94	−0.37	....					
6.....	.99	.53	.96	.49	−0.96	....				
7.....	.45	.68	.55	.73	− .50	0.49	....			
8.....	− .38	− .80	− .51	−. .78	.47	− .45	−0.88	......		
9.....	− .63	− .46	− .57	− .41	.45	− .64	.05	0.08	....	
10.....	−0.76	−0.44	−0.73	−0.52	0.72	−0.72	−0.61	0.39	0.14	....

TABLE 4

REVISED SIMILARITY MATRIX (SAMPLES)

	1	2	3	4	5	6	7	8	9	10
1......	....									
2.....	0.50	....								
3.....	.96	0.57	....							
4.....	.47	.93	0.49	....						
5.....	− .95	− .46	− .94	−0.37	....					
6......	....	....	....	....	....	....				
7.....	.47	.68	.55	.73	−0.50	....	....			
8.....	− .41	− .80	− .51	− .78	.47	....	−0.88	....		
9.....	− .64	− .46	− .57	− .41	.45	....	.05	0.08	....	
10.....	−0.74	−0.44	−0.73	−0.52	0.72	....	−0.61	0.39	0.14	....

data for these two samples are combined and averaged, using a weighting factor of 2 for sample 1 because it contains two of the original samples. Sample 3 is deleted and the correlation coefficients of the new sample 1 with the remaining samples are calculated and inserted in the matrix. This procedure is continued until all samples have been paired. The pairing sequence is shown in table 5.

From the data in table 5 a two-dimensional hierarchical cluster diagram is constructed shown in figure 1, *B*. The abscissa is the value of the correlation coefficient, from +1.0 on the left to −1.0 on the right. The clustered samples are arranged along the left margin, according to the pairing sequence of table 5. The degree of similarity between the two samples in a pair is the *X*-axis value of the vertical line joining the horizontal lines representing the two samples.

The "natural breaks" between groups are readily apparent on this cluster diagram. There are two main groups or clusters; the first composed of samples 1, 6, 3, 2, 4, and 7, and the second group composed of samples 5, 10, 8, and 9. These two groups are quite different, with a correlation coefficient of −0.90. Within the first group there are two subgroups, samples 1,

TABLE 5

PAIRING SEQUENCE (SAMPLES)

Retained Sample	Deleted Sample	Correlation Coefficient
1........	6	0.99
1........	3	.96
2........	4	.93
5........	10	.72
2........	7	.71
1........	2	.56
5........	8	.47
5........	9	.23
1........	5	−0.90

6, and 3; and samples 2, 4, and 7, with the correlation between these subgroups being 0.56. The observer may choose his groups at any desired level of similarity, depending upon the nature of the problem.

When Imbrie's cosine θ similarity function is used and when the same clustering procedures are followed, the results are only slightly different. In figure 1, *C*, the major groups and subgroups are composed of the same samples, but the values of the connectors between samples and between groups are different. In larger problems this can lead to slight differences in the clusters.

The simple distance function is the distance between the ends of the sample vectors, rather than a measure of the angle between the vectors as is the case with cosine θ and the correlation coefficient. Only slight modifications to the general clustering procedure are needed in order to cluster a distance function matrix. Small distances mean a high degree of similarity, and the greater the distance, the greater the difference between samples. The distance function similarity matrix is searched for the smallest entries. The distance-function cluster diagram for the example problem is shown in figure 1, *D*, where the major groups are the same as for the two previous functions but the subgroups are paired differently.

CLUSTERING THE VARIABLES

One of the objectives of a multivariate study is to eliminate redundant or highly correlated variables. This can be achieved by clustering the variables on the correlation coefficient. For this purpose, an *R*-type similarity matrix is constructed, comparing each variable with

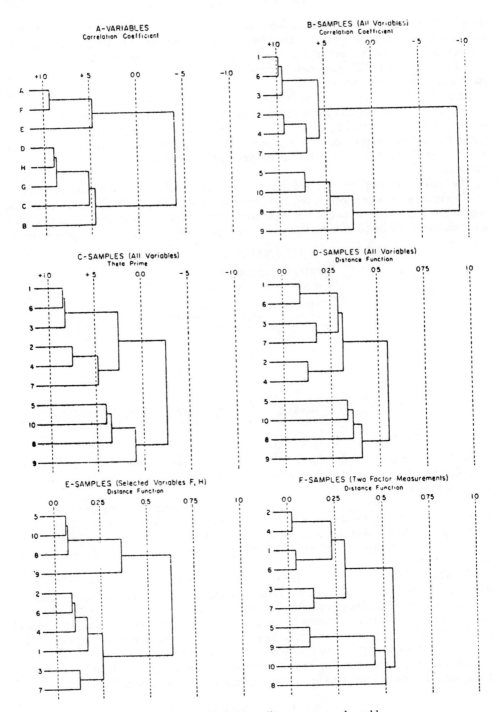

FIG. 1.—Hierarchical cluster diagrams—example problem

134

all other variables over the range of all the samples. The same example problem (table 1) is used for purposes of illustration. The raw data are again normalized columnwise (table 2). A similarity matrix is calculated, using the product-moment correlation coefficient, between all possible pairs of variables. For this problem involving eight variables, this will be an 8 × 8 matrix, shown in table 6.

Following the same clustering procedures as outlined for samples, the cluster diagram of figure 1, A, is constructed. Variables A and F are highly correlated, and either one could be eliminated without much loss of information. The same holds true for the three highly correlated variables D, H, and G. Which variable or group of variables to eliminate in future studies would depend upon external influences, such as the relative cost or ease of measuring each variable, the precision or reliability of the measurements, etc.

As shown in figure 1, A, there are two very different groups of variables in this problem. The variables within each group are fairly well correlated. One or more variables from each group could be chosen to best represent the average or centroid of that group for clustering the samples, and the remaining variables could be left out of the calculations in clustering the samples. This results in using variables that are only poorly correlated or uncorrelated

TABLE 6

CORRELATION COEFFICIENT MATRIX (VARIABLES)

	A	B	C	D	E	F	G	H
A.......	1.00							
B.......	−0.15	1.00						
C.......	−0.59	0.46	1.00					
D.......	−0.69	0.35	0.52	1.00				
E.......	0.50	−0.31	0.09	−0.16	1.00			
F.......	0.93	−0.40	−0.79	−0.71	0.39	1.00		
G.......	−0.52	0.53	0.35	0.87	−0.32	−0.53	1.00	
H.......	−0.80	0.29	0.59	0.88	−0.36	−0.76	0.82	1.00

with each other, thus approaching orthogonality. Figure 1, E, shows the results of clustering the samples by distance function using only variables F and H.

If it is desired to use all the variables, still achieve orthogonality, and not sacrifice much information, an R-type principal factor analysis can be performed on the variables, a limited number of factors representing most (perhaps 80–90 per cent)) of the total variance can be selected, and factor measurements for each of the samples can be estimated by least-squares regression techniques (Harman, 1960, p. 338–345). It is not necessary for this purpose to perform any arbitrary rotation of the factor solution; the distances remain the same whatever the rotation. Figure 1, F, shows the results of clustering the samples by distance function on factor measurements, using two factors representing 75 per cent of the total variance.

USING THE CLUSTER PROGRAM ON LARGER PROBLEMS

A study on the constituent particle composition of recent Bahamian bottom sediment samples (Purdy, 1960) resulted in twelve variables measured on each of 200 samples; data for the first forty samples are shown in table 7. The computer output of clustering these twelve variables by correlation coefficient is shown in figure 2. Using five composite variables chosen from figure 2—one (1, 2, 5, and 6); two (3, 4, 8, and 12); three (7); four (9 and 11); and five (10)— the forty samples were clustered by distance function, with the results shown in figure 3. At a distance similarity level of about 0.8 (where 1.0 equals exact similarity), the five major groups derived by factor analysis by Imbrie and Purdy (1962, p. 264)

TABLE 7

PURDY (1960) 40-SAMPLE PROBLEM (RAW DATA)

Sample	No.	A*	B	C	D	E	F	G	H	I	J	K	L
1......	1	2.0	3.4	10.0	8.8	3.2	2.2	1.0	3.8	14.6	0.4	24.8	23.1
2......	31	0.0	0.6	0.6	0.2	0.0	0.4	3.6	3.0	12.8	59.0	17.8	7.0
3......	43	0.2	1.0	0.4	1.4	0.0	1.8	7.4	1.8	25.8	38.8	16.2	6.5
4......	59	0.0	1.6	0.8	2.8	0.4	3.8	3.4	1.4	44.2	22.0	12.4	2.5
5......	171	0.0	5.2	4.8	3.0	0.4	9.8	61.0	6.8	0.0	2.6	0.6	31.3
6......	174	0.0	11.2	5.8	1.2	0.0	14.2	47.6	6.2	0.4	0.6	3.0	40.1
7......	380	0.0	16.2	4.2	1.8	0.0	10.4	47.0	3.0	0.4	3.0	1.8	51.9
8......	360	0.0	0.2	2.4	1.8	0.0	4.8	68.8	8.4	0.0	4.2	3.0	58.8
9......	366	0.0	0.8	3.4	1.2	0.0	3.0	76.8	8.4	0.0	2.8	0.4	35.2
10......	367	0.0	0.2	2.8	1.2	0.0	2.2	61.8	4.2	0.0	20.0	3.2	31.8
11......	368	0.0	2.0	6.2	2.0	0.0	4.2	67.4	8.8	0.0	4.2	0.4	51.9
12......	369	0.0	3.0	4.2	2.6	0.0	5.0	66.8	7.4	0.0	3.2	1.4	63.7
13......	370	0.0	3.4	10.0	5.4	0.2	5.4	43.0	19.6	0.0	1.6	1.2	90.7
14......	371	0.0	5.4	15.2	4.2	0.0	10.4	32.0	17.0	0.2	0.4	2.4	87.2
15......	372	0.0	10.2	12.8	7.2	0.0	11.8	22.4	16.8	0.0	1.2	2.0	85.7
16......	373	0.0	25.6	1.8	6.4	0.0	12.2	25.8	7.4	0.0	5.2	0.8	89.2
17......	374	0.0	13.8	0.2	3.6	0.0	2.8	22.0	9.2	0.2	32.8	9.4	5.0
18......	207	3.4	12.6	1.4	1.8	7.6	8.0	4.2	8.4	4.8	8.0	11.2	2.2
19......	328	0.0	0.4	0.2	0.6	0.0	1.6	18.6	3.6	35.0	16.0	20.2	7.6
20......	330	0.0	1.0	0.8	0.2	0.0	1.2	26.2	4.0	10.6	44.6	9.4	5.6
21......	334	0.0	0.8	0.0	0.0	0.0	1.2	20.0	1.8	4.8	63.8	6.2	4.3
22......	310	0.2	2.8	0.0	0.6	0.0	1.6	11.6	2.6	3.2	36.4	37.6	8.3
23......	312	0.0	1.0	0.0	0.4	0.0	2.8	25.6	2.4	12.8	36.8	15.0	5.5
24......	502	0.0	3.6	0.6	2.8	0.0	9.0	4.4	1.8	14.4	13.6	45.6	4.4
25......	504	3.6	26.4	1.0	1.8	12.8	12.0	0.2	10.4	2.8	0.0	7.8	2.4
26......	279	2.0	15.8	2.8	5.4	6.2	5.2	1.8	6.4	1.8	1.4	35.2	18.9
27......	512	2.8	1.6	0.2	0.2	0.6	4.0	6.4	1.4	38.4	18.0	22.2	2.5
28......	514	0.0	1.0	2.2	2.6	0.0	1.4	3.2	3.4	49.0	2.2	32.0	4.5
29......	317	0.0	0.8	0.4	1.8	0.0	1.8	3.0	1.2	48.6	7.8	30.6	4.6
30......	537	3.8	9.4	1.6	1.2	6.8	10.4	2.2	1.8	3.2	4.4	17.2	2.0
31......	37	0.2	39.8	2.0	3.2	1.0	10.4	0.8	2.6	3.2	0.0	19.8	4.1
32......	7-47	2.8	14.2	7.2	2.2	13.8	6.8	0.0	3.8	5.6	0.0	27.2	2.5
33......	409	0.0	5.2	5.5	0.7	0.8	5.1	0.6	1.2	36.7	24.9	15.8	2.9
34......	411	0.0	0.9	0.8	0.5	0.0	1.6	0.5	1.6	11.4	68.8	11.9	5.9
35......	418	0.0	0.0	0.0	0.0	0.0	0.3	0.0	0.0	0.3	96.8	2.1	1.4
36......	417	0.0	0.0	0.0	0.0	0.0	0.0	0.0	0.0	0.0	98.2	1.4	1.3
37......	414	0.0	0.0	0.0	0.0	0.0	0.0	0.0	0.0	0.0	97.6	2.3	2.3
38......	413	0.0	0.0	0.0	0.0	0.0	0.0	0.0	0.1	0.2	96.4	2.9	1.5
39......	438	0.0	0.0	0.0	0.0	0.0	0.1	0.0	0.0	0.5	97.9	1.3	1.3
40......	250	1.5	22.4	0.6	0.9	13.5	10.6	2.1	3.3	14.2	13.1	11.7	0.6

* A, Coralline algae; B, *Halimeda*; C, Peneroplidae; D, other Foraminifera; E, Corals; F, Molluscs; G, fecal pellets; H, mud aggregates; I, grapestone; J, oölites; K, cryptocrystalline grains; L, weight per cent < ⅛ mm.

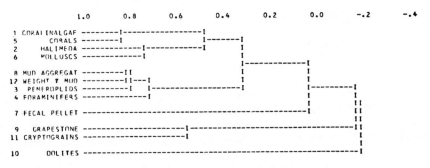

FIG. 2.—Cluster diagram for variables using correlation coefficient: Imbrie and Purdy, Bahamas, recent sediment data for 40 samples.

are apparent. However, at this level of similarity, some samples do not fit into any cluster, such as the samples shown in parentheses in table 8.

If the alternative procedure of clustering by distance function on factor measurements (three factors representing 75 per cent of the total variance) is used, the forty samples cluster as shown in figure 4. Generally similar groupings are evident. The comparison of the groupings derived from the two cluster techniques and the Q-type factor analysis technique (Imbrie and Purdy, 1962, p. 263) is shown in table 8.

When the variables for the entire sample problem are clustered, the groupings are somewhat different (fig. 5). Clustering the samples for the entire 200-sample problem, which takes about 28 min. on the IBM 704 from raw data through the cluster diagram, allows some of the samples that did not fit well into any group on the 40-sample problem to pick up other similar samples and form subgroups of their own. By cluster analysis there are more than the six facies used by Imbrie and Purdy (1962, fig. 10).

DISCUSSION

There are two principal restrictions in conventional statistical analysis that do not apply to cluster and factor analysis. First, that the experimental sample should be representative

```
                    1.0        0.9         0.8         0.7         0.6

  36           417 II
  37           414 III
  39           439 -III
  38           413 --II----------------------I
  35           419 ---I                       I
                                              I
   2            31 ----I-I                    I
  34           411 ----I I---I                I--------I
  21           334 ------I   I--I             I        I
  20           330 ----I---I I  I             I        I
  23           312 ----I   I-I  I-----I       I        I
  22           310 --------I    I     I I     I        I
   3            43 ------------I      I--I     I        I
                                      I         I
  17           374 ------------------I          I------I
                                                I      I
   4            59 ------I----I                 I      I
  33           409 ------I    II                I      I
  28           514 ---I-------II------------I   I      I
  29           317 ---I        I            I   I      I
  19           328 -----------I            II   I      I
                                           II   I      I
  24           502 ----------------------II I---I      I
                                          I           I
  27           512 -------------------------I          I
                                                       I
   6           174 -----------II                       I
   7           380 -----------II-----I                 I
   5           171 ------------I     I---I             I
  16           373 ----------------I     I             I
                                          II           I
  14           371 ------I----------I     II           I
  15           372 ------I          I----II            I
  13           370 ----------------I     I             I
                                          I-----------I I
  11           368 --II                   I            I I
  12           369 --II---I               I            I I
   8           360 ---I   I---I           I            III
   9           366 -------I   I-----------I            III
  10           367 ----------I                         III
                                                       III
  31            37 ------------------------------------III
                                                       II
  18           207 -----------I----I                   II
  30           537 -----------I    II                  I
  25           504 -----------------II--I              I
  26           279 --------I-------I    I              I
  32          7-47 ---------I           I----------I   I
                                        I          I   I
  40           250 -----------------I   I----------I
                                                   I-----I
   1             1 ------------------------------I
```

FIG. 3.—Cluster diagram for samples using distance function and five composite variables: Imbrie and Purdy, recent sediment data for 40 samples.

TABLE 8

COMPARISON OF FACIES GROUPINGS

| Cluster Analysis | | | | Imbrie-Purdy (1962) Q-Type Factor Analysis | |
| Selected Variables | | Factor Measurements | | | |
Group	Samples	Group	Samples	Facies	Samples
A	35	A	35	Oölite	35
	36		36		36
	37		37		37
	38		38		38
	39		39		39
			(21)		
B	2	B	2	Oölitic	2
	3		(17)		17
	(17)		20		20
	20		23		21
	21		34		22
	22				23
	23				34
	34				
C'	8	C'	8	Pelletal mud	8
	9		9		9
	10		10		10
	11		11		11
	12		12		12
C''	5	C''	5	Skeletal mud	5
	6		6		6
	7		7		7
	16		16		13
					14
C'''	13	C'''	13		15
	14		14		16
	15		15		
D	4	D	(1)	Grapestone	3
	19		3		4
	(24)		4		19
	(27)		19		24
	28		22		27
	29		24		28
	33		27		29
			28		33
			29		
			33		
E	(1)	E	18	Coralgal	1
	18		25		18
	25		26		25
	26		30		26
	30		31		30
	32		32		31
	(40)		40		32
					40
(F)	(31)				

of the general population or be a carefully randomized sample is unnecessary (Thurstone, 1947, p. xii). Rather than select a random group of subjects, it is better to select subjects so that their attributes are as diverse as possible within the domain to be studied. Second, it is not necessary that anything have a normal or known frequency distribution (Thurstone, 1947, p. xii). It is better to measure something significant without any sampling distribution than to measure something trivial or irrelevant because its sampling distribution is known. Conventional statistical tests of significance are difficult to apply to cluster and factor analysis because of this lack of restrictions.

Correlational analysis carries no implication of cause and effect, nor does it imply the relationship of dependent and independent variables, as is necessary for regression analysis.

```
                   1.0        0.9        0.8        0.7        0.6
    36          417 II
    39          438 II
                    II
    35          418 II------I
    37          414 II       I--------I
    38          413 -I       I         I
    21          334 --------I          I-------------I
     2           31 ---II               I               I
    34          411 ---II--------I     I               I
    20          330 ---II          I--I               I
    23          312 ---I           I                  I
    17          374 -------------I                    I
     4           59 ---II                            I-I
    19          328 ---II---I                        I I
    33          409 ----I   I--I                     I I
     3           43 ---I----I  I-----I               I I
    22          310 ---I       I     I               I I
    24          502 --------I-I    I----I            I I
    27          512 ---------I     I    I            I I
    28          514 ----I-----------I    I--------I I
    29          317 ----I               I           I
     1            1 --------------------I           I
    11          368 -I                               I
    12          369 II--I                            I
     8          360 -I  I----I                       I
     9          366 ----I    I---I                   I
    10          367 --------I   I                    I
     6          174 --I-I      I-------I             I
     7          380 --I I-----I I      I             I
     5          171 ----I      I--I    I--------I I
    16          373 ---------I      I          I I
    14          371 ----I--I        I          I I
    15          372 ----I  I------------I      I I
    13          370 -------I                   I-I
    30          537 --I-I                       I
    40          250 --I I----I                  I
    18          207 ----I    I----I             I
    32         7-47 ---------I   I----I         I
    26          279 ----------I---I  I----------I
    31           37 ----------I      I
    25          504 ------------------I
```

Fig. 4.—Cluster diagram for samples using distance function and three factor measurements: Imbrie and Purdy, recent sediment data for 40 samples.

In factor and cluster analysis, all variables are treated in the regression sense of independent variables. Cause and effect are looked for in the grouping of highly correlated variables (cluster analysis); or in meaningful rotations of R-type solutions (factor analysis).

CONCLUSIONS

1. Cluster analysis is a useful technique for analyzing large tables of data where many different measurements are made on each of many samples.

2. Cluster analysis results are easily understood and interpreted because the results are in the form of a two-dimensional hierarchical diagram.

3. Cluster analysis reveals "natural groupings," and further allows the observer to pick off groups at any desired level of similarity.

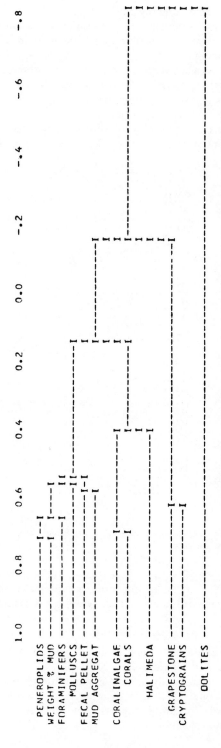

FIG. 5.—Cluster diagram for variables using correlation coefficient: Imbrie and Purdy, recent sediment data for 200 samples

4. The distance function is a useful measure of similarity for cluster analysis when the data are orthogonal or near-orthogonal.

5. Using only specified uncorrelated or poorly correlated variables can help achieve near orthogonality of data.

6. An R-type factor analysis on the variables, with least-squares regression estimates of factor measurements on each sample, achieves orthogonality with little loss of information and without arbitrary rotation.

7. Clustering the variables allows elimination of redundant or highly correlated variables.

8. Variables known to be of prime importance can be weighted.

REFERENCES CITED

BONHAM-CARTER, G. F., 1965, A numerical method of classification using qualitative and semi-quantitative data, as applied to the facies analysis of limestones: Bull. Canadian Petroleum Geol., v. 13, p. 482–502.

BONNER, R. E., 1964, On some clustering techniques: IBM Jour., v. 8, p. 22–32.

CASETTI, E., 1964, Classification and regional analysis by discriminant iterations: Tech. Rept. 12, ONR Task no. 389-135, 95 p.

EDWARDS, A. W. F., and CAVALLI-SFORZA, L. L., 1965, A method for cluster analysis: Biometrics, v. 21, p. 362–375.

HARBAUGH, J. W., 1964, BALGOL programs for calculation of distance coefficients and correlation coefficients using an IBM 7090 computer: Kansas Geol. Surv., Spec. Dist. Pub. 9, 32 p.

HARMAN, H. H., 1960, Modern factor analysis: Chicago, University of Chicago Press, 469 p.

HOWD, F. H., 1964, The taxonomy program—a computer technique for classifying geologic data: Quart. Colo. Sch. Mines, v. 59, no. 4, p. 207–222.

IMBRIE, J., 1963, Factor and vector analysis programs for analyzing geologic data: Tech. Rept. 6, ONR Task no. 389-135, 83 p.

——— and PURDY, E. G., 1962, Classification of modern Bahamian carbonate sediments: Am. Assoc. Petroleum Geologists, Mem. 1, p. 253–272.

JIZBA, Z. V., 1964, A contribution to a statistical theory of classification: Stanford Univ. Publ., Geol. Sci. v. 9, no. 2, p. 729–756.

MAHALANOBIS, P. C., 1936, On the generalized distance in statistics: Proc. Natl. Inst. Sci. India, v. 12, p. 49–55.

PURDY, E. G., 1960, Recent calcium carbonate facies of the Great Bahama Bank: Ph.D. thesis, Columbia Univ., 174 p.

RAO, C. R., 1952, Advanced statistical methods in biometric research: New York, John Wiley & Sons, 390 p.

SOKAL, R. R., 1961, Distance as a measure of taxonomic similarity: Syst. Zoölogy, v. 10, no. 2, p. 70–79.

——— and MICHENER, C. D., 1958, A statistical method for evaluating systematic relationships: Univ. Kansas Sci. Bull., v. 38, pt. 2, no. 22, p. 1409–1438.

THURSTONE, L. L., 1947, Multiple factor analysis: Chicago, University of Chicago Press, 535 p.

WARD, J. H., JR., 1963, Hierarchical grouping to optimize an objective function: Jour. Am. Stat. Assoc., v. 58, p. 236–244.

13

Reprinted from *J. Geol.* 74:786–797 (1966)

THE STATISTICS OF ORIENTATION DATA

GEOFFREY S. WATSON[1]

The Johns Hopkins University

ABSTRACT

This paper provides the key references for, and a summary of, statistical methods for describing and analyzing orientation data. Most of the paper is concerned with the author's approximate analysis of variance procedures for unit vectors in two and three dimensions. The procedures are based on probability density proportional to exp $(\kappa \cos \theta)$, where κ is an accuracy parameter and θ is the angle between the true and the observed directions. A brief account of the theory for exp $(\kappa \cos^2 \theta)$ and its generalization is given; this may be useful for axial data.

I. INTRODUCTION

Geologists have suggested one statistical problem that is novel to statistics—the problem of how to handle data in which the basic observation is a direction. Data in which each observation is a vector are common in all fields. A direction is a vector of *unit* length, and this restriction changes the problem. In recent years biologists interested in animal behavior (especially in bird navigation) have raised similar questions.

Many people contributed to the construction of distributions to represent directional data. No real progress was made with the inference questions until the problem was drawn to the attention of the late Sir Ronald Fisher (1953) by geophysicists interested in paleomagnetism. His paper showed how progress toward methods of statistical analysis could be made. The writer had the great good fortune to be introduced, at that happy moment, to this subject by E. Irving, whose book (1964) could serve as a model for the application of statistics to the earth sciences. As a result, a series of simple practical methods were devised whose exposition is the subject of this paper.

No further effort will be made to trace the history of this topic on which so many have written. Steinmetz (1962) gives many older references. New references occur below in the text; Batschelet (1965) gives an extensive list. Two recent books (Miller and Kahn, 1962; Krumbein and Graybill, 1965) on the application of statistics to geology refer only briefly to the analysis of directional data. The scientific data-analysis point of view, so effectively used by Irving and evident in his many charts and elementary statistical calculations, is not stressed in these texts, which concentrate on elaborate formal methods.

Since we are about to give a brief account of formal methods, it is worthwhile reiterating that they will give results that are only as sound as the assumptions they rest on. *Thus the most important part of the analysis precedes the application of these methods.* This initial phase includes such matters as the precise formulation of the problem, decisions about what populations are really being sampled, whether the samples are random, whether the observations are numbers, vectors (in two[?] or three[?] dimensions), directions (i.e., unit vectors), or just axes (i.e., directions without sense like the orientation of a plane), thoughts about the relative magnitudes of variability due to measurement errors, site-to-site and exposure-to-exposure variation, etc. This paper begins when the first phase has led to the question of a simple mathematical description of individual populations. Later sections then deal with the comparison of different populations by statistical arguments.

[1] This research was sponsored by the Office of Naval Research under contract Nonr 4010(09) awarded to the Department of Statistics, The Johns Hopkins University. This paper in whole or in part may be reproduced for any purpose of the United States Government.

II. DESCRIPTION OF DIRECTIONAL DATA

We will consider only unit vectors in two and three dimensions. *All* such data should first be examined graphically. Two-dimensional data can be represented by angles θ or as points on the circumference of a circle. The points may cluster about a single point suggesting a single preferred direction, or a *unimodal* distribution. The scatter about some central direction (or angle) may be symmetrical or asymmetrical. In the former case a single number might be sought to describe the scatter, and this, along with the central direction, forms a rough summary of the data. It may be possible to go further and show that the density of the data around the circle is described, except for sampling fluctuations, by

$$f(\theta) = \frac{1}{2\pi I_0(\kappa)} \, e^{\kappa \cos(\theta - \theta_0)}, \tag{2.1}$$

where $I_0(\kappa)$ is a tabulated (Bessel) function of a positive parameter κ. Equation (2.1) is variously called the "circular normal" or "von Mises" distribution. The meaning of equation (2.1) is that the expected or average fraction of data in an angular sector (θ_a, θ_n) is $\int_{\theta_a}^{\theta_n} f(\theta) d\theta$. When $\kappa = 0$, $f(\kappa)$ is a constant and the data density should be constant, approximately, for all angles; i.e., there is no preferred or mean direction. When $\kappa > 0$, $f(\theta)$ has a maximum at $\theta = \theta_0$ and falls off symmetrically as θ goes away from θ_0 to a minimum at $\theta = \theta_0 + \pi$ radians (or $\theta_0 + 180°$). The fitting of equation (2.1) to data is described, with the necessary tables, in Gumbel, Greenwood, and Durand (1953), Gumbel (1954), and Batschelet (1965). If the N observed unit vectors or directions are tightly clustered about some direction, the length R of the vector resultant of the N vectors will be nearly as large as N. If, however, the vectors are greatly dispersed, R will be small. *Thus $N - R$ is a measure of dispersion of the vectors.* If the data indicates several modes or shows marked skewness, then the case for using the distribution (2.1) to represent it is slight and the geologist should be content to show his data graphically. It may often be the case that the deviations from a symmetric unimodal distribution indicate some complicating phenomena—non-random sampling, a mixture of populations, etc.—of either conceptual or technical interest.

In three dimensions, the data must be shown by some projection. The possible distributions of the points on the surface of a unit sphere are more diverse than in two dimensions. In the simplest case, the generalization of (2.1) may be a useful description. This is called, then, the "spherical normal" or Fisher distribution and is defined by the density

$$f(\theta, \phi) = \frac{\kappa}{4\pi \sinh \kappa} \, e^{\kappa \cos \theta}, \tag{2.2}$$

where κ is a non-negative constant controlling the scatter (the larger κ is, the less the data should be scattered—see fig. 1) and θ is the angle between the preferred direction and an observation. If axes are set up so that the preferred direction is the north pole, θ is the colatitude and ϕ is the longitude of a typical data point. For equation (2.2) to fit, the data must have a circularly symmetric distribution about a single point on the sphere. Also the density of the observations should not depend on ϕ but must fall off as θ goes from 0 to π. The fraction of the data making angles between θ_a and θ_b with the preferred or mean direction should be $(\theta_a < \theta_b)$

$$\frac{e^{\kappa \cos \theta_a} - e^{\kappa \cos \theta_b}}{e^{\kappa} - e^{-\kappa}}, \tag{2.3}$$

and it should be equally dense around these bands except for sampling fluctuations. A χ^2 goodness-of-fit test may be made to check the fit; examples are given in Watson and Irving

143

(1957). In rock magnetism, Fisher's distribution is entirely satisfactory, and lack of fit often means there that the rock is unstably magnetized or that the data are in some other way (e.g., mixed populations have been sampled) unsatisfactory for further work. Steinmetz (1962) assumes that his samples come from the distribution (2.2), but an inspection of the figures displaying his data shows, without a goodness-of-fit test, that his data are clustered not about a pole but about a great circle.

One basic problem always arises: Does the distribution have any preferred directions at all? If the points are fairly uniformly scattered over the sphere, this will be strongly suggested. If N observations, regarded as unit vectors, are added vectorially to get a vector of length R, say, its length often gives us a good clue. The length R must lie between zero and N, i.e., $0 \le R \le N$. We can only have R nearly equal to N if the points cluster about a single direction. However, R small, the converse, does *not* imply a uniformity of distribution; it could happen with a bipolar distribution or an equatorial distribution. Nonetheless, for distributions like (2.2) $N - R$ provides an excellent measure of the scatter or dispersion of the sample of directions.

For large bodies of data where some automatic processing is available the following procedure has been found useful (Watson 1965). With respect to any set of rectangular coordinate axes, each direction or unit vector may be described by the components of the vector along these three axes. These are called the *direction cosines* of the vector. We denote the ith of N observations by (l_i, m_i, n_i) and form the sums as indicated in the 3 × 3 array or matrix

$$M = \begin{vmatrix} \Sigma l_i^2 & \Sigma l_i m_i & \Sigma l_i n_i \\ \Sigma m_i l_i & \Sigma m_i^2 & \Sigma m_i n_i \\ \Sigma n_i l_i & \Sigma n_i m_i & \Sigma n_i^2 \end{vmatrix} = \begin{vmatrix} M_{11} & M_{12} & M_{13} \\ M_{21} & M_{22} & M_{23} \\ M_{31} & M_{32} & M_{33} \end{vmatrix}, \quad \text{say}. \quad (2.4)$$

Symbols in boldface type stand for matrices and vectors. All computing centers have a routine program for computing the eigenvalues and vectors of a matrix. The three eigenvalues of M are always positive and add up to N. The three eigenvectors are always perpendicular to each other. The physical significance of these eigenvectors and eigenvalues may be explained as follows: The eigenvectors of equation (2.4) are related to the moments of inertia of a set of particles of unit mass placed at each point on the sphere. It may be easily shown that the moment of inertia of this set of N particles about an axis through the center of the sphere of direction d is, in matrix notation, given by

$$\text{Moment of inertia about } d = N - d'Md = N - \sum_{i=1}^{3} \sum_{j=1}^{3} d_i d_j M_{ij}, \quad (2.5)$$

where d_1, d_2, d_3 are the direction cosines of the direction d. The axis that produces the greatest moment of inertia is that which minimizes $d'Md$. The minimum value of $d'Md$ is the least eigenvalue of M and the direction d is then the eigenvector associated with the least root. Similarly the least moment of inertia is associated with the eigenvector belonging to the greatest eigenvalue. Physical intuition then enables us to interpret the computational results in terms of a picture of the distribution of the results. For example, one small root and two fairly similar large roots indicate a set of points distributed about a great circle —because the moment of inertia about an axis perpendicular to the great circle would then be the largest while the moments of inertia about any diameter of the great circle would be nearly the same and smaller than that about an axis perpendicular to the plane of the great circle. For a tight cluster of points the moment of inertia about an axis through the points would be small while those about any axis perpendicular to this direction would be similar and large. Points equally distributed over the sphere would lead to nearly equal roots because the moment of inertia would be much the same about any axis. If the points cluster at the two *opposite* ends of a diameter, the moment of inertia about that axis would be small while it would be nearly constant about any axis perpendicular to this diameter. Thus this analysis will not separate unipolar and bipolar distributions. However, one large and two small eigenvalues and a large value of R definitely indicate a unipolar distribution. Numerical examples from geology are given in Watson (1965) and Bingham (1964).

In summary, for the automatic sorting of a large body of data, the computation of R and of the eigenvectors and values of M gives a good indication of the distributions of the samples. This method may also be applied to two-dimensional data where the matrix

$$m = \begin{vmatrix} \Sigma l_i^2 & \Sigma l_i m_i \\ \Sigma m_i l_i & \Sigma m_i^2 \end{vmatrix}$$

takes the place of M.

It is important to realize that it is harmful to group data. Information is lost, and there is no corresponding reduction in the cost of computing with the data.

III. FISHER-DISTRIBUTED DATA

Suppose that the methods of Section II have led us to suppose that our three-dimensional unit vectors follow Fisher's distribution (2.2) which we will now write as

$$\frac{\kappa}{4\pi \sinh \kappa} e^{\kappa(l\lambda + m\mu + n\nu)}, \tag{3.1}$$

where κ, as before, is our positive accuracy or concentration parameter, (l,m,n) is an observation, and (λ,μ,ν) is the vector of the *true* mean direction. From a *random* sample of data from this distribution we need to estimate κ and (λ,μ,ν). From Section II we have intuitively arrived at the notion that (λ,μ,ν) ought to be estimated by the direction of the vector resultant of the sample $(\Sigma l_i, \Sigma m_i, \Sigma n_i)$ and that the κ-estimator, k, say, should be inversely proportional to $N - R$, where

$$R^2 = (\Sigma l_i)^2 + (\Sigma m_i)^2 + (\Sigma n_i)^2. \tag{3.2}$$

The method of maximum likelihood verifies both these intuitions. The estimator k of κ satisfies the equation

$$\coth k - \frac{1}{k} = \frac{R}{N}, \tag{3.3}$$

which, for $k > 3$, has, quite accurately, the solution

$$k = \frac{N-1}{N-R}. \tag{3.4}$$

This approximation has proved adequate in paleomagnetism. Many attempts have been made to convert k into a standard deviation. *This is a mistake*—it leads one away from the vectorial treatment, and the natural parameter κ is much easier to handle statistically as we show below. To understand the meaning of κ, we reprint with correction a figure from Watson and Irving (1957) (fig. 1).

In order to show how the standard techniques of analysis of variance may be applied to directional data to provide solutions for the main statistical problems, it is necessary here to quote several theoretical results that have been proved to be good approximations (Watson, 1956a, b, 1960). They are

$$2\kappa(1 - \cos\theta) \approx \chi_2^2, \quad 2\kappa(N - X) \simeq \chi_{2N}^2, \quad 2\kappa(N - R) \simeq \chi_{2(N-1)}^2, \tag{3.5}$$

where θ is the angle between an observation and the true mean direction (λ,μ,ν), X is the length of the projection of the vector sum R of the sample upon (λ,μ,ν), R is the length of R, and χ_ν^2 is the standard statistical random variable χ^2 with ν degrees of freedom. The third equation in (3.5) follows from the second, intuitively, by the statistical rule that efficient fitting reduces the degrees of freedom of a χ^2 by the number of parameters fitted —in this case two because, although (λ,μ,ν) appears to contain three, there is the constraint $\lambda^2 + \mu^2 + \nu^2 = 1$.

Given a sample $x_1, \ldots, x_N$ from a normal distribution with mean μ and variance σ^2, the basic results given in all statistics texts follow from the algebraic identity

$$\Sigma(x_i - \mu)^2 = \Sigma(x_i - \bar{x})^2 + N(\bar{x} - \mu)^2 . \tag{3.6}$$

The left-hand side measures the dispersion or scatter of the sample about the true mean. $\Sigma(x_i - \bar{x})^2$ measures the dispersion of the sample about the estimated mean $\bar{x}$, while the

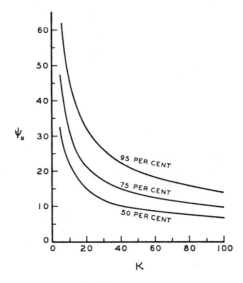

FIG. 1.—Curves giving the angle ψ_0 as a function of κ, such that 50, 75, and 95 per cent of the observations fall within ψ_0 degrees of the mean direction of the distribution (2.2). Thus ψ_0 is the semi-angle of a cone of concentration about the mean direction.

last term measures the dispersion of the sample mean about the true mean. Equation (3.6) shows the dispersion of the sample is the sum of two sources of dispersion, the first part reflecting the size of σ^2 and the second part the accuracy of $\bar{x}$. We have already seen how $N - R$ represents the dispersion of the sample of directions about the estimated mean direction. Equally $N - X$ represents the dispersion of the sample about the true mean direction; thus the algebraic identity

$$N - X = (N - R) + (R - X) \tag{3.7}$$

in our analogue of (3.6) (see fig. 2, a, below). But more is true. It is shown in all statistics texts that

$$\Sigma(x_i - \mu)^2 \simeq \sigma^2\chi_N^2 , \quad \Sigma(x_i - \bar{x})^2 \simeq \sigma^2\chi_{N-1}^2 , \quad N(\bar{x} - \mu)^2 \simeq \sigma^2\chi_1^2 , \tag{3.8}$$

146

so that equation (3.6) gives us the χ^2 resolution

$$\frac{1}{\sigma^2} \Sigma(x_i - \mu)^2 = \frac{1}{\sigma^2} \Sigma(x_i - \bar{x})^2 - \frac{1}{\sigma^2} n(\bar{x} - \mu)^2,$$

(3.9)

$$\chi_N^2 = \chi_{N-1}^2 + \chi_1^2.$$

But from equation (3.5), we see that equation (3.7) admits the parallel χ^2 resolution

$$2\kappa(N - X) = 2\kappa(N - R) + 2\kappa(R - X), \qquad \chi_{2N}^2 = \chi_{2(N-1)}^2 + \chi_2^2.$$

(3.10)

Proceeding from here we attempt to find an analogue of the t-statistic which is used in normal populations to test a hypothetical mean or to obtain confidence limits for it. Now

$$t^2 = \frac{n(\bar{x} - \mu)^2}{\Sigma(x_i - \bar{x})^2} = \frac{\chi_1^2}{\chi_{N-1}^2},$$

and our analogue suggests we use the ratio

$$\frac{R - X}{N - R} = \frac{\chi_2^2}{\chi_{2(N-1)}^2}.$$

Now statisticians have tabulated

$$F_{\nu_1, \nu_2} = \frac{\chi_{\nu_1}^2 / \nu_1}{\chi_{\nu_2}^2 / \nu_2},$$

(3.11)

so that we will naturally use

$$F_{2, 2(N-1)} = (N - 1)\frac{(R - X)}{N - R}$$

(3.12)

to test a given mean direction or to obtain a confidence cone for it. Examples of the use of equation (3.12) are given in Watson (1956a).

To examine the data to check whether they support a prescribed value of κ, or to obtain a confidence region for κ, we consider, by equation (3.5),

$$2\kappa \frac{(N - R)}{2(N - 1)} \approx \frac{\chi_2^2(N - 1)}{2(N - 1)}$$

or, using equations (3.4) and (3.11),

$$\frac{\kappa}{k} \approx F_{2(N-1), \infty}.$$

(3.13)

Immediately the comparison of the κ's of several populations becomes possible by the standard techniques used for comparing variances of samples from normal populations. The special case of testing for purely random directions is a test of $\kappa = 0$ if equation (2.2) is the alternative hypothesis. A small value of R suggests that $\kappa = 0$. Critical values for k are given in Watson (1956b); more discussion will be found in Stephens (1964).

To compare the mean directions of several populations with the same κ, we refer to figure 2, b. We will suspect that the mean directions are different if $R_1 + R_2$ is much greater than R. The dispersion within samples (i.e., of the samples from the sample mean directions) is $(N_1 - R_1) + (N_2 - R_2)$. This suggests the analysis of dispersion

$$2\kappa(N - R) = 2\kappa(N_1 - R_1 + N_2 - R_2) + 2\kappa(R_1 + R_2 - R),$$

(3.14)

$$\chi_{2(N-1)}^2 = \chi_{2(N-2)}^2 + \chi_2^2,$$

where the sample sizes are N_1 and N_2, $N = N_1 + N_2$. The second equation of (3.14) may be deduced as before. Thus the ratio of between samples dispersion to within samples dispersion is

$$F_{2,2(N-2)} = \frac{R_1 + R_2 - R}{(N_1 - R_1) + (N_2 - R_2)} \frac{2(N-2)}{2},\qquad(3.15)$$

where the second factor divides the χ^2's by their degrees of freedom to obtain an F-distribution. Equation (3.15) then provides a standard test of the equality of the mean directions of two populations. It is immediately extended to q populations in

$$F_{2(q-1),2(N-q)} = \frac{R_1 + R_2 + \ldots + R_q - R}{\sum_{i=1}^{q}(N_i - R_i)} \frac{2(N-q)}{2(q-1)},\qquad(3.16)$$

where

$$N = \sum_{1}^{q} N_i,$$

R is the length of the resultant of all the N observations, and $R_1, \ldots, R_q$ are the lengths of the sample resultants.

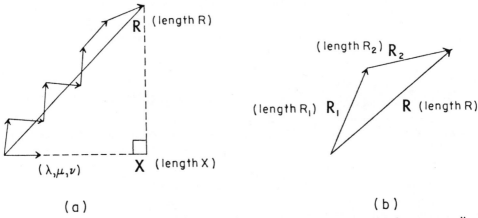

(a)   (b)

FIG. 2.—(a) The vector sum or resultant R of seven vectors and its component X on the true mean direction whose direction cosines are (λ,μ,ν). If the unit vectors have nearly similar directions, it is clear that R is longer than if their directions are very different. (b) The addition of the resultants R_1 and R_2 of two samples to the combined resultant R. If R_1 and R_2 have almost the same direction, it is clear that R will be only a little less than $R_1 + R_2$.

A very important problem arises when one wishes to estimate a mean direction of some property (e.g., the c-axis of quartz grains) of a large geological formation. It is very likely that, if w_i samples are taken at the ith of B sites, the sites being spread evenly across the formation, that there will be variation within and between sites. An erroneous idea of accuracy will thus be obtained unless several sites are examined. This suggests an analysis that is analogous to the variance-components analysis in normal-theory statistics. If it is assumed that, within the ith site, the variation is described by a Fisher distribution with $\kappa = \omega$ and an ith site mean, and that the variation of the site means is Fisher-distributed about the formation mean with $\kappa = \beta$, an analysis of variance may be used to estimate ω and β. This is described in Watson and Irving (1957). Let the site resultants have lengths

$R_1, \ldots, R_B$ and the $N = \Sigma W_i$ observations have a vector resultant of length R. Then the analysis of variance table is as shown in table 1. Here

$$\bar{W} = \frac{1}{B-1}\left(N - \frac{\Sigma W_i^2}{N}\right)$$

and is the weighted average of the W_i; if all $W_i = W$, $\bar{W} = W$. The significance of the between-site variation may be judged by an F-test. By equating mean squares and their expectations, estimates $\hat{\beta}$ and $\hat{\omega}$ may be obtained. The precision k of the estimator of the mean direction of the formation (the direction of the resultant of all the N observations) is given by

$$k = \frac{1}{(\hat{\omega}N)^{-1} + (\hat{\beta}B)^{-1}}.$$

A cone of confidence may be obtained by assuming the estimate has a Fisher distribution with this value of κ. Watson and Irving (1957) give a numerical example.

TABLE 1

ANALYSIS OF VARIANCE

Source	Degrees of Freedom	Sum of Squares	Mean Square	Expectations of Mean Squares
Between sites...	$2(B-1)$	$\Sigma R_i - R$	$(\Sigma R_i - R)/2(B-1)$	$\frac{1}{2}(1/\omega + \bar{W}/\beta)$
Within sites....	$2[\Sigma(W_i - 1)]$	$\Sigma(W_i - R_i)$	$\Sigma(W_i - R_i)/2\Sigma(W_i - 1)$	$1/2\omega$
Total.........	$2(N-1)$	$N - R$		

The above results are the spherical analogues of the elementary normal-theory problems concerning means and variances. Many other problems may, however, arise. The allocation of specimens to one of two known populations—for which the discriminant function is used in normal theory—is performed by using

$$l(\kappa_1\lambda_1 - \kappa_2\lambda_2) + m(\kappa_1\mu_1 - \kappa_2\mu_2) + n(\kappa_1\nu_1 - \kappa_2\nu_2), \qquad (3.17)$$

where $(\lambda_1, \mu_1, \nu_1)$ and $(\lambda_2, \mu_2, \nu_2)$ are the mean vectors of the two populations which have accuracies κ_1 and κ_2. The writer would be interested to hear from geologists who have problems of this kind. With three or more populations the question sometimes arises of whether the mean vectors are coplanar. The analysis of variance solution of this is given in Watson (1960). All linear hypotheses about mean directions can be tested by estimating the dispersion on the null H_0 and non-null H_a hypotheses, just as is done in normal-theory analysis of variance. The difference (dispersion) H_0-(dispersion)H_a is then divided by (dispersion) H_a to form an F-statistic.

The approximate treatment above has been examined for accuracy in various ways in a doctoral thesis and subsequent papers by Stephens (1962a, b, c, 1964). Precise methods have been suggested by Watson and Williams (1956), but they are not always practical. The results seem fairly robust against small deviations from the Fisher form of distribution, if used with the moderation with which all statistical methods should be used.

IV. VON MISES–DISTRIBUTED DATA

If a study of the two-dimensional data by the methods of Section II suggests that the von Mises distribution (2.1) will be adequate, κ and the true mean direction ($\lambda = \cos\theta_0$,

$\mu = \sin \theta_0$) will be estimated by maximum likelihood as shown in the references mentioned in Section II. Because Batschelet's monograph (1965) is available and written for the non-mathematically trained user, much less need be said here than in Section III.

It is, however, worthwhile to stress the simple analysis of variance approach, since it is rather lost in the many details of Batschelet's account. Before doing so, we introduce figure 3, which shows the meaning of the accuracy parameter κ for the circular case, just as figure 1 shows it for the spherical case. It shows an angle ψ_0 such that, for a given κ, 95 per cent (75 per cent, 50 per cent) of the observations should fall in the interval $(\theta_0 - \psi_0, \theta_0 + \psi_0)$, where θ_0 is the true mean direction; e.g., for $\kappa = 3$, 95 per cent of the observations should lie within 75° of the true mean.

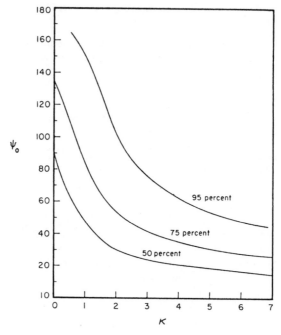

FIG. 3.—Curves giving the angle ψ_0 as a function of κ, such that 50, 75, and 95 per cent of the observations fall within ψ_0 degrees of the mean direction of the distribution (2.1). Thus ψ_0 is the semi-angle of a sector of concentration about the mean direction.

In place of equation (3.5), we have

$$2\kappa(1 - \cos \theta) \approx \chi_1^2, \quad 2\kappa(N - X) \approx \chi_N^2, \quad 2\kappa(N - R) \approx \chi_{N-1}^2 .\tag{4.1}$$

Then equation (3.12) is replaced by

$$F_{1,N-1} = (N-1)\frac{(R-X)}{N-X},\tag{4.2}$$

equation (3.13) by

$$\frac{\kappa}{k} = F_{N-1,\infty},\tag{4.3}$$

equation (3.15) by

$$F_{1,N-2} = \frac{R_1 + R_2 - R}{N_1 - R_1 + N_2 - R_2} \frac{N-2}{1},\tag{4.4}$$

150

and equation (3.16) by

$$F_{q-1,N-q} = \frac{R_1 + \ldots + R_q - R}{\sum_1^q (N_i - R_i)} \frac{N-q}{q-1}. \tag{4.5}$$

The use of the relations (4.2)–(4.5) becomes more and more accurate as κ becomes larger. If only equation (4.2) has to be used and κ is small, the exact procedures of Watson and Williams (1956) may be used with the help of nomograms derived by Stephens and quoted by Batschelet.

V. FURTHER DISTRIBUTIONS ON THE SPHERE

Data are often obtained that cluster about both ends of a diameter or that cluster around a great circle. Various suggestions have been made for distributions that might fit. However, it is only recently that statistical procedures have been suggested for them (Selby, 1964; Bingham, 1964; Watson, 1965). The latter two authors both give the theory for the most amenable distribution whose density is proportional to

$$e^{\kappa \cos^2 \theta}. \tag{5.1}$$

If $\kappa > 0$, the density is a maximum at the poles $\theta = 0°$ and $\theta = 180°$, so it represents a bipolar distribution. If $\kappa < 0$, the density is least at the poles and a maximum around the equator, so it represents an equatorial distribution. As before, $\cos \theta = \lambda l + \mu m + \nu n$ when (λ,μ,ν) and (l,m,n) are, respectively, the true axial and the observation directions.

TABLE 2*

Ratio	k	Ratio	k	Ratio	k
0.351	0.2	0.643	3.2	0.862	8.0
.370	0.4	.660	3.4	.871	8.5
.389	0.6	.676	3.6	.879	9.0
.409	0.8	.690	3.8	.886	9.5
.429	1.0	.705	4.0	.892	10.0
.450	1.2	.718	4.2	.903	11.0
.470	1.4	.731	4.4	.912	12.0
.491	1.6	.743	4.6	.919	13.0
.511	1.8	.754	4.8	.925	14.0
.531	2.0	.764	5.0	.930	15.0
.551	2.2	.788	5.5	.935	16.0
.571	2.4	.808	6.0	.939	17.0
.590	2.6	.825	6.5	.942	18.0
.608	2.8	.839	7.0	.946	19.0
0.626	3.0	0.851	7.5	0.948	20.0

* The ratio on the left-hand side of eq. (5.2) has been computed by R. J. Beran for various values of k.

It is easily shown that the estimator of (λ,μ,ν) is the eigenvector of M given in equation (2.4), corresponding to its largest root if $\kappa > 0$ and its smallest root if $\kappa < 0$. Denote these roots by $\tau_1 \leq \tau_2 \leq \tau_3$. The estimator of κ, k, may be found for the equatorial case from a table in Watson (1965) once the least eigenvalue or a latent root τ_1 of M is known. The estimated density is proportional to $\exp(-k \cos^2 \theta)$. For the bipolar form, the estimated density has the form (5.1) with κ replaced by k, the solution of the equation

$$\frac{\int_0^1 t^2 e^{kt^2} dt}{\int_0^1 e^{kt^2} dt} = \frac{\tau_3}{N}, \tag{5.2}$$

where τ_3 is the largest eigenvalue of M.

It may be shown that

$$\frac{\int_0^1 t^2 e^{kt^2} dt}{\int_0^1 e^{kt^2} dt} \begin{cases} \to \frac{1}{3}(1 + \frac{4}{15}k), & \text{as } k \to 0, \\ \to 1 - \frac{1}{k}, & \text{as } k \to \infty. \end{cases} \qquad (5.3)$$

The left-hand side of equation (5.2) was computed for values of k between $k = 0.2$ and $k = 20$ (table 2). For $k < 0.2$ and $k > 20$ the formulae (5.3) are correct to two decimal places. In the range of the table, linear interpolation may be used to solve equation (5.2) to an accuracy of at least two significant figures.

The development of statistical procedures for the distribution (5.1) is begun in Watson (1965) for the case $\kappa < 0$, and the same techniques may be adopted to the case $\kappa > 0$. A more general theory that subsumes the above has been given by Bingham (1964). It provides, for example, a mathematical framework for the work of Loudon (1964) which is very similar to the general discussion in Section II. More effort is, however, required to make Bingham's work easily available to geologists. Because of limitations of time, it was impossible to clear up this important area in the present paper. In particular the writer regrets that there has been no explicit discussion of the analysis of axial data. This omission will be made good in a future paper.

ACKNOWLEDGMENTS.—The writer is very grateful to R. J. Beran for computing table 2, and to Dr. Felix Chayes, who suggested that the paper be written and later made most helpful comments.

REFERENCES CITED

Arnold, K. J., 1941, On spherical probability distributions: Unpublished Ph.D. thesis, Massachusetts Institute of Technology.

Batschelet, E., 1965, Statistical methods for the analysis of problems in animal orientation and certain biological rhythms: AIBS Monograph.

Bingham, C., 1964, Distributions on a sphere and on the projective plane. Ph.D. dissertation, Yale University.

Fisher, R. A., 1953, Dispersion on a sphere. Proc. Roy. Soc. London. Ser. A. v. 217, p. 295–305.

Gumbel, E. J., 1954, Applications of the circular normal distribution: Jour. J. Am. Statis. Assoc., v. 49, p. 267–297.

——, Greenwood, J. A., and Durand, D., 1953, The circular normal distribution: Theory and tables: Jour. Am. Statis. Assoc., v. 48, p. 131–152.

Irving, E., 1964, Paleomagnetism and its applications to geological and geophysical problems: New York, John Wiley & Sons.

Krumbein, W. C., and Graybill, F. A., 1965, An introduction to statistical models in geology: New York, McGraw-Hill Book Co.

Loudon, T. V., 1964, Computer analysis of orientation data in structural geology: Tech. Rept. no. 13. Northwestern University, Evanston, Ill.

Miller, R. L., and Kahn, J. S., 1962, Statistical analysis in the geological sciences: New York, John Wiley & Sons.

Selby, B., 1964, Girdle distributions on a sphere: Biometrika, v. 51, nos. 3 and 4, p. 381–392.

Steinmetz, R., 1962, Analysis of vectorial data: Jour. Sed. Petrology, v. 32, p. 801–812.

Stephens, M. A., 1962a, The statistics of directions—the Fisher and von Mises distributions: Unpublished Ph.D. dissertation, University of Toronto.

—— 1962b, Exact and approximate tests for directions. I: Biometrika, v. 49, p. 463–477.

—— 1962c, Exact and approximate tests for directions. II: Ibid., p. 547–552.

—— 1964, The testing of unit vectors for randomness: Jour. Am. Statis. Assoc., v. 59, p. 160–167.

Watson, G. S., 1956a, Analysis of dispersion on a sphere: Roy. Astron. Soc. Monthly Notices Geophys. Supp., v. 7, no. 4, p. 153–159.

———— 1956b, A test for randomness of directions: *Ibid.*, p. 160–161.

———— 1960, More significance tests on the sphere: Biometrika, v. 47, p. 87–91.

———— 1961, Goodness of fit tests on the circle: *Ibid.*, v. 48, p. 109–114.

———— 1962, Goodness of fit tests on the circle. II: *Ibid.*, v. 49, p. 57–63.

———— 1965, Equatorial distributions on a sphere: *Ibid.*, v. 52, p. 193–201.

———— and Irving, E., 1957, Statistical methods in rock-magnetism: Roy. Astron. Soc. Monthly Notices Geophys. Supp., v. 7, no. 6, p. 289–300.

Watson, G. S., and Williams, E. J., 1956, On the construction of significance tests on the circle and the sphere: Biometrika, v. 43, p. 344–352.

Part II

STATISTICAL METHODS–
A SECOND LOOK

Editors' Comments
on Papers 14 Through 18

Even before the advent of computers, geologists had begun to question the basic statistical assumptions of methods they wished to use. A prime example was the investigation of transformations that could be applied to data of non-Gaussian probability distributions, as studied independently by Razumovsky (1940, 1941) and Ahrens (1953, 1954a, b, 1957). However, their application of the logarithmic normal (or log-normal) distribution to geochemical data was seen only as an empirical tool for "massaging" the data. Nevertheless, throughout the 1950s and 1960s papers were published in which the lognormal distribution of trace elements was assumed, with the graphical support of skewed histograms. Indeed, Ahrens (1953, 1954a, b, 1957) elevated this distribution to the status of a law. During the 1960s the greatly increasing volume of data from instrumental analysis led to rapid demotion of this "law," with the realization (Krige, 1960; Shaw, 1961) that trace-element distributions followed no unique statistical law but were controlled by complex geological factors (Krige, 1951; Vistelius, 1960, 1967; McCammon, 1969; Ondrick and Griffiths, 1969; Harbaugh and Bonham-Carter, 1970; Burch and Murgatroyd, 1971; Burch, 1972; Gates and Ethridge, 1972; Clark, 1976).

This admitted complexity results in insuperable problems for statistical analysis of data: An undefined mixed distribution and many of the simpler distributions (e.g., beta and gamma) cannot be transformed to a Gaussian distribution by taking logarithms. Tests can tell us whether a particular distribution is approximately Gaussian, within given confidence limits. If it is, or if it can be transformed readily to one that is, then statistical operations may give valid results. However, if we have a data set that cannot be transformed, the confidence placed in any parametric statistical analysis may be severely reduced. Richard Link from Artronic Information Systems in New York and George Koch, Jr., of the University of Georgia demonstrate the effect of automatic application of transformations that are inappropriate to the data—that is, simple statistical tests are completely invalidated (Paper 14).

As statistical analysis is extended to study the correlation between a pair, or among several variables, another problem can develop. Frequently, geological data is collected and recorded in the form of ratios, such as percentage data, where each constituent is expressed as a proportion of the whole: All geochemical and petrographic (modal) mineralogical data are recorded in this form, and frequently other data may be transformed to ratios for ease of interpretation. However, when one consitutent in a percentage analysis is dominant, its correlations with other variables tend to be negative—that is, as the dominant component increases, all other components appear to decrease. This negative correlation is a result of the closure of a data set adding to a constant sum. Felix Chayes of the Carnegie Geophysical Laboratory (1960, 1962, 1964, 1967, 1971, 1972) examined the mathematics of correlation as it applied to ratio data and together with William Kruskal, a statistician at the University of Chicago famed for developing multidimensional scaling, devised a method of statistically testing the significance of correlations between proportional data (Paper 15). Chayes and Kruskal also give details of how to determine the null value for correlation and for computing hypothetical variances of variables with the closure effect removed.

Chayes and Kruskal's conclusions are repeated and elaborated upon in subsequent papers (Chayes, 1970, 1971, 1972, 1975; Zodrow, 1975, 1976). However, there seems to be little prospect of further dramatic advance in this statistically intractable field, because the essential problem is that of compensating for an original lack of information. Data collected as ratios or percentages contain no information on absolute quantities.

With the advent of computers, it became practical to apply multivariate techniques such as principal components analysis to large data sets. However, because of the complexity of the methods and the lack of general statistical knowledge in geology, there was little discussion of the appropriateness of multivariate techniques to geological applications of the assumptions made, or of the statistical confidence that could be placed in the results. It was not until the late 1960s that geologists became aware of these problems, and Miesch of the U.S. Geological Survey, in parallel with similar studies in polynomial trend surface analysis, reviewed principal components and factor analysis by using Monte Carlo simulation (Paper 16). Al Miesch shows that the techniques are mathematically effective, but require a knowledge of their strict limitations and assumptions before any useful application to geological data should be attempted.

Robert Blackith of Trinity College, Dublin, and Richard Reyment of the Paleontological Institute in Uppsala also warned geologists of the flaws in factor analysis (Paper 17). They quote Ehrenburg's comment that after fifty years of the practice of factor analysis, it was uncertain whether anything of value had emerged. The same comment would probably be true today, for there is no rigorous statistical test for the results of factor analysis. However, the cautious approach advocated by Blackith and Reyment implies that there remains a place for factor analysis in the armory of (?quasi) statistical methods available to the geoscientist.

The reappraisal of statistical methods continues as the complexities of real data are seen to conflict with the idealized assumptions required for rigorous application of classical statistics. In the collection of data, for example, John Griffiths of Pennsylvania State University has thoroughly examined the complex pitfalls of statistical sampling methods and the meaning of a geological sample (Paper 18). A similar study initiated by Krige in South Africa has been developed into the field of geostatistics by Matheron in France. Zodrow and Sutterlin (1971) have adopted an alternative approach based on set theory in an attempt to define samples and the interrelationships between them.

It seems therefore that the re-examination of statistical assumptions in recent years has led to two divergent trends. In the Anglo-American school, sampling methods, data transformations, and classical statistics have been defined rigorously, and a variety of new statistical (and non-statistical) approaches have been investigated. Alternatively, in the French school the statistics themselves have been adapted to fit the peculiarities of spatially distributed geological data, and a family of related statistical methods have been created by using the concept of the regionalized variable.

In the present-day application of statistics to geology, it is recognized that the second look at established methods has substituted a balanced, skeptical approach to the results of statistical analysis for the frequent earlier attitude of total reliance on significance tests or the acceptance of inappropriate or unproven methods.

14

Copyright © 1975 by Plenum Publishing Corporation

Reprinted from *Math Geol.* 7(2):117–128 (1975)

Some Consequences of Applying Lognormal Theory to Pseudolognormal Distributions[1]

Richard F. Link[2] and George S. Koch, Jr.[3]

A logarithmic transformation may be used to improve the efficiency of estimates of the mean when observations follow the lognormal distribution. But if this transformation is applied to observations that follow another distribution, bias may be introduced. We consider some consequences of erroneously applying lognormal estimation theory and demonstrate that biased estimates may be obtained for certain classes of distributions. Illustrations of bias obtained in gold sampling are given. KEY WORDS: data processing, lognormal theory, sampling, simulation, statistics, transformations, economic geology, mining, sedimentology.

INTRODUCTION

Frequency distributions that mimic the lognormal distribution by being skewed also to the right may be named pseudolognormal. We consider some consequences of erroneously applying lognormal-estimation theory to several of these pseudolognormal distributions.

Both the lognormal and the pseudolognormal distributions are followed by many sets of geological data, which characteristically contain many relatively small observations but only a few large ones. Examples are the usual distributions of most substances found in nature in small or trace amounts (including diamonds, gold, other precious metals, uranium, many distributions in exploration geochemistry, and many sediment-size distributions). Because geological experience or theory seldom provides a guide to determine whether an observed distribution is lognormal or not, a risk is run when lognormal theory is applied to an empirical distribution. The purpose of this paper is to investigate the consequences of running this risk.

A familiar geological problem is to estimate a population mean, perhaps the average grade of a gold ore body, or the average grain size of a sandstone. If an empirical distribution is in fact lognormal, the efficiency of estimate of the population mean can be improved by applying lognormal theory.

---

[1] Manuscript received 24 December 1973.
[2] Artronic Information Systems, New York (USA).
[3] Department of Geology, University of Georgia (USA).

The improvement is substantial if the coefficient of variation is large (say larger than 1.2). However if the distribution is not lognormal, a mean estimated through lognormal theory is generally biased. We investigate the size of this bias by applying lognormal theory to several pseudolognormal theoretical distributions. The work is done by mathematical simulation, because a purely analytic treatment is mathematically intractable.

Investigators of trace-element distributions in geology have studied the lognormal distribution for several years, following the first work done on gold data by Sichel (1947) and Krige (1952). Recently, we (Link, Koch, and Schuenemeyer, 1971) studied the distribution, also with particular regard to gold data. Among other subjects, we reviewed a method used by Krige (1960) to remove positive bias from estimates of population means made through lognormal theory by adding a constant to each gold assay value. Through mathematical simulation, we investigated the consequences of adding a constant to observations that were already lognormal, in knowledge of the fact that it was theoretically inappropriate to add a constant. In the experiment, 1000 random samples of size 10 were drawn from a population that was modified from a lognormal population. The modification was made by subtracting a constant (0.255) from all those observations larger than the constant and by replacing all those observations smaller than the constant with the observation 0.005 (an arbitrary small number). The resulting censored distribution is of the type postulated by Krige for gold observations that were originally lognormal but from each of which a constant amount of gold was

Table 1. Statistics Calculated in Sampling Experiment to Test Effect of Adding Constant to Modified Lognormal Population (Recalculated from Link, Koch, and Schuenemeyer, 1971, p. 43)

Constant	Average of 1000 estimated means, assuming lognormality	Percentage bias
0	1.77	77
0.026	1.17	17
0.051	1.07	7
0.128	0.98	−2
0.255	0.95	−5
0.510	0.94	−6
2.511	0.96	−4
5.102	0.97	−3

removed, with the smallest observations, which the assayer would record as trace or 0, being replaced by the arbitrary small number.

Table 1 gives the results of this simulation. (The numerical values of the original table have been rescaled for consistency with the results of the present study.) The first column lists the constants; the second column lists the arithmetic means of the 1000 estimates of the population mean; the third column lists the estimated percentage biases obtained by subtracting the theoretical mean of 1.0 from the means of column 2 and multiplying by 100. The table shows that, whether or not a constant is needed, adding too large a constant does not on the average lead to much bias, but if a constant is needed, adding too small a constant leads to large positive bias. This simulation does in fact reproduce the phenomenon discovered by Krige.

PRESENT INVESTIGATION

In this paper, we change the fundamental shape of the distribution of the logarithms in order to consider a different type of deviation from lognormality than that discussed by Krige. In particular, 14 theoretical frequency distributions of logarithms, corresponding to 14 original skewed distributions, are

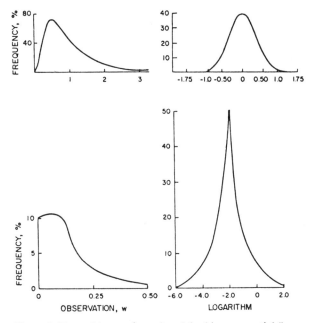

Figure 1. Normal (upper figures) and double exponential (lower figures) distributions of logarithms.

161

studied. These original distributions are alike or similar to empirical distributions followed by many sets of geological data. Figures 1 to 3 illustrate several of these distributions. The distributions of observations are on the left, and those of the logarithms of the observations are on the right. Figure 1 shows normal and double exponential distributions of logarithms. Figure 2 shows rectangular, triangular, and quadratic distributions of logarithms; these nonnormal distributions are included because the corresponding distributions of observations are skewed more or less similar to the others. Figure 3 shows two lambda distributions. Not illustrated, but also discussed are chi-square distributions. The chi-square distribution with 25 degrees of freedom resembles the normal distribution; chi-square distributions with fewer degrees of freedom are increasingly skewed.

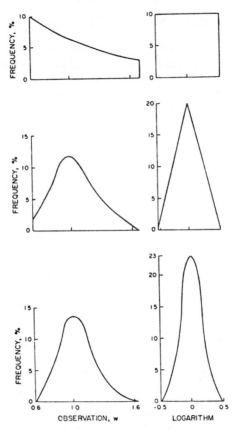

Figure 2. Rectangular (upper figures), triangular (middle figures), and quadratic (lower figures) distributions of logarithms.

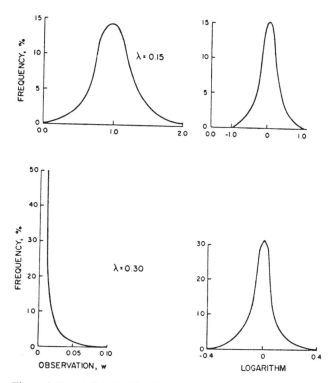

Figure 3. Low-tailed lambda distribution (upper figures) and high-tailed lambda distribution (lower figures) of logarithms.

Calculations

The method of calculating is explained by giving the procedure for the analysis of observations for random samples of size 5 drawn from a chi-square distribution with 1 degree of freedom. The method for other distributions and sample sizes differs only in the details of drawing the observations. Most of the calculations are done with published computer programs (Koch, Link, and Schuenemeyer, 1972).

Steps are as follows:

(1) Generate a random number between 0 and 1 through use of a pseudo-random number generator (Koch, Link, and Schuenemeyer, 1972, p. 12).

(2) From this number, generate a random normal number (Koch, Link, and Schuenemeyer, 1972, p. 6).

(3) Square the normal number and thereby obtain a random number from a chi-square distribution with 1 degree of freedom (Koch and Link, 1970, p. 199).

(4) Exponentiate this number. Multiply the result by a constant to get an observation from a distribution with a mean of 1 (the value of the constant is found by a preliminary analysis up to and including step 8).

(5) Repeat steps 1 to 4 five times to get a random sample of size 5.

(6) Apply lognormal theory (Koch, Link, and Schuenemeyer, 1972, p. 46) to calculate a sample mean from logarithms, also calculate the arithmetic sample mean.

(7) Steps 1 to 6 are repeated 1000 times, to obtain 5000 observations, 1000 sample means based on logarithms, and 1000 arithmetic sample means.

(8) Calculate the means of the 1000 arithmetic and lognormal sample means, and calculate their ratio to estimate the percentage bias.

(9) Calculate frequency distributions of the 5000 observations and of their logarithms (Koch, Link, and Schuenemeyer, 1972, p. 8).

Results of Calculations

In order to interpret the results, the concept of the shape of the distribution tails must be understood. This shape may be characterized with the terms "low-tailed distribution" or "high-tailed distribution." Consider symmetric, unimodal distributions with the same mean and variance. A distribution is *low-tailed* if, as the absolute values of the observations get large, the ratio of $f_1(x)/f_2(x)$ approaches zero, where $f_1(x)$ is the frequency distribution of the low-tailed distribution and $f_2(x)$ is the frequency distribution of the normal distribution (Fig. 4). A distribution is *high-tailed* if, as the absolute values of the observations get large, the ratio of

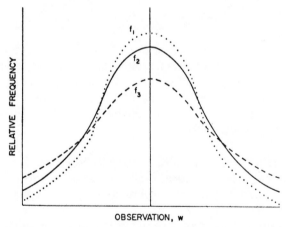

Figure 4. Low-tailed (f_1), normal (f_2), and high-tailed (f_3) distributions.

$f_3(x)/f_2(x)$ approaches infinity, where as before $f_2(x)$ is the frequency distribution of the normal distribution and $f_3(x)$ is the frequency distribution of the high-tailed distribution (Fig. 4).

In many scientific investigations, variables whose distributions might reasonably be expected to be normal, have turned out to have high tails, because extreme observations were introduced by causes such as more or less random analytical blunders. In geology, one also must expect to have to deal with distributions of logarithms of observations with high tails. Other examples may come from situations where extreme observations come from mixed distributions like so many sediment-size distributions.

Percentage bias, rather than *bias*, is the measure used in this paper to compare the arithmetic means with those estimated with logarithms. Percentage bias is equal to

$$[(\mu' - \mu)/\mu]100$$

where μ is the mean of the observations w, and μ' is the mean estimated with logarithms. Percentage bias has the desirable property of being essentially independent of the size of μ, whereas the size of the bias δ, equal to $\mu' - \mu$ (Koch and Link, 1970, p. 80), is nearly directly proportional to the size of μ.

The results of the present investigation are summarized in Tables 2, 3, and 4. The distributions are grouped into three classes. Given first is a class of four chi-square distributions, because these distributions have basically the same type of tail as a normal distribution. The second class contains several distributions which are studied as alternatives to the normal distribution; three of these have low tails, and one has a high tail. The third class consists of several examples of the lambda distribution, which has a parameter that raises or lowers the tail. The lambda distribution of a variable w is defined by the formula

$$w = [1/(1-u)^\lambda] - 1/u^\lambda \qquad (\lambda > 0)$$

where u has a uniform distribution $(0, 1)$. The parameter controls the height of the distribution tails. This distribution was introduced by Hastings and others (1947) in an investigation simulating the behavior of statistics from other than normal distributions. For easy comparison of the percentage biases, the tabled results are scaled to make the arithmetic means equal to 1.

For samples of sizes 5 and 10 drawn from chi-square distributions with various numbers of degrees of freedom, Table 2 compares arithmetic means with those estimated from logarithms. The sign of the bias is negative if the mean estimated from logarithms is less than 1 and positive if it is greater than 1. For the samples of size 5, the chi-square distribution with 1 degree of freedom has the largest percentage bias. It decreases as the degrees of freedom increase, because for many degrees of freedom the chi-square

Table 2. Percentage Biases of Lognormal Estimates for Samples of Sizes 5 and 10 from Chi-Square Distribution

Distribution	Sample size	Degrees of freedom	Arithmetic mean	Lognormal mean	Percentage bias
		1	1	0.89	−11
	5	5	1	0.97	−3
		10	1	0.99	−1
		25	1	1.0	0
Chi-square					
		1	1	0.19	−81
	10	5	1	0.95	−5
		10	1	0.98	−2
		25	1	1.0	0

distribution closely approximates the normal distribution. It also increases as the sample size increases from 5 to 10.

Table 3 compares percentage biases for selected low-tailed and high-tailed distributions. For the low-tailed distributions, lognormal estimation introduces no bias for samples of either size 5 or 10. However, for the high-tailed distribution, lognormal estimation introduces a severe negative bias.

Table 3. Percentage Biases of Lognormal Estimates for Selected Low-Tailed and High-Tailed Distributions

Type of distribution	Sample size	Distribution	Arithmetic mean	Lognormal mean	Percentage bias
		Uniform	1	1.00	0
	5	Triangular	1	1.00	0
		Quadratic	1	1.00	0
Low-tailed					
		Uniform	1	1.00	0
	10	Triangular	1	1.00	0
		Quadratic	1	1.00	0
	5	Double exponential	1	0.66	−34
High-tailed					
	10	Double exponential	1	0.56	−44

Table 4. Percentage Biases of Lognormal Estimates for Several Examples of Lambda Distribution

Type of distribution	Sample size	Value of lambda	Arithmetic mean	Lognormal mean	Percentage bias
		0.05	1	1.0	0
		0.10	1	1.0	0
Low-tailed	5	0.15	1	1.0	0
		0.25	1	0.85	−15
High-tailed	5	0.30	1	0.19	−81
		0.35	1	0.15	−85
		0.05	1	1.0	0
Low-tailed	10	0.10	1	1.0	0
		0.15	1	1.0	0
		0.25	1	0.06	−94
High-tailed	10	0.30	1	0.03	−97
		0.35	1	0.10	−90

Table 4 compares percentage biases for selected values of lambda, the parameter that controls the height of the tail. Small values of lambda produce low-tailed distributions; high values produce high-tailed distributions. (The entries for a value of lambda of 0.20 are not tabulated, because in this region the distribution shifts from low-tailed to high-tailed.) No biases are obtained by lognormal estimation for the low-tailed distributions, but considerable bias is obtained for the high-tailed distributions. For both Tables 3 and 4, the bias from samples of size 10 drawn from high-tailed distributions is more severe than from those of size 5.

The logarithmic transformation introduces bias in high-tailed distributions because of the shape of the tails. In a high-tailed distribution, an observation can be farther out in the tail, as measured in units of standard deviation away from the mean, than in a normal distribution. Because the logarithmic transformation assumes a normal distribution of logarithms, it will excessively reduce observations from a high-tailed distribution. For larger samples the effect of this reduction is more severe than for smaller samples, because on the average the large observations in larger samples are more extreme than those in smaller samples. The numerical example of Table 5 illustrates this phenomenon. For three samples drawn from the double exponential distribution, the table compares the arithmetic and lognormally estimated means for samples A and B of size 5 with those of sample C of

Table 5. Comparison of Means Calculated for Samples of
Sizes 5 and 10 from Double Exponential Distribution

Item	Sample name		
	A	B	C
	0.70	0.33	0.70
	0.32	688.33	0.32
	1.63	0.94	1.63
	6.09	8.89	6.09
	0.99	0.75	0.99
			0.33
			688.33
			0.94
			8.89
			0.75
Arithmetic mean	1.94	139.85	70.90
Lognormal mean	1.83	61.30	16.68

size 10; sample C is a composite of samples A and B. Samples B and C include
the observation 688.33, one of the larger but not the largest observation obtained in simulating 10,000 observations from the double exponential distribution. For sample A, the arithmetic and lognormal means are nearly
the same, but for sample B, the arithmetic mean is larger than the lognormal
mean. For sample C, the arithmetic mean 70.90 is necessarily the mean of the
arithmetic means 1.94 and 139.85 of samples A and B. But the lognormal
mean 16.68 is only about half the mean 31.57 of the lognormal means 1.83
and 61.30 of samples A and B. For a sample of size 5, the observation 688.33
is not recognized by the transformation as being so extreme (because there
are so few observations) as in the sample of size 10.

Application

These results may be related to empirical gold data from mines in South
Africa and Kolar, India. The South African data, obtained by Krige (1960)
and later summarized by us (Koch and Link, 1971, p. 384–385), are evidently
censored and yield positive bias. In 21 of 28 samples, the estimates based
on logarithms are substantially higher than the arithmetic means. (Krige
compensated for this overestimate by adding a constant to each original
observation.) One set of Kolar data (Sarma, 1969, p. 104) yields negative
bias, suggesting high-tailed distributions. A later set of Kolar data (Sarma,

1970, p. 26) yields both positive and negative bias, suggesting both high- and low-tailed distributions.

SUMMARY

If lognormal estimation is used for observations that are not in fact lognormally distributed, either positive or negative bias, or a combination of the two can occur.

Positive bias can occur if a significant fraction of the observations in a sample are censored (Table 1). If only positive bias occurs, owing to a censoring effect on a lognormal distribution, in practice it can be corrected by adding a constant to the logarithms of the original observations.

Negative bias can occur if the tails of the distribution of logarithms are higher than the tails of a normal distribution. There is no straightforward manner to correct for negative bias.

One cannot determine easily, if at all, whether a given distribution of logarithms has tails that are high, low, or normal. Although only the high-tailed distributions lead to bias, one must be cautious, because high-tailed distributions may be common in geology. Sources of both positive and negative bias may operate to produce a net positive or negative bias.

One employs the lognormal transformation when data exhibit extreme variability as reflected by a large coefficient of variation (Link, Koch, and Schuenemeyer, 1971). Using the transformation obtains more efficient estimates of the mean; however, the ordinary arithmetic mean is always unbiased. The methods of correcting for bias in log-transformed data in the past have grown from experience in using the transformation and comparing results with untransformed analysis. This growth of understanding will continue in the future as experience is gained about the behavior of different types of geological data. The purpose of this paper is to suggest that, until experience is obtained with a substantial amount of data, caution should be exercised in applying the lognormal distribution without examining consequences that may be undesirable.

ACKNOWLEDGMENTS

We acknowledge, with thanks, the advice of Frederick Mosteller. We are grateful to the American Philosophical Society for financial support.

REFERENCES

Hastings, C., Mosteller, F., Tukey, J. W., and Winsor, C. P., 1947, Low moments for small samples: a comparative study of order statistics: Ann. Math. Stat., v. 18, p. 413–426.

Koch, G. S., Jr., and Link, R. F., 1970, Statistical analysis of geological data, v. 1: John Wiley & Sons, New York, 375 p.

Koch, G. S., Jr., and Link, R. F., 1971, Statistical analysis of geological data, v. 2: John Wiley & Sons, New York, 438 p.

Koch, G. S., Jr., Link, R. F., and Schuenemeyer, J. H., 1972, Computer programs for geology: Artronic Information Systems, Inc., New York, 142 p.

Krige, D. G., 1952, A statistical analysis of some of the borehole values in the Orange Free State goldfield: Jour. Chem. Met. and Min. Soc. of South Africa, v. 53, p. 47–70.

Krige, D. G., 1960, On the departure of ore value distributions from the lognormal model in South African gold mines: Jour. South African Inst. Mining Metall., v. 61, no. 4, p. 231–244.

Link, R. F., Koch, G. S., Jr., and Schuenemeyer, J. H., 1971, Statistical analysis of gold assay and other trace element data: U.S. Bur. Mines Rept. Invest. 7495, 127 p.

Sarma, D. D., 1969, A preliminary statistical study on the basic mine valuation problems at Kolar goldfields, Mysore state (India): Geoexploration, v. 7, no. 2, p. 97–105.

Sarma, D. D., 1970, A statistical study in mining exploration as related to the Kolar gold fields (India): Geoexploration, v. 8, no. 1, p. 19–35.

Sichel, H. S., 1947, An experimental and theoretical investigation of bias error in mine sampling with special reference to narrow gold reefs: Trans. Inst. Min. and Met. (London), v. 56, p. 403–474.

15

Reprinted from *J. Geol.* 74:692–702 (1966)

AN APPROXIMATE STATISTICAL TEST FOR CORRELATIONS BETWEEN PROPORTIONS

FELIX CHAYES[1] AND WILLIAM KRUSKAL[2]

ABSTRACT

The observed means and variances of data occurring as proportions or percentages may be used to estimate analogous parameters of a theoretical open array, X, which, on closure, yields a new array, Y, whose means and variances are exactly those of the observed data, but in which the covariances have been generated entirely by closure. The correlations in Y, found directly from the means and variances of X, are appropriate null values against which the observed correlations may be tested. A testing procedure is outlined, and a practical example is given.

INTRODUCTION

The examination of closed arrays—tables of chemical analyses, modes, or norms, for instance—is an essential part of the daily work of the descriptive petrographer, and the interpretation of correlations found in such tables is a continuing source of difficulty. Even in the absence of any specifically petrographic relation between them, variables whose variances are substantial relative to the sum of all variances of a closed array must exhibit appreciable and sometimes very strong negative correlation (see, e.g., Chayes, 1960). The appropriate null value against which correlations observed in closed arrays are to be tested is very clearly not the zero of ordinary statistical experience, but just what it is or should be no one seems to know.

In the absence of some well-defined null state, in fact, there can be no well-defined null value. Attempts to define such a state usually involve constructing a set of (theoretical) open variables and then imposing closure upon them by dividing the amount of each variable in an item by the sum of all. Throughout this paper the non-negative open variables of such theoretical populations are denoted by x and the theoretical closed variables formed from them by y. Mosimann (1962) examined relations between proportions assumed to follow one of the multinomial-related distributions—i.e., those in which the proportions y_i, with expectations p_i, have variances proportional to $p_i(1 - p_i)$—and found that the correlation between pairs of such variables "independent except for the constraint" of closure is simply $- [p_ip_j/(1 - p_i)(1 - p_j)]^{1/2}$.

A binomial or "sufficiently" binomial relation between means and variances is certainly far from rare, and Mosimann's result should find wide application. In that part of chemical petrography which deals with the major constituents of igneous rocks, however, there is little reason to suppose—and, in fact, considerable reason to doubt—that means and variances are related in this fashion. On the whole, there seems to be little consistency of any sort between them, and for our purposes null correlations presuming no relation between means and variances would appear to be more realistic. The only procedure so far suggested for the generation of such null values relies heavily on simulation techniques both for assigning variances to the theoretical open array and for estimating the expected correlation imposed by closure (Chayes, *op. cit.*). In the present note we show that both these operations can be carried through without simulation and suggest use of the resulting expectations as null values in an approximate statistical test of correlations between proportions.

[1] Geophysical Laboratory, Carnegie Institution of Washington, Washington, D.C. Some of Chayes's work on this paper was carried out while he was visiting professor of geology at Northwestern University.

[2] University of Chicago. Kruskal's participation was partly sponsored by the Army Research Office, Office of Naval Research, and Air Force Office of Scientific Research by contract Nonr 2121(23), NR 324-043.

171

THE UNDERLYING RELATIONS BETWEEN OPEN AND CLOSED
MEANS, VARIANCES, AND COVARIANCES

One of the first and most obvious questions raised by a closed array is whether the covariances it contains could have been generated by the closure of an open array consisting of uncorrelated variables. In the context of interest the open variables cannot be isolated and separately measured, if indeed they exist; closure is a condition of observation. With no information about the parent array, the question is too broad for a specific answer. The variables of the parent open array are to be uncorrelated, however, so that for the purpose at hand we need specify only their means and variances. Narrowing the inquiry, we may construct the open array (or random vector), X, from which the operation of closure will generate a new array, Y, whose elements have the same means and variances as those of the columns of an observed array, U, and then ask whether a covariance observed in U differs significantly from its analogue in Y, generated completely by closure. With some reservations and approximations, this question can be answered.

Starting from a hypothetical random vector $(x_1, x_2, \ldots, x_m)$ whose elements are uncorrelated, and unconstrained except that each $x_i \geq 0$, let the elements of the corresponding *closed* vector be

$$y_i = x_i \Big/ \sum_{j=1}^{m} x_j = x_i/T, \qquad\qquad T = \Sigma x_j .$$

The y_i are then proportions, closed variables, the correlations between which have been generated entirely by the closure.[3]

If μ_i and τ are the expected values of x_i and T in X, $\Delta_t = T - \tau$, and $\Delta_i = x_i - \mu_i$, we may write

$$y_i = \frac{x_i}{T} = \frac{\mu_i + \Delta_i}{\tau + \Delta_t} \cong \frac{\mu_i + \Delta_i}{\tau}\left(1 - \frac{\Delta_t}{\tau}\right), \qquad\qquad (1)$$

where the approximation consists in using only the first two terms of the Taylor expansion of $1/(\tau + \Delta_t)$. (That is, we are going to use the so-called delta method. For expositions of this method, see chap. 10 of Kendall and Stuart (1963) or Appendix A3 of Goodman and Kruskal (1963).)

From (1), continuing the linear approximation,

$$y_i \cong \frac{\mu_i}{\tau} + \frac{\Delta_i}{\tau} - \frac{\mu_i \Delta_t}{\tau^2} \cong p_i + \frac{\Delta_i}{\tau} - p_i \frac{\Delta_t}{\tau}, \qquad\qquad (2)$$

where $p_i = \mu_i/\tau$, the parent proportion of x_i. Taking expectations,

$$E(y_i) \cong p_i \qquad\qquad (3)$$

since $E(\Delta_i) = E(\Delta_t) = 0$. Evidently the desired equivalence of means in the observed and hypothetical closed arrays, U and Y, requires that in the hypothetical open array, X, we set $p_i = \bar{u}_i$ for all i, that is, each parent proportion of the open array is equal to the average of the corresponding column of U.

To find the variances of Y we rearrange (2) to obtain

$$\tau \Delta_{y_i} = \tau(y_i - p_i) \cong \Delta_i - p_i \Delta_t = (1 - p_i)\Delta_i - p_i(\Delta_t - \Delta_i) . \qquad\qquad (4)$$

---

[3] The distribution of the x's is presumed to be such that the distribution of Fisher z for any pair of y's will be approximately normal. Some simulation experiments described elsewhere (Chayes, 1962, p. 445–446) suggest that this is hardly a stringent requirement.

Squaring, taking expectations, and recalling that the x's are uncorrelated, we have

$$\tau^2 s_i^2 \cong (1 - p_i)^2 \sigma_i^2 + p_i^2 \sum_{j \neq i} \sigma_j^2, \tag{5}$$

where var $(y_i) = s_i^2$, var $(x_i) = \sigma_i^2$. Restating (5) in a form which will be convenient later,

$$\tau^2 s_i^2 \cong p_i^2 \sigma_t^2 + (1 - 2p_i)\sigma_i^2, \tag{6}$$

where $\sigma_t^2 = \Sigma \sigma_i^2$. Although for completeness the constant τ has been retained throughout, it would involve no loss of generality to take $\tau = 1$, since the hypothetical open variables can be multiplied by any positive constant without affecting either the means or the relations between variances and covariances of the closed variables.

From (6) we readily obtain

$$\sigma_i^2 \cong \frac{\tau^2 s_i^2}{1 - 2p_i} - \frac{p_i^2}{1 - 2p_i} \sigma_t^2. \tag{6'}$$

Summing (6') over i, rearranging terms, and changing the subscript i to j ($j = 1, \ldots, m$) for future convenience, we get

$$\sigma_t^2 \cong \tau^2 \frac{\Sigma[s_j^2/(1 - 2p_j)]}{1 + \Sigma[p_j^2/(1 - 2p_j)]},$$

in which τ may be assigned any arbitrary positive value. Substituting this expression for σ_t^2 into the right side of (6'), we have, again after some rearrangement of terms, that

$$\sigma_i^2 \cong \frac{\tau^2}{1 - 2p_i} \left\{ s_i^2 - p_i^2 \frac{\Sigma[s_j^2/(1 - 2p_j)]}{1 + \Sigma[p_j^2/(1 - 2p_j)]} \right\}, \tag{6''}$$

subject only to the mild restriction that no $p_i = 0.5$. Of course, if $\Sigma[p_j^2/(1 - 2p_j)] = -1$ the fraction inside the braces is undefined. It is shown in an appendix, however, that this occurs if and only if $m = 2$, whereas the proposed test is meaningful only if $m \geq 4$. Accordingly, in arrays of practical interest, explicit solutions for σ_i^2 will always be possible unless, for some i, p_i is exactly equal to $\frac{1}{2}$.

If some p, say p_1, is exactly equal to $\frac{1}{2}$, the right side of (6'') is of course undefined, but in this event explicit solutions for the σ_i^2 may be found directly from (6'). For, clearing (6') of fractions, we have at once that $\sigma_t^2 = 4\tau^2 s_1^2$, and it follows readily from substitution of this value for σ_t^2 in (6') that

$$\sigma_i^2 \cong \frac{\tau^2}{1 - 2p_i}(s_i^2 - 4p_i^2 s_1^2) \tag{6'''}$$

for all $i > 1$. Then σ_1^2 may be obtained by subtracting the sum of the $\sigma_i^2 (i > 1)$ from σ_t^2.

Alternatively, entering (6) for successive values of $i = 1, 2, \ldots, m$, we have, in matrix notation, that

$$A\delta = \tau^2 s, \tag{7}$$

where δ and s are column vectors of length m whose elements are, respectively, the variances of arrays X and Y, the hypothetical open parent and its closed equivalent, and A is the non-singular $m \times m$ coefficient matrix

$$\begin{pmatrix}
(1 - p_1)^2 & p_1^2 & p_1^2 & . & . & . & p_1^2 \\
p_2^2 & (1 - p_2)^2 & p_2^2 & . & . & . & p_2^2 \\
p_3^2 & p_3^2 & (1 - p_3)^2 & . & . & . & p_3^2 \\
. & . & . & . & . & . & . \\
. & . & . & . & . & . & . \\
. & . & . & . & . & . & . \\
p_m^2 & p_m^2 & p_m^2 & . & . & . & (1 - p_m)^2
\end{pmatrix}. \tag{8}$$

Non-singularity of A follows from the specific solution found. Left-multiplying (7) by A^{-1}, we obtain

$$\delta = A^{-1} \tau^2 s \tag{9}$$

in which the elements of δ and s will be in the same scale if τ is taken as unity. We have already seen that $\bar{y}_i$ will be equal to $\bar{u}_i$ if we set $p_i = \bar{u}_i$. There is of course not the same simple equivalence between variances, but the analogy is evident; if we solve (6″) or (9) with the sample statistics, taking $p_i = \bar{u}_i$ for all i and the observed variances of the u's as elements of s, the resulting elements of δ will be such as to ensure that Y formed by closure of X will have the same variances as U up to the linear approximation.

The approximate covariances which would appear in Y as a result of closure of X are readily obtained in similar fashion; the product of any pair of Δ_y's is

$$\begin{aligned}
\tau^2 \Delta_{v_i} \Delta_{v_j} &\cong (1 - p_i)(1 - p_j)\Delta_i\Delta_j - p_i(1 - p_j)\Delta_j(\Delta_t - \Delta_i) \\
&\quad - p_j(1 - p_i)\Delta_i(\Delta_t - \Delta_j) + p_i p_j(\Delta_t - \Delta_i)(\Delta_t - \Delta_j) .
\end{aligned} \tag{10}$$

Taking expectations, and remembering once more that the x's are uncorrelated, so that expectations of all cross-products in the Δ's vanish, we have at once that

$$\begin{aligned}
\tau^2 \operatorname{cov}(y_i, y_j) &\cong p_i p_j \Sigma \sigma_k^2 - p_i(1 - p_j)\sigma_j^2 - p_j(1 - p_i)\sigma_i^2 \qquad \begin{cases} k \neq i \\ k \neq j \end{cases} \\
&\cong p_i p_j \sigma_t^2 - p_i \sigma_j^2 - p_j \sigma_i^2 .
\end{aligned} \tag{11}$$

The required null value of the correlation between y_i and y_j is then

$$\rho_{v_i v_j} = \frac{\operatorname{cov}(y_i, y_j)}{\sqrt{\operatorname{var}(y_i) \cdot \operatorname{var}(y_j)}} = \frac{p_i p_j \sigma_t^2 - p_j \sigma_i^2 - p_i \sigma_j^2}{\sqrt{[p_i^2 \sigma_t^2 + (1 - 2p_i)\sigma_i^2][p_j^2 \sigma_t^2 + (1 - 2p_j)\sigma_j^2]}}, \tag{12}$$

where cov (y_i, y_j) is obtained from (11) and var (y_i), var (y_j) from (6).

Recapitulating, X being the conceptual open array of uncorrelated variables which yields, by closure, an array, Y, having the same means and variances as the observed closed array, U, the correlations found in Y, *and generated entirely by closure*, would have the values given by (12); these are appropriate null values against any one of which, by means of Fisher z for example, the analogous observed correlation may be tested.[4]

The test, which is tedious but not impossible to apply by hand calculation, would hardly tax the resources of even a rather small electronic computer. Although the general path to be followed by the calculations will no doubt be clear to readers who have some experience in machine computation, it may be useful to review it in outline form:

1. The raw data are read in, the covariance matrix of the observed (U) array is computed and stored, together with the means for all variables.

2. With the stored means, as proportions, either

    a) the open variances are found directly, from (6″), or

    b) the coefficient matrix A is formed and inverted, and (9) is solved with the variances found in step 1. (If the δ vector found in step 2*a* or 2*b* contains negative elements the test terminates, and the appropriate interpretation is that described in the following section.)

---

[4] Provided that $m > 3$. If the array contains only three variables, the correlations in Y will be exactly equal to those in U, since in any three-variable closed array the covariances may be obtained directly from the variances (Chayes, 1960, p. 4186–4187), and we have purposely taken var (y_i) = var (u_i). The "test" will therefore always indicate that the observed correlations are exactly equal to the "expected" ones. In any two-variable closed array, further, the correlation between the variables will be exactly equal to -1. Thus, any test is trivial if $m = 2$, and the test described here is inapplicable if $m = 3$.

3. With the elements of δ from step 2 as variances, and sample averages from step 1 as means, (12) is solved for all pairs of variables (i,j), $i \neq j$, yielding the required null values, ρ_{ij}.

4. A Fisher z-test is performed on any r_{ij} from step 1 against ρ_{ij} from step 3. The statistic of interest is $(z_r - z_\rho)/(n - 3)^{-1/2}$, which is normal $(0,1)$ to a very good approximation. Here n is sample size, the number of rows of U.

Except for assignment of degrees of freedom (see below), the testing of any particular correlation is straightforward. But if the objective is to test a number of correlations—as, for instance, the correlations of other essential oxides with silica in a Harker array—the situation is far from simple. Presumably one may look at the correlations "one at a time," but we know of no practicable and fully understood procedure by means of which two or more of the correlations in U may be simultaneously tested for significance. An approximate test is described on page 120 of Bartlett and Rajalakshman (1953) and on page 297 of Bartlett (1954).

Indeed, even the initial test in step 4 is open to some question. No account has been taken of the fact that the proposed null correlations are sensitive to sampling fluctuations in the observed means and variances. At the present writing the correct adjustment is not known; if we are to regard both r and ρ as sample statistics, it might be argued that although $n - 3$ d.f. remain after the calculation of r in the usual fashion, only $n - 2m - 1$ survive the calculation of ρ. Whatever the final solution, it is clear that n should be rather large if the proposed test is to have sufficient power to be useful. In any event, although the test outlined above provides considerable protection against attributing significance to "closure correlation"—far more than any test that ignores closure—it still leaves something to be desired on this score.

More basically, the approach of this paper is somewhat ad hoc. At a more general level, the problem is of the following form: observations are made on a random vector Y, which, under the null hypothesis, is a given linear transformation of another random vector X, whose components are uncorrelated. Under some distributional assumption (presumably the simplest is that of normality) how should one estimate the first and second moments of Y? How should one test the null hypothesis that the given linear structure holds? At a still more basic level, one would prefer to avoid leaning on the linear approximations (delta method) that are central in the approach of the present paper. We know of no satisfactory treatments at these levels.

In the practical example described later the significance levels mentioned are those appropriate to $n - 3$ d.f., so that the test is only of whether a correlation observed in U might have arisen in random sampling of a Y array having exactly the means and variances of U. Correlations regarded as not significant by this test will also be rejected by any test that makes allowance for sampling variance in Y.

ZERO AND NEGATIVE ELEMENTS IN THE VARIANCE VECTOR

Solutions of (6″) or (9) may sometimes yield a δ vector containing elements that are either zero to within the rounding error of the calculation or definitely negative. A δ vector containing one or more zero elements evidently corresponds to the so-called "concretionary" model of Sarmanov and Vistelius (1958); that $s_i^2 \neq 0$ if $\sigma_i^2 = 0$, although perhaps at first sight a little unreasonable, is in fact unavoidable. A certain amount of the variance of any closed variable arises from interaction between it and T in the course of closure; if $\sigma_i^2 = 0$, all the variance of the closed variable is introduced via T.

Negative elements in δ require a quite different interpretation, for variance is by definition non-negative. If δ contains one or more negative elements, null values cannot be obtained either by simulation or by calculation but, in a rather crude and unspecific sense, they are not necessary. The occurrence of a negative element in σ is the strongest possible indication that there is no set of open *uncorrelated* variables with appropriate means whose closure would generate the observed covariance matrix. It is tempting to suppose that a variable whose open variance appears to be negative is in some way particularly responsible for the failure, but this is an oversimplified view of the situation, and quite possibly a misleading one. Unless the negative elements in σ are all small enough to be accounted for by rounding errors or sampling fluctuation, the only permissible conclusion is the very general one that the observed covariance matrix could not have been generated by closure of a set of open uncorrelated variables with appropriate mean values.

TWO SPECIAL CASES AND A POSSIBLE GENERALIZATION OF (12)

It was shown earlier (Chayes, 1960) that, in any m-variable closed array in which all parent variances are equal, all expected correlations have the value $(1 - m)^{-1}$. This result is simply an algebraic consequence of the definitions of variance and covariance, and thus provides a useful check of (12); if the parameters of X are such as to generate equal variances in Y, the expected correlations in Y ought to be $(1 - m)^{-1}$, and in fact they are. Entering (12) with, for instance,

$$p_1 = p_2 = \ldots = p_m = 1/m \quad \text{and} \quad \sigma_1 = \sigma_2 = \ldots = \sigma_m = \theta,$$

so that $\sigma_t^2 = m\theta^2$, we have at once that

$$\rho_{y_i v_j} = \frac{\theta^2/m - \theta^2/m - \theta^2/m}{\theta^2/m + \theta^2 - 2\theta^2/m} = \frac{1}{1 - m}.$$

If the p's are not equal, however, the equal variances of the open array will become unequal on closure, and the expected correlations will not be $(1 - m)^{-1}$. It is to be stressed, further, that the requirement of equal variance extends to *all* the variables, not merely to the pair whose correlation is of immediate concern.

If, rather less restrictively, the parameters of X are so chosen that variances are proportional to means, viz., $\sigma_i^2 = kp_i$ for all i, (12) reduces to

$$\rho_{y_i v_j} = -\sqrt{\frac{p_i p_j}{(1 - p_i)(1 - p_j)}},$$

which is the Mosimann relation; similar substitution reduces (6) to

$$\text{var}(y_i) \sim p_i(1 - p_i);$$

that is, the relation between means and variances in Y will be multinomial if in X the variances are proportional to the means. Since, as shown in (6''), σ_i^2 may be found explicitly from the means and variances of Y, it also follows that $\sigma_i^2 \sim p_i$ if, for all i, $s_i^2 \sim p_i(1 - p_i)$.

Finally, although a null state in which correlation is completely lacking prior to closure and a test of observed correlation against that arising solely from closure have perhaps the most immediate intuitive appeal, more complicated null hypotheses could be constructed and tested. If, for instance, $\text{cov}(x_i, x_j) \neq 0$ for some specific (i,j), $i \neq j$, but all other covariances in X remain zero, expectations taken from the square of (4) lead to

$$\text{var}(y_i) = p_i^2 \sigma_i^2 + (1 - 2p_i)\sigma_i^2 - 2p_i\sigma_i\sigma_j\rho_{ij}', \tag{13a}$$

a similar statement for $\text{var}(y_j)$, and

$$\text{var}(y_k) = p_k^2 \sigma_i^2 + (1 - 2p_k)\sigma_k^2, \tag{13b}$$

where $k \neq i,j$ and the prime on ρ indicates that the correlation in question is between (x_i,x_j).

The covariances in Y, obtained by taking expectations in (10) after each rotation of subscripts, are now of three types, depending on whether both, either or neither of the y's in question is the closed equivalent of either of the correlated x's. Specifically,

$$\text{cov } (y_i,y_j) \cong p_i p_j \sigma_i^2 - p_i \sigma_j^2 - p_j \sigma_i^2 + (1 - p_i - p_j)\sigma_i \sigma_j \rho'_{ij}, \qquad (14a)$$

$$\text{cov } (y_i,y_k) \cong p_i p_k \sigma_i^2 - p_i \sigma_k^2 - p_k \sigma_i^2 - p_k \sigma_i \sigma_j \rho'_{ij}, \qquad (14b)$$

a similar statement for cov (y_j,y_k), and

$$\text{cov } (y_k,y_l) \cong p_k p_l \sigma_i^2 - p_k \sigma_l^2 - p_l \sigma_k^2, \qquad (14c)$$

where $k \neq l$ and $k,l \neq i,j$.

From equations (13a), (13b) and (14a)-(14c) one could now extend the preceding argument to provide a test of the hypothesis that some one of the other observed correlations might have resulted from random sampling of a Y array in which cov $(y_i,y_j) = $ cov (u_i,u_j) for some particular pair of variables and var $(y_i) = $ var (u_i) for all i.

It is surprising and rather alarming, considering the elaborate associations implied in much petrochemical speculation, how much even this minimal departure from zero correlation in X complicates both the arithmetic and the interpretation of the test.

AN APPLICATION OF THE TEST

The initial Monte Carlo experimentation in this field was an attempt to "model" a set of modes of the Bellingham, Minnesota, granite, chosen as an example of data in which there seemed no reason to attribute substantive significance to strong negative correlations between major constituents. The simulation was reasonably successful (see Chayes, 1960, p. 4192, table 3), but no satisfactory test of the results was available. The observed means and variances, the means and variances of U in the present discussion, are shown here in table 1. The variances of X, found from (9), are given in the last column; the mean proportions of X are estimated directly from the observed means. An X having these variances and mean proportions would yield, on closure, a Y whose means and variances were those of U. (The variances of Y may in fact be found by substitution in (6) and of course agree with those of U.)

In table 2 the correlations of Y found from (12), the null values of the proposed test, are juxtaposed to the observed correlations. The observed correlation of quartz with muscovite appears to be significant at the 0.10 level, but the considerably larger correlations of quartz

TABLE 1
DATA FROM MODAL ANALYSES OF 15 THIN SECTIONS
OF BELLINGHAM GRANITE

MINERAL		OBSERVED		OPEN VARIANCES CALCULATED FROM EQ. (6) OR (9)
No.	Name	Mean	Variance	
1.......	Quartz	29.37	19.5021	21.8262
2.......	Microcline	34.19	35.7878	68.1950
3.......	Plagioclase	29.85	20.5712	24.1408
4.......	Biotite	4.50	6.1757	6 8.5157
5.......	Muscovite	2.08	1.0189	1.0082

TABLE 2

OBSERVED AND PROPOSED NULL CORRELATIONS FOR BELLINGHAM MODAL DATA

Variable Pairs	Observed Correlations	Null Values Calculated from Eq. (12)	Variable Pairs	Observed Correlations	Null Values Calculated from Eq. (12)
12........	−0.6466	−0.5781	24........	+0.1302	−0.2303
13........	+0.0086	−0.1466	25........	−0.4782	−0.1487
14........	−0.4492	−0.1173	34........	−0.2529	−0.1239
15........	+0.5243	−0.0015	35........	−0.0303	−0.1014
23........	−0.6543	−0.5967	45........	−0.1316	−0.0267

with microcline and of microcline with plagioclase, the major element correlations which first attracted suspicion, fail entirely of significance. The Mosimann test gives the same result. Against $\rho = 0$, however, the null value appropriate to open variables, both correlations involving microcline are easily significant at the 0.01 level.

ESTIMATES OF CORRELATION BASED ON SIMULATION

Given the means and variances of X, the required null values, the correlations of Y can, of course, be approximated by simulation experiments. Actual estimates of correlation based upon simulation, however, are subject to considerable uncertainty even when the simulated arrays are rather large. Table 3 summarizes new data for twelve "mock-Bellingham" simulations in each of which the X array contained four hundred items.

From the last two columns of the table it is evident that correlations that are averages of results found in rather large numbers of large simulations will indeed agree well with those obtained from (12) by direct calculation. From the column headed "Range," however, it is equally clear that results for a single simulation may fall wide of this mark, even if the simulated array is large. It is to be remembered, further, that errors in the correlations found in a particular simulated array are not independent. In any closed array the algebra of the situation requires (see Chayes, 1962) that

$$\mathrm{var}\,(y_i) + \sum_j \mathrm{cov}\,(v_i, y_i) = 0. \qquad\qquad j \neq i.$$

TABLE 3

SUMMARY OF CORRELATION DATA IN 12 "MOCK-BELLINGHAM" SIMULATIONS, EACH BASED ON 400 ITEMS*

Variable Pair	Range of Simulated Correlations	r Calculated from Average z	Correlation Calculated from Eq. (12)
12.........	−0.6140 ... −0.4968	−0.5717	−0.5781
13.........	−0.2439 ... −0.0356	−0.1412	−0.1466
14.........	−0.2137 ... −0.0320	−0.1255	−0.1173
15.........	−0.0583 ... +0.1473	+0.0321	−0.0015
23.........	−0.6619 ... −0.5346	−0.6138	−0.5967
24.........	−0.2710 ... −0.1541	−0.2204	−0.2303
25.........	−0.2930 ... −0.0691	−0.1635	−0.1487
34.........	−0.2008 ... −0.0409	−0.1172	−0.1239
35.........	−0.1203 ... +0.0824	−0.0162	−0.0104
45.........	−0.1028 ... +0.1188	−0.0180	−0.0267

* Means, variances, and subscripts as in table 1.

Thus, sampling error in the estimate of one covariance must be compensated by errors of the same sign in the variances, or of opposite sign in the other covariances, of both variables. Similarly, error in the estimated variance of a simulated variable must be compensated by a net error of opposite sign in its estimated covariances.

POSITIVE CORRELATION ARISING FROM CLOSURE

It is clear even from a cursory examination of the situation that variables that are major contributors to the total variance of a closed array must be negatively correlated, and from the preceding discussion it is evident that the net covariance of every variable in a closed array must be negative. Mosimann (1962) has shown, furthermore, that if the relation between parent closed means and variances is multinomial, i.e., if $s_i^2 \sim p_i(1 - p_i)$ for all i, all expected correlations between variables "independent except for the constraint" are necessarily negative; if open variances are proportional to means, as implied by Mosimann's argument, substitution in (11), above, shows immediately that for their closed equivalents

$$\operatorname{cov}(y_i, y_j) = -k p_i p_j,$$

which is always negative, since by definition k, p_i, and p_j are positive.

Positive correlations were nevertheless noted between minor variables in some early simulation experiments, and a rather limited rationalization of this curious phenomenon has already been presented (Chayes, 1960). A more general statement of the condition under which correlation imposed by closure *must* be positive is readily obtained from (11), for positive correlation between variables y_i and y_j clearly occurs if and only if the quantity $p_i p_j \sigma_i^2 - p_j \sigma_i^2 - p_i \sigma_j^2 > 0$. This, in turn, requires that

$$p_i p_j \sigma_i^2 > p_i \sigma_j^2 + p_j \sigma_i^2$$

or, dividing both sides by $p_i p_j$,

$$\sigma_i^2 > \frac{\sigma_j^2}{p_j} + \frac{\sigma_i^2}{p_i}.$$

Whenever the sum of the variances in X is greater than the sum of the ratios (σ^2/p) for any two open variables, the appropriate null correlation for the analogous closed variables must be positive. This condition will be satisfied most readily by variables whose means and variances are both small; null correlations between variables present in accessory or trace amounts will often, perhaps usually, be positive.

In terms of the model proposed here, the presumed effects of closure on major and trace constituents are thus in a sense complementary. Between major constituents a weak positive correlation is far more likely to indicate a substantively meaningful association than a strong negative one. Between trace constituents, on the other hand, weak negative correlations are more likely to be of substantive interest than strong positive ones. This rather paradoxical result suggests that in principle the effect of closure on correlation requires consideration whatever the tenor of means and variances of the variables involved. If the absolute value of a null correlation is small—as for ρ_{15} in the example discussed above—it will usually make little practical difference whether the significance of correlations based on small samples is tested by departures from it or from the zero appropriate to the testing of data free of the closure restriction. It is only in this strictly practical sense, however, that the effects of closure may be ignored.

Although positive null correlations are perhaps most likely to be encountered in arrays composed of variables differing widely in mean value, this is not a necessary condition for their emergence. If, for instance, $p_i = 1/m$ for all i in X, (11) reduces to

$$\operatorname{cov}(y_i, y_j) = \frac{\sigma_i^2}{m^2} - \frac{\sigma_i^2}{m} - \frac{\sigma_j^2}{m},$$

so that $\rho_{ij} > 0$ if

$$\sigma_i^2/m > \sigma_i^2 + \sigma_j^2.$$

Thus, if the mean proportions are equal for all members of an open array, closure will impose positive correlation on any two uncorrelated open variables the sum of whose variances is less than the mean variance of the array.

SUMMARY

The means and variances of an indefinitely extended open array, X, which, on closure, yields a new array, Y, having the means and variances of an observed array, U, are

$$p_i = \bar{u}_i, \qquad i = 1, 2, \cdots, m,$$

and

$$\boldsymbol{\delta} = A^{-1} s,$$

where the variance vector of X is

$$\boldsymbol{\delta} = \{\sigma_1^2, \sigma_2^2, \cdots, \sigma_m^2\},$$

the variance vector of Y is

$$s = \{s_1^2, s_2^2, \cdots, s_m^2\},$$

and the ith element of the ith row of A is $(1 - \bar{u}_i)^2$, all other elements in this row being simply $\bar{u}_i^2$. Alternatively, the elements of $\boldsymbol{\delta}$ may be found directly by means of (6″).

The correlations in Y, generated entirely by closure and thus suitable for use as null values in a test which will often seem intuitively reasonable, are

$$\rho_{ij} = \frac{\operatorname{cov}(y_i, y_j)}{\sqrt{\operatorname{var}(y_i) \cdot \operatorname{var}(y_j)}} = \frac{p_i p_j \sigma_t^2 - p_j \sigma_i^2 - p_i \sigma_j^2}{\{[p_i^2 \sigma_t^2 + (1 - 2p_i)\sigma_i^2][p_j^2 \sigma_t^2 + (1 - 2p_j)\sigma_j^2]\}^{1/2}},$$

where $\sigma_t^2 = \Sigma \sigma_k^2$ and $m > 3$. The test is easily programmed for machine computation.

If, in X, $p_i = 1/m$ and $\sigma_i^2 = \theta$ for all i, ρ_{ij} reduces to $(1 - m)^{-1}$, a value well known from earlier work as a direct algebraic consequence of the definitions of variance and covariance. If, for all i, $\sigma_i^2 \sim p_i$ in X, the analogous variances in Y are proportional to $p_i(1 - p_i)$, and

$$\rho_{ij} = -\sqrt{\frac{p_i p_j}{(1 - p_i)(1 - p_j)}},$$

a relation previously found by Mosimann. If there is reason a priori to suppose that the parent closed variances are proportional to $p_i(1 - p_i)$ *for all i*, there is no need for the more extended calculation, the simpler one actually being preferable. It might also be preferable where the sample is too small to yield sound estimates of variance, but in such cases it would probably be better to make no test at all.

The basic formula for ρ_{ij} may be expanded to include the effect of non-zero correlation between any pair of the variables in X, as indicated in (13a) and (13b) and (14a)–(14c), and the accompanying discussion.

ACKNOWLEDGMENTS.—We are indebted to J. M. Cameron, H. H. Ku, J. Mosimann, and P. Nemenyi for extensive discussion and criticism.

APPENDIX

LIMITING VALUES OF $P_m = \Sigma \left[p_i^2/(1 - 2p_i) \right]$ FOR $\Sigma p_i = 1$ AND ALL $p_i \neq \frac{1}{2}$

If $p_i < \frac{1}{2}$ for all i, it is obvious that $P_m > 0$. Negative values of P_m may occur only if, for some i, $p_i > \frac{1}{2}$. If $m = 2$,

$$P_2 = \frac{p_1^2}{1 - 2p_1} + \frac{p_2^2}{1 - 2p_2} = \frac{p_1^2}{1 - 2p_1} - \frac{(1 - p_1)^2}{1 - 2p_1} = -1, \tag{A1}$$

and of course one of the p's is greater than $\frac{1}{2}$. Denoting this large p by subscript 1, we examine the effect on P_m of subdividing p_2. Suppose first that $m = 3$ and $p_2 = p_a + p_b$, so that

$$\frac{p_a^2}{1 - 2p_a} + \frac{p_b^2}{1 - 2p_b} = \frac{p_a^2 + p_b^2 - 2p_a p_b p_2}{1 - 2p_2 + 4p_a p_b} < \frac{p_2^2}{1 - 2p_2}, \tag{A2}$$

and it follows that $P_3 < P_2 = -1$.

Since any $p < \frac{1}{2}$ may be subdivided with similar effect on its contribution to P_m, extension to the general case is immediate, and

$$P_k < P_{k-1} \ldots < P_2 = -1, \qquad 2 < k \leq m. \tag{A3}$$

Thus the quantity $1 + P_m = 0$ if and only if $m = 2$, a case of no concern here, since the proposed test is meaningful only if $m \geq 4$. To summarize,

$$1 + P_m > 0 \qquad \text{if} \qquad p_i < \tfrac{1}{2} \text{ for all } i,$$

$$1 + P_m = 0 \qquad \text{if} \qquad m = 2,$$

$$1 + P_m < 0 \qquad \text{if} \qquad m > 2 \text{ and } p_i > \tfrac{1}{2} \text{ for some } i.$$

REFERENCES CITED

BARTLETT, M. S., 1954, A note on the multiplying factors for various χ^2 approximations: Jour. Royal Stat. Soc., Ser. B, v. 16, p. 296–298.

———, and RAJALAKSHMAN, D. V., 1953, Goodness of fit tests for simultaneous autoregressive series: Jour. Royal Stat. Soc., Ser. B, v. 15, p. 107–124.

CHAYES, F., 1960, On correlation between variables of constant sum: Jour. Geophys. Res., v. 65, p. 4185–4193.

——— 1962, Numerical correlation and petrographic variation: Jour. Geology, v. 70, p. 440–452.

GOODMAN, L. A., and KRUSKAL, W. H., 1963, Measures of association for cross classifications. III. Approximate sampling theory: Jour. Am. Stat. Assoc., v. 58, p. 310–364.

KENDALL, M. G., and STUART, A., 1963, The advanced theory of statistics, v. 1 (2d ed.): New York, Hafner Publishing Co.

MOSIMANN, J., 1962, On the compound multinomial distribution, the multivariate β-distribution, and correlations among proportions: Biometrika, v. 49, p. 65–82.

SARMANOV, O. V., and VISTELIUS, A. B., 1958, On the correlation of percentage values: Doklady Akad. Nauk SSSR, v. 126, p. 22–25 (in Russian).

ERRATA

AN APPROXIMATE STATISTICAL TEST FOR CORRELATIONS BETWEEN PROPORTIONS (VOLUME 74, 1966, PAGES 692-702): SOME CORRECTIONS[1]

FELIX CHAYES AND WILLIAM KRUSKAL

Geophysical Laboratory, Washington, D.C. 20008, and Department of Statistics,
University of Chicago, Chicago, Illinois 60637

On page 694, the subscript of p should be i throughout equation (6'); in the first occurrence of p only the dot over the i is printed.

In the second paragraph of page 697, the word "expected," which appears three times, should be omitted or replaced by "anticipated." The point is that "expected" was used here in a psychological sense, not in the statistical sense of expected value.

In the bottom paragraph of page 697 and the material of page 698 through the end of the section, σ_i^2 should be interpreted as the variance of Σx_h, which is there *not* $\Sigma \sigma_h^2$ but rather

$$\Sigma \sigma_h^2 + 2\sigma_i \sigma_j \rho'_{ij}$$

because of the nonzero covariance term that is assumed. With this interpretation (which was intended, but unfortunately not stated), the material of pages 697–698 seems to us correct.

Note also that in this section the scaling factor τ^2 has been omitted from (13) and (14), in which it plays the same role as in (6) and (11). If, as in our paper, only the correlations are of interest, τ^2 may be ignored, since it appears as a multiplier in both numerator and denominator of ρ.

Finally, on page 698, the last entry in row 4 of table 1 should be 6.5157, not 5.5157.

16

Reprinted from *Math Geol.* 1(2): 171–184 (1969)

Critical Review of Some Multivariate Procedures in the Analysis of Geochemical Data[1]

A. T. Miesch[2]

Simulation experiments have been conducted to examine the potential usefulness of R-mode and Q-mode factor methods in the analysis and interpretation of geochemical data. The R-mode factor analysis experiment consisted of constructing a factor model, using the model to generate a correlation matrix, and attempting to recover the model by R-mode techniques. The techniques were successful in determining the number of factors in the model, but the factor loadings could not be estimated even approximately on the basis of mathematical procedures alone. Q-mode factor methods were successful in recovering all of the properties of a model used to generate hypothetical chemical data on olivine samples, but it was necessary to use a correction previously regarded as unimportant.

INTRODUCTION

Most investigations in field geochemistry are directed simultaneously at a number of chemical elements and involve assessments of either the relations among the elements or the classification of samples or sampling localities on a multivariate basis. A large number of useful or potentially useful multivariate techniques are available and an increasingly large number of them are being applied in geologic and geochemical investigations. Some of the more interesting methods involve more or less intricate mathematical procedures or are based on intricate mathematical and statistical reasoning. Evaluation of the methods is commonly difficult, and their merits usually are judged by the geologic reasonableness of the results they yield. The difficulty here is that geologically reasonable results or answers are no assurance that the results or answers approach those being sought. Each of the methods is based on the initial assumption of a statistical model, and attempts are made to estimate the parameters of the model and to test its fit to the observed data. Again, the difficulty in evaluating a method by applying it to a geologic problem is that both the appropriateness of the model for the particular problem and the parameters of the model are unknown. It is impossible, therefore, to evaluate a multivariate method on the basis of application alone. The purpose of this paper is to describe a supplementary approach involving simulation as applied in an examination of certain procedures in both R- and Q-mode factor analysis.

---

[1] Manuscript received 6 August 1969; publication approved by Director, U.S. Geological Survey.
[2] U.S. Geological Survey (USA).

SIMULATION

Two types of simulation have been employed. Both begin with the construction of a multivariate model. In one type of simulation, the model is used to generate a *data* matrix, which then is processed through the multivariate procedure in an attempt to recover the model. In the other type of simulation, the model is used to generate statistical parameters which then are processed through the multivariate procedures, again in an attempt to recover the known model. One advantage of the simulation approach is that the methods are examined under optimum conditions; all prerequisite assumptions demanded by the method can be satisfied. If the method fails under ideal conditions, it cannot be expected to be successful in application to real problems. The principal advantage of this approach, of course, is the fact that the correct model and its essential parameters are known and so the estimates provided by the multivariate procedure can be properly evaluated. Simulation methods were used previously in an examination of factor analysis methods by Imbrie (1963) and Imbrie and van Andel (1964).

The simulation procedures described here were designed to examine the effectiveness of the basic mathematical techniques contained in the multivariate methods that were examined. No examinations were made of the effects of deviations of the model used to generate the *data* from the model assumed in the multivariate method, and no consideration was given to the effects of various types of error that are present in actual data. The simulation procedures, however, may be extended easily to examine effects of these types. Some generally applicable techniques are described by Miesch, Connor, and Eicher (1964), and a computer program for investigating the effects of sampling errors is given by Eicher and Miesch (1965).

R-MODE FACTOR ANALYSIS

R-mode factor analysis is a multivariate method used to resolve an array of correlation coefficients, or other product-moment measures of association, into a form that can be interpreted more easily in terms of processes thought to have been responsible for the correlations. A geometric explanation of the procedures was given by Miesch, Chao, and Cuttitta (1966), but they are discussed more fully by Imbrie (1963), and in great detail by Harman (1967). The model assumed in R-mode factor analysis has the general form

$$X_{ij} = \alpha_{j1}F_{i1} + \alpha_{j2}F_{i2} + \cdots + \alpha_{jm}F_{im} \tag{1}$$

where X_{ij} is the element in the ith row and jth column of the normalized data matrix, the α's are factor loadings, and the F's are normalized factors or factor scores. Dropping the i subscript for convenience and replacing X_j by the conventional symbol for a chemical element, we may write the following hypothetical model for use in the simulation examination:

$$Cu = 0.2500\ F_1 + 0.9682\ F_2$$
$$Pb = 0.6024\ F_1 - 0.7982\ F_2$$
$$Zn = 0.4344\ F_1 + 0.9007\ F_2$$

$$Ag = 0.4025\ F_1 + 0.9154\ F_2$$
$$As = 0.1000\ F_1 + 0.9950\ F_2$$
$$Cd = 0.5750\ F_1 + 0.8182\ F_2 \tag{2}$$
$$Sb = 0.0500\ F_1 + 0.9987\ F_2$$
$$Au = 0.4216\ F_1 - 0.9068\ F_2$$

Using two pseudorandom and uncorrelated normal deviates ($\eta = 0$, $\sigma^2 = 1$) for F_1 and F_2 (which may represent normalized measured values of temperature, pressure, concentration, etc.), we may generate the first row of a normalized data matrix, Cu through Au. Subsequent rows are generated in a similar manner and the final *data* matrix then may be used to derive an 8×8 correlation matrix. Alternatively, the same correlation matrix that would be obtained if a large number of *data* rows were generated may be obtained from the sums of squares and cross products of the loadings in eqs (2) using

$$r_{jk} = \sum_{p=1}^{m} \alpha_{jp}\alpha_{kp} \tag{3}$$

where j and k refer to two chemical elements. The loadings in eqs (2) are fixed so that where $j = k$, $r_{jk} = 1.00$. The upper triangle of the complete correlation matrix is

Pb	Zn	Ag	As	Cd	Sb	Au	
−0.62	0.98	0.99	0.99	0.94	0.98	−0.99	Cu
	−0.46	−0.49	−0.73	−0.31	−0.77	0.47	Pb
		1.00	0.94	0.99	0.92	−1.00	Zn
			0.95	0.98	0.93	−1.00	Ag
				0.87	1.00	−0.95	As
					0.85	−0.99	Cd
						−0.93	Sb

$$(4)$$

The task of R-mode factor analysis is to use the correlations in (4) to learn as much as possible about the model in (2) from which they were derived.

The eigenvalues of the complete correlation matrix, with unities in the diagonal, are

$$7.1 \quad 0.9 \quad 0.0 \quad 0.0 \quad 0.0 \quad 0.0 \quad 0.0 \quad 0.0 \tag{5}$$

As is standard practice in factor analysis procedures, the number of factors required to account for all of the variance and covariance in the data is interpreted to be equal to the number of eigenvalues greater than zero. Where less than all of the total variance and covariance is to be explained, some critical value greater than zero is used. The eigenvalues in (5) show that the data from which the correlations were derived could have resulted from a two-factor model, as indeed they did. The power of eigenvalue analysis, just one of a number of procedures used in factor analysis, is evident. Most correlation matrices, as that in (4), are difficult to interpret by inspection alone. It seems likely that eigenvalue analysis can be used to determine at least the complexity of the model required to account for the correlation among chemical or other types of variables.

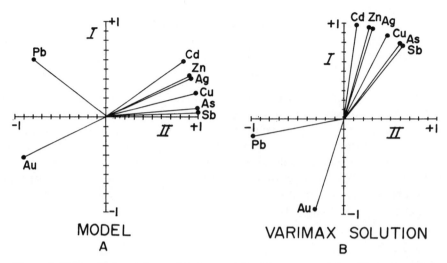

Figure 1. Vector representations of factor model used to generate *data* (A) in simulation examination of R-mode factor analysis and varimax factor solution (B).

We have recovered one important aspect of the model from which the *data* and the correlation matrix were derived, specifically the fact that $m = 2$ (the model contains two factors). The next objective in the procedures is to estimate the factor loadings.

By using mathematical procedures described by Harman (1967, p. 137), the correlation matrix in (4) is represented by a vector cluster wherein each vector represents a chemical variable, and the cosine of the angle between any two vectors is equal to the correlation between the variables they represent. Although the procedures allow the vector cluster to occupy as many dimensions as may be required to maintain the proper angles between vectors, the cluster actually will occupy m dimensions only. As $m = 2$ in the present example, the vector cluster can be displayed in a two-dimensional graph; a graph was constructed using the factor loadings in (2) plotted with respect to two orthogonal axes. The cosines of the angles between the vectors are found to be equal to the respective correlations between the variables represented (Figure 1A). In a real problem, of course, the factor loadings are unknown and are the object of estimation; the vector cluster can be formed from the correlations, but no means are available for placing the reference axes correctly within the vector cluster or for obtaining consistent estimates of the factor loadings. The positions of the reference axes shown in Figure 1B were determined from a varimax rotation of previous axes derived from a principal components factor solution.

If the correct factor loadings of $m - 1$ of the variables were known, it would be possible to rotate the entire vector cluster into its correct position with respect to the reference axes; the correct factor loadings for all of the variables then would be known. In the present example, because $m = 2$, all factor loadings could be determined correctly if any one of them were known. For example, if it were known that the variability in antimony was determined almost entirely by a single factor, rotation of

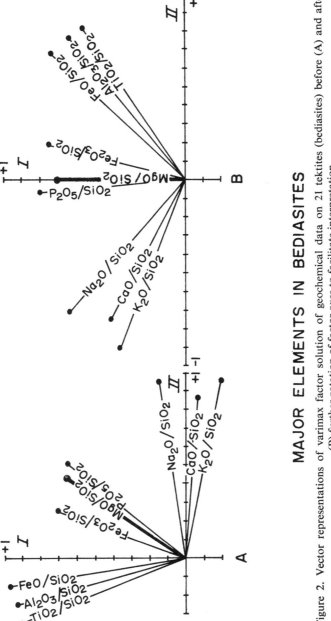

MAJOR ELEMENTS IN BEDIASITES

Figure 2. Vector representations of varimax factor solution of geochemical data on 21 tektites (bediasites) before (A) and after (B) further rotation of factor axes to facilitate interpretation.

the vector cluster in Figure 1B so that the antimony vector coincided with one of the reference axes would result in a vector solution, and factor loadings, similar to those displayed in Figure 1A. A varimax solution alone, or any other solution based only on mathematical criteria, can lead to consistent estimates of the true factor loadings only by chance.

In a factor analysis examination of some compositional data on 21 tektites from Texas, the correlation matrix was derived from the ratios of various major element oxides to SiO_2; this was done to reduce some of the effects of the constant sum problem described by Chayes (1960, 1962) and to facilitate interpretation (Miesch, Chao, and Cuttitta, 1966). The varimax factor solution shown in Figure 2A was not amenable to any interpretation consistent with accepted theory of tektite formation. However, two factors were thought to be of particular importance in controlling the tektite compositions, assuming that the tektites were derived from the lunar surface (Chapman, 1964; Chao, Dwornik, and Littler, 1964). One of these factors is considered to be the amount of chondritic meteoritic material incorporated into the tektite relative to the amount of principal parent material. As MgO is considered the best indicator of chondritic material, the vector representing MgO was rotated to coincide with one of the reference axes (Figure 2B). This axis, then, is interpreted to represent the incorporation of meteoritic debris into the tektite; the second axis can now be interpreted in a manner consistent with prevailing theory of tektite formation (Miesch, Chao, and Cuttitta, 1966); prior to the rotation performed to position the MgO vector, no acceptable interpretation was apparent.

Simulation of the R-mode factor analysis problem, therefore, demonstrates that the procedures can be used successfully to determine the complexity, the number of factors, required in the factor model. Determination of the correct factor loadings, or consistent estimates of the correct factor loadings, however, requires an additional subjective rotation of the vector cluster after the varimax solution has been obtained. The correctness of the final solution depends on the correctness of the assumption on which the subjective rotation is based. Similarly, the validity of estimates of factor scores [the F's in eq (1)] rests on the validity of the same assumption.

The simulation examination of R-mode factor analysis techniques described here did not include examination of the effects of data errors or of the effects of the constant sum. Other experiments in progress suggest that the constant sum effects on correlations between two minor chemical constituents are small, although they may be significant where either one of the constituents is present in major amounts.

Q-MODE FACTOR ANALYSIS

Q-mode factor analysis is essentially a classification technique and is used in geologic and geochemical studies principally in attempts to place rock samples, stratigraphic sections, wells, or other objects on which observations are made into objectively determined or natural categories. The method, as used here, was first described by Imbrie (1963), who provided a computer program which was revised later by Manson and Imbrie (1964). Variations of the method have been applied to petrographic and stratigraphic problems by a number of workers, including Krumbein and Imbrie (1963), Imbrie and van Andel (1964), Harbaugh and Demirmen (1964), and Griffiths (1966).

The simulation procedures followed in the examinations of Q-mode factor analysis are essentially the same as used by Imbrie (1963, p. 23–27) and Imbrie and van Andel (1964) and were designed to test the feasibility of the method for describing a seemingly complex matrix of compositional data as a mixture of various end members, or of various items of extreme composition. For example, according to Imbrie (1963, p. 23), Q-mode factor analysis of compositional data on plagioclase could be used in an academic exercise, of course, to show that all of its compositional variations can be explained by mixing of two compositional end members, in this situation albite $(NaAlSi_3O_8)$ and anorthite $(CaAl_2Si_2O_8)$. It also could be used to determine the proportions of each end member in each plagioclase specimen; this is not a remarkable feat where only two end members are involved, but could be difficult otherwise where there are more than two or where the fundamental structure of the data is masked by errors. Similarly, the method might be used to determine that olivine is a two end-member series ranging from forsterite (Mg_2SiO_4) to fayalite (Fe_2SiO_4).

Although these mineralogical examples serve as convenient and familiar subjects for examining Q-mode methods, the purpose of the methods is not to identify and describe solid-solution series. Rather, their use in geologic and geochemical problems is intended to resolve complex matrices of multivariate data into reduced forms. If we return to the plagioclase example again, it is more meaningful to describe the variation in the albite content among a series of plagioclase specimens than it would be to describe the variations in all four of its oxide components. Similarly, it seems more meaningful to describe the variation among a number of stratigraphic sections in terms of a few sections that are extreme lithologically (Krumbein and Imbrie, 1963).

The model assumed in Q-mode factor analysis is similar to the R-mode model and is written as

$$X_{ij} = \alpha_{i1}F_{j1} + \alpha_{i2}F_{j2} + \cdots + \alpha_{im}F_{jm} \tag{6}$$

where X_{ij}, in the Q-mode situation, is the percent concentration of the jth constituent in the ith sample; the α's are the true factor loadings, and the F's are percent concentrations of the jth constituent in m end-member samples. In contrast to the R-mode model, the factor loadings differ across samples rather than across variables; it is required also that the sum of the loadings, α_{i1} through α_{im}, must equal unity.

Hypothetical data simulating a series of olivine analyses were generated from the following *olivine model*:

$$
\begin{aligned}
Si &= \alpha_1(19.9) + \alpha_2(13.8) \\
O &= \alpha_1(45.5) + \alpha_2(31.4) \\
Mg &= \alpha_1(34.6) + \alpha_2\ (0.0) \\
Fe &= \alpha_1\ (0.0) + \alpha_2(54.8)
\end{aligned}
\tag{7}
$$

The row subscript, i, has been dropped for convenience, and X_j is replaced by chemical symbols for the elemental constituents in olivine. The percentage values replacing F_1 represent the composition of forsterite; those replacing F_2 represent fayalite. The

A. T. Miesch

first row of the data matrix is generated using any two fractions that sum to unity in place of α_1 and α_2. Subsequent rows are generated by using any other pairs of α's with unit sum. As an example, if α_1 is taken as 0.60 and α_2 is 0.40, the row generated in the *data* matrix represents a sample consisting of 60 percent forsterite and 40 percent fayalite. The data matrix used in the present examination was derived by using pairs of pseudorandom normal numbers after adjusting them to sum to unity and has constant row sums. The population from which α_1 was derived had a mean of 0.60 and variance of 0.04; α_2 was derived from a population with a mean of 0.40 and vari-

Table 1. Part of an Hypothetical *Data* Matrix Representing Compositions of Olivine *Samples* Generated from the Q-Mode Factor Analysis Model

Sample number	Si (percent)	O (percent)	Mg (percent)	Fe (percent)
A. Raw *data* matrix				
1 (forsterite)	19.9	45.5	34.6	0.0
2 (fayalite)	13.8	31.4	0.0	54.8
3	17.7	40.5	22.3	19.4
4	17.2	39.2	19.2	24.4
5	17.1	38.9	18.5	25.5
6	18.1	41.3	24.3	16.3
.	.	.	.	.
.	.	.	.	.
13	16.1	36.7	13.0	34.2
.	.	.	.	.
15	18.2	41.6	25.0	15.2
.	.	.	.	.
.	.	.	.	.
50	17.5	40.1	21.2	21.1
B. Transformed *data* matrix				
1 (forsterite)	1.00	1.00	1.00	0.00
2 (fayalite)	0.00	0.00	0.00	1.00
3	0.65	0.65	0.65	0.35
4	0.55	0.55	0.55	0.45
5	0.54	0.54	0.54	0.46
6	0.70	0.70	0.70	0.30
.	.	.	.	.
.	.	.	.	.
13	0.38	0.38	0.38	0.62
.	.	.	.	.
15	0.72	0.72	0.72	0.28
.	.	.	.	.
.	.	.	.	.
50	0.61	0.61	0.61	0.39

ance of 0.01. Forty-eight pairs of α_1 and α_2 were drawn, yielding a *data* matrix with four columns, representing the four elemental constituents, and 50 rows including those representing pure forsterite and pure fayalite. Part of this matrix is given in Table 1A.

The underlying structure of the data can be seen from direct examination of the data in either its raw or transformed form (Table 1) because only two factors were used to generate the data and the data contain no error whatsoever. It is expected that Q-mode analysis can be used to reveal the structure in cases where direct examination of the data cannot, due to three or more factors and imperfect chemical analyses.

The *data* matrix was transformed by scaling each column from zero to one, using

$$Y = X - X_{min}/X_{max} - X_{min} \tag{8}$$

where Y is the transformed value of X, and X_{min} and X_{max} are, respectively, the minimum and maximum values in the column of X. The purpose of the transformation is to give approximately equal weight to each elemental constituent in the determination of similarities among *samples*. As the abundances of the four elemental constituents in olivine are not extraordinarily different, the transformation is not especially important in this exercise. Other problems to which Q-mode factor methods are applied, however, include data pertaining to both major and minor chemical elements. Without a transformation such as this, the minor element contents of the samples would exert little influence on measures of similarity relative to that of the major elements. The transformation was included to determine any possible effects it may have on estimations of the loadings in the initial factor model. As will become apparent, its effect in this situation is significant. Part of the transformed data matrix is given in Table 1B.

The similarities among *samples* were measured using cos θ as defined by Imbrie and Purdy (1962, p. 257) as follows:

$$\cos \theta_{pq} = \sum_{i=1}^{M} Y_{ip} Y_{iq} / (\sum_{i=1}^{M} Y_{ip}^2 \sum_{i=1}^{M} Y_{iq}^2)^{\frac{1}{2}} \tag{9}$$

where the subscript i refers to the M chemical variables, and p and q refer to two *samples*. Part of the upper triangle of the 50×50 matrix of cos θ is given as follows:

2	3	4	5	6 ...	13 .	15 ..	50		
0.00	0.95	0.91	0.89	0.97	0.72	0.96	0.94	1	
	0.30	0.42	0.45	0.24	0.69	0.22	0.34	2	
		0.99	0.99	1.00^3	0.90	1.00	1.00	3	
			1.00	0.98	0.95	0.98	1.00	4	
				0.98	0.96	0.97	0.99	5	
					0.87	1.00	0.99	6	(10)
					.	.	.	.	
					.	.	.	.	
					.	.	.	.	
						0.86	0.92	13	
						.	.		
							0.99	15	

---
[3] All correlations of unity result from rounding upward to two places.

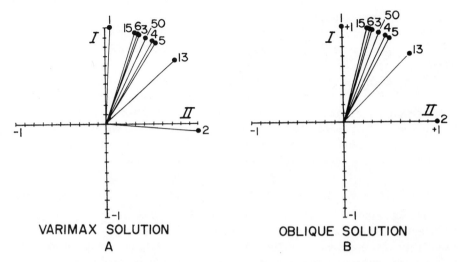

Figure 3. Partial vector representations of varimax (A) and oblique (B) factor solution from Q-mode analysis of olivine *data*.

The objective of the Q-mode analysis is to determine as much as possible about the model in (7) from the cos θ matrix in (10). The first two eigenvalues of the complete cos θ matrix, with unities in the diagonal, are 48.35 and 1.65; all 48 of the other eigenvalues are zero. The eigenvalues correctly indicate, therefore, that the model from which the *data* were generated contains two factors.

In Q-mode factor analysis the samples rather than the variables are represented by vectors, with the angles between any two vectors equal to the corresponding value of θ as defined in (9). The initial estimates of the factor loadings were determined by the principal components method (Harman, 1967, p. 137) and consist of vector coordinates with respect to principal components axes. Rotation of the axes using the varimax criterion (Harman, 1967, p. 304) provides a first approximation to estimates of the factor loadings used in the model in (7). The positions of 9 of the 50 vectors with respect to the axes determined by the varimax rotation are shown in Figure 3A. The vectors representing the forsterite and fayalite end members lie close to the orthogonal reference axes, their coordinates being close to the correct values (1, 0 and 0, 1, respectively). Further rotation of the reference axes using the method of oblique vector resolution described by Imbrie (1963) places the reference axes through the extreme vectors representing the end members (Figure 3B); the new vector coordinates are, respectively, 1, 0 and 0, 1 exactly. However, the coordinates of the vectors representing samples of composition intermediate to forsterite and fayalite depart significantly from the loadings in the factor model from which the olivine *data* were generated. The true factor loadings, the α's in (7), are listed for comparison with the estimates, the a's, derived from the oblique vector resolution in Table 2; the derived loadings on F_1 (forsterite) are plotted against the true loadings on the same factor in Figure 4.

Table 2. Comparison of Selected Factor Loadings (α) in Model Used to Generate Olivine *Data* with Uncorrected and Corrected Estimates of Loadings (*a*) from Oblique Vector Matrix

Sample number	Loadings in model used to generate the *data*		Uncorrected loadings from the oblique vector matrix		Corrected loadings from the oblique vector matrix[a]	
	α_1	α_2	a_1	a_2	a_1	a_2
1	1.0000	0.0000	1.0000	0.0000	1.0000	0.0000
2	0.0000	1.0000	0.0000	1.0000	0.0000	1.0000
3	0.6453	0.3547	0.9532	0.3024	0.6454	0.3546
4	0.5541	0.4459	0.9069	0.4214	0.5541	0.4459
5	0.5353	0.4647	0.8940	0.4481	0.5353	0.4647
6	0.7018	0.2982	0.9712	0.2383	0.7018	0.2982
.	.	.	.	.	.	.
.	.	.	.	.	.	.
.	.	.	.	.	.	.
13	0.3764	0.6236	0.7226	0.6913	0.3764	0.6236
.	.	.	.	.	.	.
15	0.7222	0.2778	0.9762	0.2168	0.7222	0.2778
.	.	.	.	.	.	.
.	.	.	.	.	.	.
50	0.6141	0.3859	0.9401	0.3410	0.6142	0.3858

[a] The correction consisted of dividing the uncorrected loadings, a_1 and a_2, by 1.73205 and 1.0, respectively, and then adjusting the quotients to sum to unity. These values are the square roots of the sums of squares of the first two rows of the transformed data matrix (Table 1).

It is apparent that the loadings as derived directly from the oblique vector resolution are sufficiently biased with respect to the true loadings to make their use in determining the correct proportions of forsterite and fayalite in olivine of doubtful value. Imbrie (1963, p. 26–27) and Imbrie and van Andel (1964, p. 1141) refer to a correction that may be applied to the loadings derived from the oblique vector resolution, but dismiss it as unimportant. However, after the correction has been made, the loadings agree with true loadings, generally to four or more significant figures (Table 2); without the correction, the estimates of the loadings in this experiment contain a variable bias of as much as 100 percent or more. The effects of variable bias in statistical analysis and interpretation of geologic data have been discussed elsewhere (Miesch, 1967).

The correction, described by Imbrie (1963), consists of dividing each loading in a column of the oblique vector matrix by the square root of the sum of squares of the corresponding column of the transformed data matrix, assuming that the data matrix has been transposed so that columns represent samples. After each column of the oblique vector matrix has been adjusted in this manner, the rows must be adjusted to

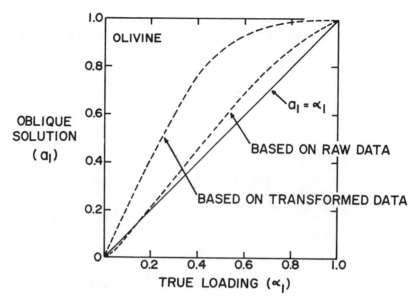

Figure 4. Graphs showing relations of estimates of first factor loading derived from uncorrected oblique vector matrix using both raw and transformed data matrices.

sum to unity. The first of these adjustments is linear, but the second is not. Used in combination, the exact factor loadings used to generate the *data* can be recovered from the oblique vector matrix.

The necessity for the correction of the loadings given in the oblique vector matrix in this experiment seems to have been caused by the transformation made on the *data* matrix prior to computation of the cos θ matrix. Imbrie (1963) and Imbrie and van Andel (1964) suggested that the transformation may be desirable in some problems, but did not use it. When the transformation was omitted in a repetition of the olivine simulation, the uncorrected loadings in the oblique vector matrix were in considerably better agreement with the true loadings, even though they were distinctly biased and related to the true loadings in a complex manner. The uncorrected loadings on F_1 (forsterite) derived using the untransformed *data* matrix are plotted against the corresponding true loadings in Figure 4. Corrections made by dividing by the appropriate square root of the column sum of squares from the original transposed *data* matrix, and adjustment to sum to unity, yield loadings correct to four or more significant figures. It seems likely that the correction is desirable where the original data are used in Q-mode analysis, but almost mandatory where transformed data are used. The transformation is extremely easy to achieve, and there is no good reason not to apply it.

The simulation experiment was repeated after omitting the *samples* representing the forsterite and fayalite end members. The 15th row of the original *data* matrix then represented the *sample* richest in forsterite, and the 13th row represented the sample richest in fayalite. After transformation of the *data* by scaling the values for each chemical element to range from zero to one, the measure of similarity, cos θ, between samples 13 and 15 was zero and the angle between the corresponding vectors was 90°. Correction of the derived oblique vector matrix yielded loadings that correctly described the remaining 46 *samples* in terms of the 2 extreme *samples*.

None of the experiments included any examination of the effects of data errors on either the determination of the number of factors from the eigenvalues of the cos θ matrix, or on the estimation of the factor loadings. Both data errors, and deviation of the model used to generate the data from the model assumed in the Q-mode method, could be included easily in further experiments. Experiments of this type may serve to determine the sensitivity of the Q-mode method to reality.

In any event, the method of Q-mode factor analysis as developed by Imbrie (1963) is seen as a method that can accomplish precisely what it is purported to accomplish. The method should serve to clarify a large number of geologic and geochemical problems in future applications.

SUMMARY AND CONCLUSIONS

Simulation is regarded as an effective supplementary technique for the evaluation of multivariate procedures for the analysis of geochemical data. It can, by no means, substitute entirely for critical mathematical examination of the methods, but is useful and informative where the mathematical and logical bases for the methods are unclear or in doubt. The advantages of simulation examinations over examinations made by applying the method to real geologic or geochemical problems are (1) the underlying model and all of its parameters are known, and (2) all assumptions demanded of the method in regard to the model and the amount and type of data errors can be satisfied. If the method is unsuccessful in estimating the model parameters under these circumstances, it cannot be applied successfully in real problems. The converse, of course, is not necessarily true, and examinations of the methods by applying them to real problems are needed. Simulation methods also may serve to obtain a better understanding of a multivariate method prior to real applications, and to become familiar with both its power and its limitations.

Simulation examination of R-mode factor analysis methods indicates that they may be useful in determining the complexity of the geochemical system required to explain the covariance in the data. Where applications of the methods show evidence that a linear factor model is sufficient and where data errors are small, the number of factors required in the model may be determined correctly from the eigenvalues of the correlation matrix, but consistent estimates of the factor loadings are impossible on the basis of the varimax or other mathematical criteria alone. These limitations in factor analysis methods, and others, have been discussed by Matalas and Reiher (1967). Further rotation of the factor axes derived from the varimax solution, based entirely on subject matter criteria, may serve to develop a factor model that is more consistent with other available geologic information—information that cannot be considered in a pure mathematical solution.

The examination of Q-mode factor analysis showed that it can accomplish exactly what it is purported to, providing that the estimates of the factor loadings from the oblique vector matrix are corrected as described by Imbrie (1963). The correction seems particularly important where the original data matrix was transformed, so that each variable is scaled from zero to one, prior to the determination of the similarity matrix. An attractive characteristic of the Q-mode method is that the results are unaffected by the constant row sum in the data matrix.

Results of the Q-mode simulation may have a bearing on the R-mode method. In brief, because of the observed success of the Q-mode experiment, it may be desirable to redefine F_{i1} through F_{im} in eq (1) as measurements on m extreme variables rather than as theoretical normalized factors. An oblique vector resolution, then, would yield estimates of the factor loadings that, after correction, would express each variable as a function of m extreme variables, the m extreme variables being the m variables that have the least correlation with each other. Imbrie (1963) proposed this method and tested it with a simulation experiment, but, to my knowledge, it has not been used in a real application.

REFERENCES

Chao, E. C. T., Dwornik, E. J., and Littler, J., 1964, New data on the nickel–iron spherules from Southeast Asian tektites and their implications: Geochim. et Cosmochim. Acta, v. 28, no. 6, p. 971–980.

Chapman, D. R., 1964, On the unity of origin of the Australian tektites: Geochim. et Cosmochim. Acta, v. 28, no. 6, p. 841–880.

Chayes, F., 1960, On correlation between variables of constant sum: Jour. Geophysical Res., v. 65, no. 12, p. 4185–4193.

Chayes, F., 1962, Numerical correlation and petrographic variation: Jour. Geology, v. 70, no. 4, p. 440–452.

Eicher, R. N., and Miesch, A. T., 1965, Computer simulation program for the investigation of geochemical sampling problems: U.S. Geol. Survey Open-File Rept., 65 p.

Griffiths, J. C., 1966, A genetic model for the interpretive petrology of detrital sediments: Jour. Geology, v. 74, no. 5, pt. 2, p. 655–672.

Harbaugh, J. W., and Demirmen, F., 1964, Application of factor analysis to petrologic variations of Americus Limestone (Lower Permian), Kansas and Oklahoma: Kansas Geol. Survey Sp. Distr. Publ. 15, 40 p.

Harman, H. H., 1967, Modern factor analysis (2nd ed.): Univ. Chicago Press, Chicago, 474 p.

Imbrie, J., 1963, Factor and vector analysis programs for analyzing geologic data: Office Naval Res., Geog. Branch, Tech. Rept. 6, ONR Task No. 389-135, 83 p.

Imbrie, J., and Purdy, E. G., 1962, Classification of modern Bahamian carbonate sediments, in Classification of carbonate rocks: Am. Assoc. Petroleum Geologists Mem. 1, p. 253–272.

Imbrie, J., and van Andel, Tj. H., 1964, Vector analysis of heavy-mineral data: Geol. Soc. America Bull., v. 75, no. 11, p. 1131–1155.

Krumbein, W. C., and Imbrie, J., 1963, Stratigraphic factor maps: Am. Assoc. Petroleum Geologists Bull., v. 46, no. 12, p. 2229–2247.

Manson, V., and Imbrie, J., 1964, FORTRAN program for factor and vector analysis of geologic data using an IBM 7090 or 7094/1401 computer system: Kansas Geol. Survey Sp. Distr. Publ. 13, 46 p.

Matalas, N. C., and Reiher, B. J., 1967, Some comments on the use of factor analyses: Water Resources Res., v. 3, no. 1, p. 213–223.

Miesch, A. T., 1967, Theory of error in geochemical data: U.S. Geol. Survey Prof. Paper 574-A, 17 p.

Miesch, A. T., Chao, E. C. T., and Cuttitta, F., 1966, Multivariate analysis of geochemical data on tektites: Jour. Geology, v. 74, no. 5, pt. 2, p. 673–691.

Miesch, A. T., Connor, J. J., and Eicher, R. N., 1964, Investigation of geochemical sampling problems by computer simulation: Colorado Sch. Mines Quart., v. 59, no. 4, p. 131–148.

17

Reprinted from pp. 201–210 of *Multivariate Morphometrics,*
Academic Press, 1971, 435 pp.

FACTOR ANALYSIS

R. E. Blackith

Department of Zoology, Trinity College, Dublin

R. A. Reyment

Department of Historical Geology, Uppsala, Sweden

It is very hard to discuss factor analyses (for there are many different kinds) without generating more heat than light. They are the most controversial of the multivariate methods. Part of the difficulty stems from the widely different terminology of factor analysts as opposed to practitioners of other forms of multivariate analyses. Partly, confusion arises because principal component analysis grew up in the same context as factor analysis and most factor analysts consider it as a special case of their techniques; this situation makes discussion exceptionally difficult. For instance, when Ehrenberg (1962) pointed out that, after 50 years of the practise of factor analysis, it was still quite uncertain whether anything of value had emerged, he was answered by other contributors to the same symposium in terms of benefits accruing from the use of multivariate analysis in general, and it remains unclear to this day whether any considerable service is performed by factor analysis (in any strict sense of the term) that is not better performed by one or other of the multivariate techniques mentioned earlier in this book. Individual psychologists would, no doubt, claim that factor analysis had made possible an orderly simplification (Burt, 1941; Thurstone, 1947; Cattell, 1965a, b) of the vagaries of the workings of the human mind: this view we can gladly accept in principle, but there is a disquieting disagreement between schools of psychologists as to the nature of the structure of the mind, which suggests that the interpretative methods are something less than objective. Could it not be that factor analysis has persisted precisely because, to a considerable extent, it allows the experimenter to impose his preconceived ideas on the raw data?

There seem to be two quite serious difficulties with factor analytical methods in the strict sense: one is that they fail to include criteria for assessing the agreement of two or more sets of results, the other that they are much less amenable than principal components to the calculation of scores for individual organisms along the relevant vectors; whereas one can easily, and advantageously, compute scores along components, one can only estimate, by methods of doubtful utility, the corresponding factor scores. The first problem has been diminished, but not eliminated, by the maximum likelihood methods associated with Lawley (1940, 1958, 1960); Lawley and Maxwell (1963) and Jöreskog (1963), and the second remains. Readers who wish to pursue the topic may peruse the symposium comprising the following references: Lindley (1962, 1964); Warburton (1962, 1964); Jeffers (1962, 1964); Ehrenberg (1962, 1963, 1964). To this rather discouraging body of opinion should be added the papers by Rasch (1962), who summarizes at least one of the systems of factor analysis, and illustrates the summary with an analysis of 13 body measurements of cattle into three factors, as well as those of Cattell (1965a, b) and some thoughtful, and probably apposite, comments on the concept of simple structure by Sokal *et al.* (1961). One of the main ways in which factor analysis differs from principal component analysis is that factors may be rotated to determinable positions in which they are not necessarily or even generally, orthogonal. One way of determining these positions is to adopt "simple structure", described by Thurstone (1947) as one of the turning points in the solution of the multiple factor problem. Other solutions have been described with unusual clarity by Gould (1967).

Where the factors are not to be rotated, there is some evidence that despite the apparently distinct mathematical models, factor analysis and principal component analysis give closely similar outcomes; indeed, Gower (1966), who specifies some of the conditions under which this statement is approximately correct, considers that the meaningful results obtained when factor analysis is used under apparently unsuitable circumstances may well stem from the extent to which this technique can simulate a principal components analysis. We may note that Dagnelie (1965b) has presented some interesting comments on the use of the technique.

Seal (1964) has distinguished factor analyses from other multivariate techniques in two respects: that the p original characters are analysable into m ($m < p$) orthogonal factors with uncorrelated residuals; and that these m orthogonal factors may be subsequently rotated to conform to a new set of factors, imposed by the experimenter, as a consequence of his theories about the natural processes underlying the measurements he has taken. In the earliest papers of Spearman (1904), the unique underlying factor (the g-factor of intelligence) was the only one considered, but, as special factors of various kinds were introduced, the distinction between imposing a factor model and allowing the data to decide how many factors were at work became blurred. Hotelling (1957) has also considered the relations between factor analysis and multivariate methods.

Far from principal component analysis being a special case of factor analysis, as one might think from the historical sequence of their respective developments, factor analysis is in fact a special (and, in Rao's (1964b) view, perhaps an unrealistic) case of principal component analysis. Rao notes that if Λ is the actual correlation matrix of the variables, we wish to choose a matrix $\mathbf{B}_q$ such that the off-diagonal elements of $\Lambda - \mathbf{B}_q \mathbf{B}_q'$ are as small as possible. Here q represents the number of factors by which the interrelationships between the variables are to be explained (there being, of course, fewer factors than variables) and $\mathbf{B}_q$ is the matrix whose i-th row is $(b_{i1}, \ldots, b_{iq})$, where b_{ik} is the standardized factor loading of the ith factor on the kth variable. As Rao remarks, the choice of $\mathbf{B}_q$ is not easy, there being still only iterative solutions.

Whether or not factor analysis really is a helpful way of viewing biological data, and in particular, morphometric data, is considered doubtful by Gower (1967a), as the same analysis of the Q-space obtained by a principal components analysis should give similar results.

The main motivation for indulging in the factor model seems to us to lie with the convictions of the would-be user: how strong his belief is in the existence of the postulated specific factors and how favourably these are interpretable in terms of his psychological theory.

Even among psychometricians, opinion is not wholly favourable to factor methods; Storms (1958) considers that where factor analysis gives results which differ from those found by multivariate methods in the sense used in this book, the factor analytical results represent a distortion, and even a falsification, of the truth in so far as this can be discovered.

The interpretation of specific factors becomes all the more difficult and tenuous when the matrix to be factorized, $\mathbf{S}$, is a Q-matrix of order n, where the elements represent comparisons between individuals of the sample, than when it is an R-matrix $(p \times p)$, where the elements are variances and covariances between the variables.

Attempts to use factor analyses have convinced us that the results rarely differ from those given by a principal components or principal coordinates analysis in any way that seriously influences a biologist or a geologist seeking to interpret a mass of data. A somewhat similar conclusion was reached by Crovello (1968c). Insofar as some differences appear, their induction and interpretation involve such a high degree of subjective judgement, often disguised from the user because he is led to believe that procedures advocated by other workers are justified mathematically when they are in fact arbitrary manipulations of the vectors, that we prefer not to give details which might be interpreted as recommendations. For the same reasons we do not give a computer programme, but refer the interested reader to Jöreskog's (1963) paper. We now offer some published examples of factor analysis that we would prefer to treat by principal components or principal coordinates analyses.

Middleton (1964) has investigated the interrelationships of ten chemical constituents in scapolites, beginning with a principal components analysis and continuing with a factor analysis, by rotation of the principal components to positions determined by the quartimax procedure (Harman, 1960); which has analogies with the least square technique for fitting a line through the points. The elements fell into two bands roughly at right angles, considered to represent the consequences of substitution in one of two series of solid solutions. The rotation of the axis does not seem to have done more than formalize conclusions apparent from inspection of the principal components analysis.

Some genera of fusulinid foraminifera were studied by a factor analysis, which involved the calculation of nine factors and the rotation of seven of these according to the varimax criterion (Pitcher, 1966). Only four of these factors were considered to be significant, and all four contributed to the separation of the various genera in successive two-dimensional charts (Fig. 23). Since the study of the characters which contributed most to the separation of the genera and species was conducted by means of factor scores, one might think that it would have been easier to interpret the results had a simple principal component analysis been performed since principal component scores are more readily computed and interpreted. Pitcher makes the useful general point that when a specimen remains within a particular grouping as the data are plotted on different axes of variation, the credibility of the proposition that the individual belongs to that group is enhanced: conversely, if replotting the data along new axes of variation shuffles the individual into another group, its classification must remain dubious.

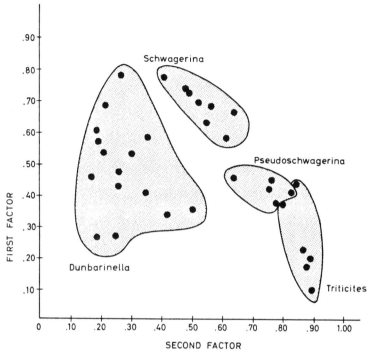

Fig. 23. Factor scores of representative species of four genera of fusulinid foraminifera showing clustering into generic groupings along the first two factors. Redrawn from Pitcher (1966).

The Case of the Bermudan Snails

Gould (1969) has uncovered a fascinating biological situation in Bermuda, where the snail *Poecilozonites bermudensis* underwent rapid evolutionary changes during the Pleistocene giving rise to four distinct paedomorphic lines, regarded by Gould as subspecies. A matrix of 33 variables and combinations of variables, measured on these pulmonate snails, was factorized, and then treated by varimax rotation. The first two axes divided the collection of shells into those of paedomorphs and those of non-paedomorphs, and separated one of the paedomorphic subspecies from the remainder. However, when the third and fourth axes of variation were made into the chart shown in Fig. 24, much finer resolution of the situation became possible. There is a general coarse division into Pleistocene paedomorphs, Recent paedomorphs, and the non-paedomorphic forms, and within the Pleistocene paedomorphs there is a distinct subdivision into the four subspecies as shown in the figure. Gould

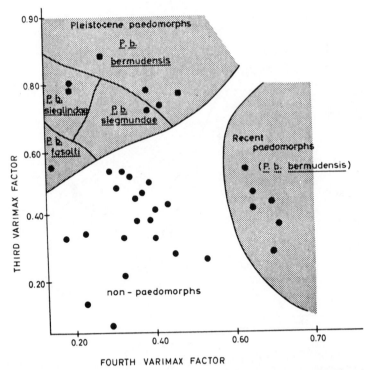

Fig. 24. Factor analysis of the paedomorphic and non-paedomorphic forms of the pulmonate snails of the genus *Poecilozonites* in Bermuda. Redrawn from Gould (1969).

took particular care to make his character combinations expressive of the shape of the shell and of its colouration.

We then tried a principal coordinates analysis on the same material as was used for Gould's factor analysis. Figure 25 shows the first two coordinate axes, taking up somewhat more than half of the "variation" in the latent roots of the association matrix. This does not show a clear separation into paedomorphic and non-paedomorphic fields [the plot of axes two and three does this (Fig. 26) well], but the various samples are clearly distinguished (Gould's axes III and IV did this); the subdivision of fossil and recent paedomorphs of *P.b. bermudaensis* stressed by Gould's third and fourth axes, is not apparent here.

We note *en passant* that Gould's specimen 05, which is on the edge of the charts for his axes I, II and III, IV, is centrally placed on all of the first four principal coordinate axes (this specimen is the Shore Hills paedomorph of Fig. 25): in view of its markedly isolated

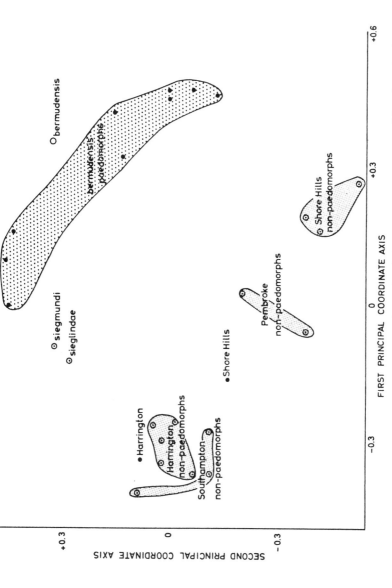

Fig. 25. A principal coordinates chart of the first and second axes for paedomorphs and non-paedomorphs of *Poecilozonites* in Bermuda.

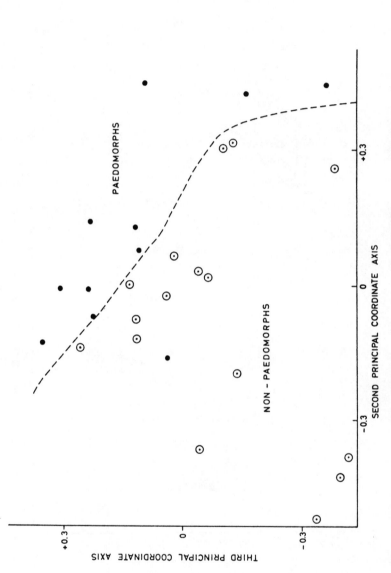

Fig. 26. A principal coordinates chart of the second and third axes for paedomorphs and non-paedomorphs of *Poecilozonites* in Bermuda.

position in Gould's plot of the first two factor axes, some mishap may have occurred in the processing by the factor programme. In Fig. 26, in which the first and third principal coordinates are plotted, it is, however, the only misplaced paedomorph.

Q- and R-Techniques of Analysis

Before we come to a discussion of the broader uses of multivariate taxonomy, a clear distinction has to be made between an analysis of the matrix of correlations between the characters (or character-states, attributes) of a set of organisms, and the analysis of the correlations between the organisms themselves. In ecological terms, to make the illustration more concrete, we always have the choice of regarding an organism as being the subject of the analysis, and analysing the quadrats in which it occurs to discover, for instance, how closely it resembles another organism ecologically, or alternatively, selecting quadrats of ground and measuring their affinities in terms of the numbers of organisms shared by two habitats, in which case the organisms are the characters of the habitat.

Q-techniques deal with the analyses in which the numbers of organisms in each habitat are studied, whereas R-techniques comprise those in which the habitats or other attributes of the organisms are correlated and analysed. To some extent these types of analysis are interchangeable. There is, however, a shift of emphasis in passing from one to the other, and most of the strictly morphometric work on shape and size described in Chapters 2-9 illustrates R-techniques, whereas a great deal of numerical taxonomic work involves examples of Q-techniques, as does the method of principal coordinates. It is often instructive to turn one's problem inside out in this way. Sokal and Sneath (1963) who discuss the differences between these two approaches at some length, point out the small amount of work, amounting at that time to a single paper by Stroud (1953), which has been performed by means of R-techniques at higher taxonomic levels, and have commented that information of interest to phylogenetically-minded taxonomists should be accessible by this route.

Orloci (1967b) discusses the degree of interchangeability between R and Q techniques in the context of data centering. He notes that when the two sets of techniques are used as alternative strategies, the various intermediate results should be directly transferable. Application of principal component analysis to a Q-matrix which has been generated by the R-expression of the correlation coefficient is inappropriate for the computation of component scores, although attempts to do this are frequently to be found in the literature.

The principal components of the R-matrix are not the same as the principal axes of the Q matrix when unstandardized data are used, but if the data are appropriately standardized, the two calculations become rearrangements of one another. There is, therefore, a real sense in which the argument about whether Q or R techniques are "better" is a non-question, a point which might be more widely appreciated than it appears to be.

Some information is lost in the course of each type of analysis, but the information is naturally qualitatively different, according to the nature of the analysis: in ecological studies, information about the relationships between the species is lost during Q-type analyses, and about relationships between the quadrats or stands during R-type analyses (Lambert and Williams, 1962).

Kendall and Stuart (1966) note that it would be desirable to distinguish between clustering of various kinds in the two cases. Clustering of individuals in respect of the characters that they possess would be called "classification analysis", whereas clustering of the characters in respect of the individuals in which they occur would be called "cluster analysis", on the system that these authors advocate. On this basis classification analysis corresponds to Q-type and cluster analysis to R-type treatments.

18

Reprinted from *Trans. Inst. Min. Met.* **80**:B346–B356 (1971)

Problems of sampling in geoscience

J. C. Griffiths M.Sc., Ph.D., D.I.C.

Synopsis

Sampling is one step in the process of problem-solving or question-answering; problem-solving presupposes an objective, explicitly defined, a universe or population of interest also clearly defined and identification of the elements of the population. The properties of the elements are used to characterize the population.

There are two kinds of problems concerning the elements of a population; the first is to estimate the proportion of some specific kind of element in the population, e.g. the amount of gold in a block of ground (equals population) or the proportion of mercury in a sample, etc. The second is to estimate the length, breadth, volume, etc., of a specific kind of element, e.g. the grain size of quartz grains in a beach sand, the size, shape and arrangement of lenses of ore in an ore deposit, etc. The first kind leads to count, the second to measurement data; in both cases the sampling problem is the same, although the model frequency distributions representing characteristics of the population may differ.

Given a specific objective, a bounded population, identified elements and the element characteristics of interest, it is then necessary to select a sample of the elements, i.e. a sample of n elements ($n = 1, 2, \ldots$) from the population and the problem of sampling requires a decision on how to select the sample. This decision is at the disposal of the investigator.

It is also necessary to establish, either by heuristic analysis, or empirically, based on long experience, the model frequency distribution of the desired element characteristics; in general, constant-probability models are used as models, the binomial or Poisson for count data and the normal distribution for measurements (or their transformed equivalents).

The objective of sampling then becomes to obtain an appropriate statistical estimate of the desired population parameter; for example, the estimators may be the mean ($\bar{X}$) and variance ($\hat{\sigma}^2$); these should be unbiased sufficient estimators of their corresponding parameters (u, σ^2, respectively). Statistical tests are performed in order to determine that the estimators are of the appropriate kind; all statistical tests are designed to test against bias of a specific kind, although they are stated as tests against randomness. Given a random sample of some population, the estimators are usually stable and adequate to solve the sampling problem.

Random samples are simple to define but exceedingly difficult to achieve in practice; in selecting a sample, if every element in the population has an equal chance of occurring, then the sample is a random sample. Interaction between the arrangement of elements in a population, i.e. its structure, and the sample selection process frequently introduces bias—that is, some elements have more chance than others of appearing in the sample and, hence, the sample is not random. An algorithm for dealing pragmatically with this aspect of the sampling problem offers one way of achieving random samples and, therefore, adequate estimators.

Examples of the use of this algorithm on various kinds of geological populations, in both field and laboratory, illustrate the requirements for solving sampling problems and the achievement of appropriate estimators.

'I contend that there is no such thing as absolute randomness, just as there is no such thing as absolute velocity. The latter has meaning only with reference to a co-ordinate framework, the former only with reference to a selector framework.'

'In any sampling inquiry it is necessary to ask oneself, Is the sampling method I am using random for the universe I am considering, for the characteristics I am discussing, and for the sampling distributions or tests of significance I am employing? Randomness is relative.'

M. G. Kendall, 1941–42

Sampling is an integral part of the general process of problem-solving; it cannot be resolved without consideration of the entire process. It is first necessary to state the question or objective of the enquiry; this leads to a definition of the universe or population of interest. It is then necessary to define the elements of which the population is composed and to decide which of the many characteristics of the elements is to be evaluated. The problem of sampling thus becomes a decision of how to select that set of elements which is to form the sample of elements whose characteristics will be evaluated.

Sampling is certainly common in the geosciences, and because the populations are quite often very large and the samples are quite small in terms of number of elements, the problem of selection becomes critical. In most cases this problem has not received the attention it requires; as an example of the present role of statistical methods, including sampling, in the mineral industries (i.e. applied geoscience) consider the articles published in the *Transactions* of the American Institute of Mining and Metallurgical Engineers from 1871 to 1968 (see Fig. 1). The total number of articles, per five-year interval, has increased from near 100 to well over 1000; during this same period the number of statistical

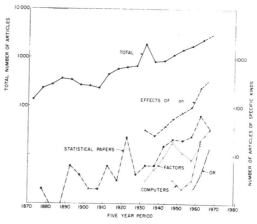

Fig. 1 Classification of articles from the *Transactions* of the American Institute of Mining and Metallurgy, 1871–1968

papers, which deal largely and directly with sampling problems, ranging from assaying a metal in a specimen of rock to estimating the reserves of petroleum in an oil field or of ore in a mine, has also increased from zero to about 100 per five-year interval. Despite both the wide range of questions, and of scale, and the obvious importance of these questions, no organized attempt has been made to resolve the problem of sampling in the mineral industries (or the geosciences).

Questions in other articles, which are statistical in nature and include problems of sampling, are entitled 'Effects of ... on ...'; these are essentially problems in simple and multiple regression, although such methods of analysis are not used in the several hundreds of articles published in the *Transactions* (see Fig. 1). Similarly, questions which are included under such titles as 'Factors which underlie ...' or 'Principles that determine ...' match in problem structure procedures of factor analysis, although, so far, these analytical procedures have not been used in this context to resolve such questions. Articles on the applications of computers and Operations Research (OR in Fig. 1) are also growing rapidly in number and many include sampling problems. By way of contrast it is interesting to note that problems of sampling which were equally critical in the area of agriculture (applied biology) were systematically analysed in the 1930s by means of uniformity trials (e.g. Cochran, 1937, Stephan and McCarthy, 1963); these extensive trials served to decide the process of selection in agricultural problems and led to square blocks and rectangular plots which cross the blocks. They are deliberate attempts to overcome soil heterogeneity (fertility gradients) by selecting size, shape and arrangement of samples to ensure that random samples from homogeneous populations are achieved.

Definition of the problem

Many statistical texts are devoted to the problem of sampling (e.g. Yates, 1949, Cochran, 1953, Stephan and McCarthy, 1963, Kish, 1967) and many publications discuss special aspects of sampling ('Student', 1931, Cochran, Mosteller and Tukey, 1954). There are very few statistical textbooks that do not include chapters on sampling and their associated experimental designs. There are also extensive discussions of problems of sampling in the geological literature (e.g. Otto, 1938, Krumbein, 1960, Milner, 1962, Krumbein and Graybill, 1965, Griffiths, 1967b, Griffiths and Ondrick, 1968, 1970, Gy, 1967, 1971, Koch and Link, 1971a); such detailed and recurrent treatment attests to a common and important problem, and its refractory nature, and implies that it has not been completely resolved.

One procedure for introducing the problem is to use an analogue model, and one common analogue is the tossing of a coin. This 'experiment' has two possible outcomes, heads or tails; all other outcomes are neglected. Given a fair toss of a fair coin, the problem of sampling is resolved and the probability (p) of either outcome is $p = 0.5$. Despite the transparent simplicity of such a model, in few cases has the result been confirmed empirically; Weldon's extensive data on dice throwing, another common analogue model used in

sampling experiments, also rarely achieved the expected model outcome (see, for example, Kendall, 1948, pp. 117 and 199). In such cases one is led to ask the following questions: was the sample random, was the toss fair, was the coin fair? It is reasonably obvious that a single set of data is unlikely to supply answers to these three separate questions. This, then, is the problem of sampling; how does one interpret the answer?

In practical experimentation in, say, the geosciences and mineral industries, one is generally faced with the option of questioning the model or the randomness of sampling or both. It is necessary to realize at this juncture that the problem largely arises from the definition of randomness—randomness, as Kendall stated, is relative (see quotation above); however, the textbook definition rarely gives one that impression. When one attempts to investigate the property of randomness and to apply the concept in practice, a paradox appears to arise;[*] one possible explanation is that the concept relies on convergence to limits at infinity, whereas all experiments are finite. One aspect of the paradox is clearly exposed in attempting to obtain a set of random numbers from a computer; it is necessary to give the computer a rule for generating the 'random' numbers, but randomness implies unpredictability, so no rule exists. Computer-generated random numbers are now recognized to be pseudo-random numbers (Tocher, 1963). Similar problems arise in testing tables of 'random numbers' for randomness (see Tocher, 1963).

This then is a brief exposé of the problem of sampling; it certainly has not been satisfactorily resolved in concept, and so in practice a pragmatic approach is necessary to achieve a 'workable' solution. Given a well defined objective, a population to which the objective is to apply and the individual units of the population, i.e. its elements, it is necessary to decide which of the many properties or characteristics of the elements is of interest (see Griffiths, 1967a, chapter 3); it is then required to select a sample of the elements, and this selection process is the component of the analytical procedure which is at the disposal of the investigator. Since statistical estimators are calculated from observations made on samples with a view to estimating parameters of the population from which the samples were drawn,[†] the samples should be representative of the population; random samples are generally representative. To obtain a random sample of elements a selection procedure is chosen such that each element has equal probability of being selected for the sample. In coin tossing, for example, let us assume that the coin and toss are fair; then the probability of appearance of heads or tails is equal at

[*]For general discussions of this problem see von Mises, 1957, Popper, 1961, Kendall, 1941–42, Tocher, 1963, Breiman, 1968, and, in a geological context, Mann, 1970, Simpson, 1970, Smalley, 1970.

[†]Given random samples, one may extrapolate by induction from sample estimators to population parameters.

$p = 0 \cdot 5$; if we assume that $p = 0 \cdot 5$ is representative, then failure to achieve this expected value during an actual experiment leads one to suspect that either the toss or the coin (or both) is (are) unfair; in sampling populations the equivalent constraints, to ensure fulfilment of a given frequency distribution model, are that samples are random and that distribution of the elements in the population is random. These constraints are expressed as, for example, 'Student', 1919, and Griffiths, 1967a, chapter 14.

The probability that an element appears in a sample is · (1) constant from trial to trial, (2) constant from element to element and (3) that the outcome of any trial is independent of any other trial. Such constraints imply that the sample selection is random *and* that the distribution of the element in the population is homogeneous.

Such constraints are exceedingly difficult to fulfil in practice when dealing with real populations; in effect, the fulfilment of these constraints demands that the population is structureless in terms of the element or at least of the characteristic of the element of interest.[*] Since most populations are structured, it is necessary to 'correct for' the structure by suitable adjustment of the selection procedure. Supposing each sample to be of the same kind, then the properties which may be varied are sample size, shape and arrangement (i.e. sample orientation and packing, see Griffiths, 1967a, chapter 3). When dealing with a population of unknown structure it is necessary to perform a series of reconnaissance sampling experiments to determine the structure and then to arrange the sampling so that the constraints are fulfilled, i.e. so that the sampling is random or so that the sample estimators are independent of the element arrangement (structure).

Given that the sampling is random, then the sample estimators will be unbiased estimators of their appropriate population parameters; it is also desirable that the estimators should be sufficient as defined by Fisher (1946). To obtain sufficient estimators constant-probability models are advantageous, so, given that the observable characteristic is measurable, the appropriate distribution is the normal or Gaussian distribution; whereas, if the observable property is countable, either the binomial or Poisson models will be appropriate. Conventional estimators such as the mean and variance for the normal distribution and the mean for the Poisson are sufficient estimators (Fisher, 1946).

It is, of course, necessary to test for the fulfilment of these constraints, and various tests of significance are available which purport to be tests for randomness; these tests are, however, tests against specific types of bias. For example, given a constant-probability model frequency distribution, one set of tests uses the Pearsonian beta statistics to evaluate specific departures from randomness in the shape of the frequency

curve. The asymmetry is tested by $\sqrt{b_1}$, an estimator of $\sqrt{B_1}$; similarly, peakedness is monitored by means of b_2, an estimator of kurtosis, B_2. An alternate test which examines the class by class frequency and is (for large numbers of samples) independent of the shape of the curve is the chi-square test. These two test procedures are independent and both require that the expected distribution be available for comparison. Still another series of tests may be performed by means of analysis of variance, which compares the relative magnitude of variation contributed by different assigned sources with that from all unassigned sources (error). To perform this test the sampling experiment is usually specifically designed and a wide variety of experimental designs exist to meet a wide range of experimental circumstances (Cochran and Cox, 1957, Scheffe, 1959).

Recent studies of pattern recognition and of trend-surface fitting (Krumbein and Graybill, 1965) are still other attempts to establish structure, or the pattern of variation in a population, on the basis of a sample of observations. There is no test for randomness and, if a structureless population exists, there does not appear to be any satisfactory means of detecting it.

One pragmatic solution

Extensive experience in sampling geological populations has led to the construction of the following algorithm to achieve adequate statistical estimators from samples which may be used to represent population parameters (see Fig. 2). It is assumed that the population, defined under a specific objective, possesses some unknown structure, and that this structure is determined by the pattern of variation in the population. The element and its characteristic (variate) are also specified in advance. From previous experience, supported by heuristic reasoning, a suitable model frequency distribution is devised; in general, if the variate is expressed on a continuous scale, then the normal model is appropriate. In some cases transformation of the variate, for example, to its logarithm (Gaddum, 1945), may be necessary for the normal model to be appropriate. If the variate is expressed on a discontinuous scale, such as when counting the proportion of elements of specific kinds, i.e. per cent composition, parts per million, etc., then the binomial or Poisson distributions are appropriate models. From the model suitable parameters are derived, such as those based on the first four moments, the mean (u), the variance (σ^2), the asymmetry ($\sqrt{B_1}$) and the peakedness (B_2).

The sample is then defined in terms of its size (number of elements), shape and arrangement, and a set of 'random' samples is collected by use of random number tables (e.g. Kendall and Babington-Smith, 1938, or the Rand tables, 1955). The data, counts or measurements, are obtained from the samples and an observed frequency distribution is constructed (see Fig. 2). From the observed distribution statistical estimators are calculated and these may be compared with their

[*]Structure may be defined in the present context as the pattern of variation of the element (characteristic or property) in the population (see Griffiths and Ondrick, 1970).

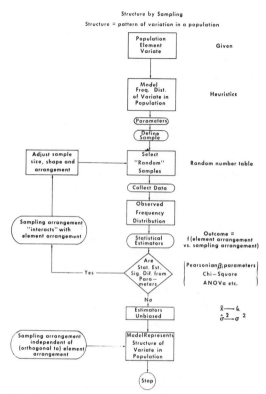

Structure by Sampling

Structure = pattern of variation in a population

Population Element Variate — Given

Model Freq. Dist. of Variate in Population — Heuristics

Parameters

Define Sample

Adjust sample size, shape and arrangement

Select "Random" Samples — Random number table

Collect Data

Sampling arrangement "interacts" with element arrangement

Observed Frequency Distribution

Statistical Estimators — Outcome = f (element arrangement vs. sampling arrangement)

Are Stat. Est. Sig. Dif. from Para-meters — Pearsonian β_i parameters Chi–Square ANOVa etc.

Yes

No

Estimators Unbiased — $\bar{x} \longrightarrow \mu$ $\hat{\sigma}^2 \longrightarrow \sigma^2$

Sampling arrangement independent of (orthogonal to) element arrangement

Model Represents Structure of Variate in Population

Stop

Fig. 2 Algorithm for evaluation of structure by sampling. Reproduced, by permission of The Pennsylvania State University Press, from Griffiths and Ondrick (1970)

corresponding population parameters. The frequency data grouped into classes in terms of the variate may be compared with the expected theoretical model by means of the chi-square statistic; and from the experimental design a suitable form of analysis of variance is performed. On the basis of these several tests of the various null hypotheses the response is either that the statistical estimators are or are not significantly different from their corresponding parameters.

If they are not, then there is no evidence against randomness of the samples and the estimators are unbiased. The model is an adequate representation of the structure of the variate in the population and, in effect, the sampling selection is independent of the structural arrangement.

If the statistical estimators are significantly different from their corresponding population parameters, then the sample selection interacts with the arrangement of the elements in the population and the sampling is biased; from the significance tests the form of the bias may be deduced and the sample size, shape and arrangement may be modified and a new set of samples selected. More data are collected and the testing is repeated; generally, with few repetitions the estimators

will converge on the parameters and the tests will yield no significant differences leading to unbiased estimators.

By way of example consider a layered or stratified population and assume the structure is known to be layered and that the layers are essentially horizontal; then, if channel samples are selected which cross the entire population perpendicular to the layers, these will be random self-weighting samples and they will yield unbiased estimators. On the other hand, such samples will not reveal the presence of layers. If it is desired to specify the kind of layering, then the sample selection must be arranged to be stratified random, i.e. each sample should be small enough to fit between the boundaries of the layers and large enough to be representative of the sub-populations of elements. In addition, to obtain unbiased estimators of population parameters, the strata must be represented by their correct proportions in the population. It is useful in such cases to arrange the sampling design as a square or rectangular grid and to use channel and stratified sampling by selecting the sampling sites at grid intersections. The channel samples (columns in the example) yield unbiased estimators of population parameters and the estimators from the stratified samples (rows) may be adjusted to match those of the channel samples, thus yielding an appropriate weighting (see Griffiths and Ondrick, 1968). This serves to reveal how many layers of different kinds occur in the population; since the strata are defined as 'greater variation among layers than within layers', the structure is defined by the sampling. Unbiased estimators from channel samples are independent of the structure; similarly, appropriately weighted, unbiased estimators from stratified random samples are also independent of the structure.

In practice the nature of the structure, i.e. the pattern of variation in the population, is often unknown, and one of the objectives of the experiment may be to define the structure as well as to obtain unbiased estimators of population parameters. Even if the structure is known to be layered, the arrangement of the layers is frequently unknown, For example. suppose the population is stratified in terms of the variate of interest and that the strata are oblique to the grid, then it commonly happens that one or more of the estimators is biased; if the experiment has been appropriately designed, an analysis of variance of the rows and columns compared with the interaction among rows and columns will exhibit the structural arrangement. If the layers and the grid are concordant, as in the initial example, the column means from channel samples will not be significantly different, whereas row means from samples within layers will be significantly different; furthermore, row by channel interaction will be effectively zero. On the other hand, if the grid is oblique to the layers, variation among column means and row means will not be significantly larger than the interaction between columns and rows.

These several outcomes emphasize the advantage of adequate experimental design; grid sampling is quite efficient in the search for structure and, if at each grid intersection replicates are taken, the interaction terms may be tested against variation within sites. This gives a powerful means of determining unknown structures, but it requires a very well specified sampling procedure. Small samples at each site are most efficient and the square or rectangular grid is equally desirable; if it turns out that the grid is oblique to the structure, then the orthogonality of rows and columns permits orthogonal rotation of the grid until it is concordant with the structure. In cases where layering is suspected and a grid is used as above described, the criterion of minimax variation may be used to define the layers, i.e. when variation is at a maximum between the rows (means) and minimum within the rows and among the columns (means) the grid may be considered concordant with the structure, and this serves to define the layering.

On the basis of extensive sampling experiments a model for sampling sedimentary populations varying from unstratified random (homogeneous) populations to finely layered deposits has been designed (see Milner, 1962, vol. 1, Fig. 93, p. 607, Griffiths, 1967a, Fig. 2.3, p. 20, and Griffiths and Ondrick, 1968, Fig. 2, p. 4). Subsequent experiments with the model and the concepts above described have illustrated the effectiveness of these grid designs in resolving the problem of defining the structure of the population and obtaining unbiased estimators of population parameters (Griffiths, 1959, 1967a, p. 416 ff, Griffiths and Ondrick, 1968). The algorithm exhibited in Fig. 2 was devised from these experiments.

Since many igneous and most metamorphic rock bodies and their contained orebodies and sediments with their contained oil and gas fields possess some structure, the model and the algorithm are equally applicable to these populations; indeed, both the model and the algorithm are sufficiently general to be applicable to populations in any field. These analytical tools are also invariant to change of scale from the optical microscope to regional exploration (Griffiths and Ondrick, 1968, 1970).

As a final example, to illustrate the versatility of these

sampling arrangements, consider three typical 'structures' which encompass a wide range of features exhibited by natural populations (Fig. 3); the three populations represented may be microscope slides or regional geological maps. The rectangular grid is composed of traverses labelled E–W and N–S, respectively; the structures are (I) homogeneous,* in which the elements are arranged at random throughout the population; (II) layered; and (III) patchy. Small samples of equal numbers of elements are collected within each traverse at grid intersections. The sources of variation are, therefore, among different directions, and among traverses within directions, and among samples within traverses. The apportionment of the relative magnitude of variation among these three sources suffices to identify each structural type. If the sampling design is arranged to differentiate directions, then the stratified population will show a large difference between north–south traverse means. This arrangement will also be effective in determining obliquity of grid and layering, and, in the case of obliquity, by rotation of the grid and use of the miximax criterion, will lead to location of the layering.

Sampling arrangement and problem structure

The above description applies essentially to problems in which estimation of the population mean (u) and variance (σ^2) are the objectives; each such problem requires repeated sampling following the loop in Fig. 2. All such problems require a relatively elaborate experimental design and are among the most difficult and expensive problems to resolve—a feature amply demonstrated by the extensive references to sampling problems and their experimental designs.

An alternative approach, which is frequently feasible, is to structure the problem in the form of a regression analysis. Suppose that it is required to study the relationship of two variables in a population and, for simplicity, suppose that they are linearly related. Then the simplest sampling design is to select the samples from the extreme values of each variable; for example, suppose that the objective is to evaluate the relationship between permeability and porosity of an oil sand. In many cases the relationship between log permeability, the dependent variable (Y), and porosity, the independent variable (X), is linear (Griffiths, 1958, 1960a, 1967a). Then, to determine the parameters of the best-fitting line, where α is the intercept on the Y axis and B is the slope of the line, a least-squares criterion is used (Griffiths, 1967a). Given a perfect linear relationship, the two extreme points of highest log permeability and porosity and lowest log permeability and porosity, respectively, will suffice to determine this line. If, as is usual, the relationship is not perfect but possesses some degree of random variation, several samples from near the extremes are the best choice.

If the form of the relationship is not known to be

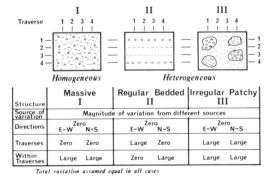

Fig. 3 Effects of structures on sampling arrangement and estimation of population parameters

	Massive I	Regular Bedded II	Irregular Patchy III
Structure	I	II	III
Source of variation	Magnitude of variation from different sources		

	Massive I		Regular Bedded II		Irregular Patchy III	
Directions	Zero		Zero		Zero	
	E–W	N–S	E–W	N–S	E–W	N–S
Traverses	Zero	Zero	Large	Zero	Large	Large
Within Traverses	Large	Large	Zero	Large	Large	Large

Total variation assumed equal in all cases

*The geological term 'massive' frequently implies a homogeneous structure.

211

linear, a few samples scattered throughout the range will suffice to yield a test for linearity (in the ideal case three samples, one at either extreme and one near the middle, are adequate). As the relationship becomes more complex more samples are needed to define the complexity. However, the sampling design for simple linear regression is frequently much easier to arrange and execute than the more elaborate requirements of estimating the mean and variance of the population. In addition, it is possible to use the regression relationship to obtain estimates of mean and variance and these estimates are often equally effective for less invest-

of the Susquehanna River at Montoursville, Pennsylvania, has been sampled by various sampling arrangements in alternate years from 1957 to 1971 by graduate students in the writer's class during the autumn of each year. This gravel deposit is almost horizontally layered and so is a close representative of population II in Fig. 3; in the intervening years the class has sampled boulders in various screes on the slopes of the Appalachian folded belt in Central Pennsylvania (see, for example, Griffiths, 1959). Since the surface boulders of the scree are essentially haphazardly arranged, they approximate population I of Fig. 3.* The effect of

Table 1   Summary of sampling experiments in Lycoming silica sand; gravel pit, Montoursville, Pennsylvania, U.S.A.

Sampling arrangement	Year	No.*	Operators	Channels or layers	Pebbles per Operator	Pebbles per Sample unit	Operator sample unit	Total	Remarks	Normality beta statistics
Channel	58–59	C_1	6	6	54	54	9	324	White Quartzite	Okay
	60–61	C_2	7	5	75	105	15	525	White Quartzite	Okay
	64–65	C_3	6	10	40	24	4	240	White Quartzite	Biased
	64–65	C_4	6	10	40	24	4	240	Red Sandstone	Okay (?)
	66–67	C_5	4	5	75	60	15	300	Red Sandstone	Biased
	66–67	C_6	4	5	75	60	15	300	Green Sandstone	Biased
	70–71	C_7	6	4	32	48	8	192	Red Sandstone	Biased
Stratified	57–58	S_1	4	8	80	40	10	319†	White Quartzite	Biased
	58–59	S_2	6	8	56	42	7	336	White Quartzite	Biased
	60–61	S_3	6	15	75	30	5	450	White Quartzite	Okay
	64–65	S_4	6	20	40	12	2	240	White Quartzite	Biased
	64–65	S_5	6	20	40	12	2	240	Red Sandstone	Okay (?)
	66–67	S_6	4	16	80	20	5	320	Red Sandstone	Okay (?)
	66–67	S_7	4	16	80	20	5	320	Green Sandstone	Biased
	66–67	S_8	4	25	75	12	3	300	Red Sandstone	Biased
	66–67	S_9	4	25	75	12	3	300	Green Sandstone	Okay
	70–71	S_{10}	6	15	45	18	3	270	Red Sandstone	Biased
Spot	58–59	Sp	5	15	60	20	4	300	White Quartzite	Biased
Field	57–58	F	1		Irregular			104	White Quartzite (talus)	Biased

*See Fig. 4.   †Missing value.   (?) Near borderline of significance at 5 per cent probability level.

ment of time and money than the alternate procedure (see Cochran, 1953, chapter 7).

When there are large numbers of variates, each with its respective estimators and each requiring a different sampling design, recourse to multivariate statistical analysis based on sophisticated regression procedures, such as covariance and factor analysis, may be the only effective recourse (see, for example, Kendall, 1958, Griffiths, 1966a).

The sampling arrangement remains relatively simple to achieve by selecting samples which cover the range of variation present in the population. Such an arrangement is frequently independent of the structure of the population.

Examples of application of the algorithm

A fluvio-glacial gravel from a terrace of the West Branch

different sampling arrangements on these populations which possess different structures then permits the evaluation of the algorithm of Fig. 2.

Table 1 is a summary of the results of these repeated sampling experiments on the gravel, the sampling arrangement being given on the left of the table. Each experiment is identified by year (column 2) and number (column 3). The number of operators in the experiment and the number of main sampling units follow in columns 4 and 5. Column 6 exhibits the number of pebbles selected by each operator and is a guide to the number which may be measured during a three- to

---

†These two populations act as analogues of populations I (the scree) and III (the gravel) in a model designed for sampling sedimentary rocks (Milner, 1962, vol. 1, Fig. 93, p. 607, Griffiths, 1967a, Fig. 2.3, p. 20, Griffiths and Ondrick, 1968, Fig. 2, p. 4).

four-hour field period. The number of pebbles per sampling unit is varied widely, as is shown in column 7; and the number of pebbles per operator per sampling unit is listed in column 8. Column 9 contains the total number of pebbles per sampling experiment.

During the experiment the length, breadth and thickness of individual pebbles of known composition (strictly colour and texture, see column 10, Table 1) are measured according to rule (see Griffiths and Ondrick, 1968, pp. 9–10); these data are recorded directly on Port-A-Punch cards in the field and, subsequently, the data are analysed by computer through standard programs (Griffiths and Ondrick, *op. cit.*).

One output from the computer program consists of the mean, standard deviation and beta statistics for each axis; these beta statistics are tested for significance by referring to tables (Pearson and Hartley, 1954, pp. 183–4). As an example the beta statistics for the long (*a*) axis of pebbles, measured during the experiments summarized in Table 1, are exhibited in Fig. 4; the ordinate scale lists the values of the statistic b_2, which varies around its parametric value of $B_2 = 3$ in a normal distribution.* The abscissa represents the values of $\sqrt{b_1}$, which varies around its parametric value of $\sqrt{B_1} = 0$ in a normal distribution. Given the sample sizes $n = 200, 300, 450$, respectively, the values, under the hypothesis of normality, which the b_i may achieve at the 5 per cent probability level, are read from tables and inserted as rectangles in Fig. 4. If the sample point falls outside the rectangle appropriate to its value of n, it is significantly different from the expected parametric value.

The results for these sampling experiments are summarized in column 11 of Table 1 and are graphically displayed in Fig. 4; discrepancies between the table and the diagram arise because the designation in column 11 is based on the results from all three axes, whereas Fig. 4 contains results from the long *a* axis only. First, it seems clear that different sampling arrangements may, and quite often do, yield different results (biased and unbiased, respectively); since it is expected, on heuristic grounds, that the expected population frequency distribution is (log) normal, then the expected parametric values are $\sqrt{B_1} = 0$, $B_2 = 3$. Three of the channel sampling experiments (C_1, C_2, C_4) achieve non-significance of their b_i—hence supplying no evidence against the null hypothesis that these are random samples from (log) normal populations. Four of the channel sampling experiments exhibit bias (C_3, C_5, C_6, C_7); the biases are not consistent and vary from significantly skewed but not significantly peaked, e.g. C_5, C_6, C_7, to significantly peaked but not skewed, C_3. It should be clear that to attribute geological implications to variation in these statistics is quite

*The frequency distribution of long *a* axis is essentially log normal and field measurements are transformed to phi units whereby the frequency distribution of *a* phi is expected to be normal, where phi is defined as phi = $-\log_2 d_{mm}$ (Krumbein, 1938).

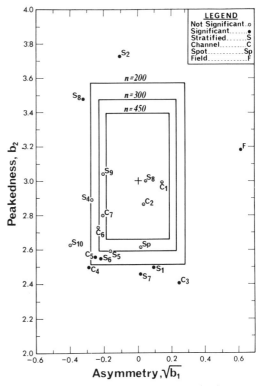

Fig. 4   Effect of different sampling selection procedures on asymmetry and peakedness: long axes of pebbles from Montoursville gravel

unjustified; they vary because of non-random sampling of the same population.

From the various channel sampling arrangements it may be deduced that to stabilize the sampling results the total number of pebbles should exceed 300; secondly, the number of channels should be at least five; thirdly, the number of pebbles from each channel should be at least 50 and preferably around 100. From these and other experiments we have learned to use between five and seven operators; less encourages bias and more introduces too much variation (per degree of freedom).

Stratified sampling is, of course, much more sensitive than channel sampling; again a total sample size of $n = 300$ is a required minimum. In this case the number of layers and the number of pebbles per layer are quite critical; in general, at least 15 layers are necessary, but 20–25 are an upper limit. In varying the number of layers the number of pebbles per layer must be adjusted; within the 15–25 layer limits the less the number of layers the more the pebbles per layer. In general, with 15 layers, 30 pebbles per layer appear adequate, whereas with 25 layers 10 pebbles per layer represent the lower limit. These figures, of course, are not exact but are approximate guidelines and we have found that any

213

serious departure leads to biased estimators. Again, the number of operators performing the experiments should be around 5–7 for best balance.

As may be seen from Table 1, in some cases 4 operators and different numbers of pebbles may achieve unbiasedness, but these are fortunate accidents and further testing by chi square reveals discrepancy. Similarly, the numbers of operators and their performance is best judged from the appropriate analysis of variance (two-way classification with replicates, see Griffiths and Ondrick, 1968) rather than from tests of the beta statistics.

Only one example of spot sampling is recorded in Table 1, but such experimental arrangements have also been used on a glacial gravel from the Livermore Valley, California,* and in an extensive set of experiments on the Hollidaysburg gravel (a fluvial terrace) at Zarah, in Kansas (1968). The sampling experiments in Kansas were first performed by a class of 25 graduate students and subsequently repeated in a short course by 64 geologists; the outcomes were similar and spot samples show no convergence even when the number of spots per population and number of pebbles per spot were widely varied.

The last sampling arrangement, labelled 'Field' in Table 1 and (F) in Fig. 4, was an attempt to sample the talus below the face of the gravel pit to see if a random sample could be easily obtained from this type of deposit; the outcome was extremely skewed and, on further testing, it became clear that in the pebbles from the surface of the talus slope a very considerable sorting on the basis of size and shape had taken place in the 30-ft drop from the gravel face!

Similar experiments, performed on scree from the slopes of the Appalachian Mountains, by means of grids of various sizes and shapes, usually resulted in unbiased estimators when the total number of boulders exceeded 200; convergence on stability was more rapid with $n = 300$. Since the experimental design was similar[†] (see Griffiths, 1959, 1967a, p. 418 ff) presence of structure in the scree could be evaluated by the analysis of variance, but none was found.

Although these experiments were performed in the field with populations ranging in area from 30 ft by 300 ft to 500 ft by 500 ft, they have also been extended to small-scale microscopic mounts of loose grains and thin sections (1 in by 1 in) and to regional scales of hundreds of square miles; limited experience with 8 in × 10 in plates of clay minerals obtained from the electron microscope further extends the scale towards hundreds of Ångström units. The algorithm and the recommendations based on these analyses appear to be invariant to scale changes of this magnitude.

Experiments described above result in data which are measurements of the dimensions of elements on a continuous scale in Ångströms, microns, millimetres, inches and feet; experiments in which an angular measure was used to characterize the elements present similar features (e.g. Griffiths, 1967a, chapter 7). When the data are counts of elements of different kinds, as in the characterization of material by its composition, chemical or mineralogical, the heuristics lead to different frequency distribution models; but, otherwise, the use of the algorithm and definition of structure again remain the same (Griffiths, *op. cit.*, chapter 14). This approach, therefore, represents a very general model for determining structure by sampling.

With this sampling experience as background we may now examine some of the literature on sampling in the geosciences; there is an extensive literature on geochemical sampling, ranging from isotope analysis (Engels and Ingamells, 1970) through trace-element data (Landy, 1961, Ondrick, 1968, Koch and Link, 1971a, Link, Koch and Schuenemeyer, 1971) to bulk chemical and mineralogical analysis (Wilson, 1964, Coatney, 1965, Kleeman, 1967). In each case the variate is discontinuous and the postulated model frequency distributions have varied from discrete distributions, such as the Poisson and binomial, to log normal (Ahrens, 1954, Chayes, 1954, Vistelius, 1960, Rodionov, 1965). The main conclusions which emerge are that different elements show differing 'types' of variability; this is usually attributed to the location of the elements of the population within the structure of other minerals, within the minerals but not as a part of their structure or outside the minerals. In terms of this article these elements possess differing structural arrangements in the population and so necessitate different sampling arrangements to yield unbiased estimators.

Similarly, in sampling ore minerals, there appears to be no consensus leading to a general recommendation, even for individual ores (Hazen, 1958, Becker and Hazen, 1960, 1961, Rowland and Sichel, 1960, Munro, 1966, Gy, 1967, Ottley, 1966, Koch and Link, 1971b). Sampling for estimation of gold ores appears to be a particularly difficult problem (see, for example, the 1966 symposium held under the auspices of the South African Institute of Mining and Metallurgy). Recurrent reviews of these sampling problems serve to emphasize that they remain unresolved (Becker and Hazen, 1961, Munro, 1966, Gy, 1967, 1971, Visman, 1969, Duncan, 1971). In many cases one is led to suspect that the initial sample size, shape and arrangement is inappropriate and the sample selection procedure represents spot sampling which fails to converge; in layered deposits, such as gold and uranium, the weighting of spot samples is inconsistent and, subsequently, no amount of manipulation of the data appears capable of removing the bias. Our own experience suggests that similar features are associated with the estimation of compositional variables from sampling in hand speci-

---

*American Geological Institute Short Course given prior to the Annual Meeting of the Geological Society of America, San Francisco, 1967.

†Actually a row by column by operator three-way classification design was used, whereas the rows (strata) and columns (channels) were treated separately for the gravel, yielding two two-way classification designs; in both cases replicates were used to test the interactions.

men, in thin section and for trace elements determined from this kind of sampling arrangement (Landy, 1961, Coatney, 1965, Ondrick, 1968).

Textural variation,· particularly grain size measurement determined in thin section, may yield different patterns of structure variation (Wood and Griffiths, 1963) and often shows the lack of convergence characteristic of spot sampling (Griffiths, 1967a, Fig. 5.5, and text pp. 95–6); in the latter case, even with relatively large $n = 15\,000$, the asymmetry remains significant, indicating the presence of bias. Similarly, in counts of heavy accessory minerals interaction of sampling arrangement with both 'natural' sampling variability (Griffiths, 1960b) and technique error contributes to sampling bias (Krumbein and Rasmussen, 1941, Griffiths and Ondrick, 1970).

Exploration for natural resources also relies on sampling (Griffiths, 1966b); given the characteristics of targets, such as their sizes, shapes and values, grid sampling may be devised to locate the targets at a specified probability level (Drew and Griffiths, 1965, Griffiths and Drew, 1966, Drew, 1967, Griffiths, 1967b). Tables are available for designing grids of various shapes, square, rectangular and hexagonal, to achieve sampling arrangements which lead to estimators for populations with different structural arrangement (Savinskii, 1965, Singer and Wickman, 1969). Since the grids possess orthogonal axes in two and three dimensions, they may be rotated to minimax solutions to define structures independent of scale (e.g. Bennion and Griffiths, 1966).

It has been suggested that in exploration sampling of regions for valuable minerals the elimination of large areas, because of alluvial cover, lease boundaries and so on, may induce an unfortunate bias; the resultant frequency distribution representing the incidence of the natural resource becomes negative binomial. Whereas if the whole region were given adequate representation in the sampling selection, the frequency distribution might then become a constant-probability model, such as a Poisson distribution. This implies that the discovery success ratio has been biased downwards and the resource may be more plentiful than the biased exploration programme has revealed (Griffiths and Ondrick, 1970).

Summary and conclusions

The place of sampling in the context of problem-solving is described and a review of articles in the area of applied geoscience indicates that sampling has not received the attention its importance appears to merit.

The problem of sampling is defined and the paradox arising from conceptual definition of randomness is discussed; a pragmatic solution is offered by means of an algorithm (Fig. 2) which leads by iteration to unbiased estimators, which are also sufficient estimators. If structure is defined as the pattern of variation of a characteristic of an element, in a population, the relationship between the sample selection procedure and structure holds the key to random sampling; essentially, the selection procedure should be orthogonal to, or independent of, the structure, to obtain unbiased estimators. Variation in sample size, shape and arrangement is used to obtain this independence; this variation in sample selection may be used to evaluate the structure.

A sampling procedure in which regression, both simple and multiple, is used is also mentioned as an alternative and sometimes more efficient approach.

Examples of application of the algorithm to sampling layered and irregularly structured populations are used to illustrate the sensitivity of balanced sampling in both defining the structure and obtaining unbiased estimators of population parameters.

On the basis of this sampling experience a number of examples of sampling experiments are reviewed, encompassing compositional characteristics from isotope, trace-element and bulk chemical and mineralogical analyses. Sampling of ore minerals suggests that it may be useful to apply the algorithm to these problems and to adjust sample size, shape and arrangement to improve the estimation procedure.

Similar features also apply to the study of textural characteristics, such as grain size and shape, and the thin section is shown to lead to a type of spot sampling which fails to yield convergence or unbiased estimators.

Finally, exploration for non-renewable natural resources also depends on sampling, and orthogonal grids are recommended as sampling designs for improving the efficiency of the search procedure.

References

Ahrens L. H. (1954) The lognormal distribution of the elements—a fundamental law of geochemistry and its subsidiary. *Geochim. cosmochim. Acta*, **5**, 49–73.

Becker R. M. and Hazen S. W. Jr. (1960) Probability in estimating the grade of ore. *Bull. Miner. Inds Exp. Stn Penn. St. Univ.* no. 72, 39–54.

Becker R. M. and Hazen S. W. Jr. (1961) Particle statistics of infinite populations as applied to mine sampling. *Rep. Invest. U.S. Bur. Mines* 5669, 79 p.

Bennion D. W. and Griffiths J. C. (1966) A stochastic model for predicting variations in reservoir rock properties. *J. Soc. Petrol. Engrs*, **6**, 9–16.

Breiman L. (1968) The kinds of randomness. *Sci. Technol.* no. 84, Dec., 34–44.

Chayes F. (1954) Discussion of **Ahrens L. H.** (1954). *Geochim. cosmochim. Acta*, **6**, 119–20.

Coatney R. L. (1965) Modal analysis of the granitic rocks of northern Sierra Nevada between Yosemite and Lake Tahoe, California. Ph.D. thesis, The Pennsylvania State University, University Park, Pennsylvania.

Cochran W. G. (1937) A catalogue of uniformity trial data. *J. R. statist. Soc. Supple.*, **4**, 233–53.

Cochran W. G. (1953) *Sampling techniques* (New York: Wiley), 330 p.

Cochran W. G. and Cox G. M. (1950) *Experimental designs* (New York: Wiley), 454 p.; (1957) *Experimental designs, 2nd edn,* 611 p.

Cochran W. G. Mosteller F. and Tukey J. W. (1954) Statistical problems of the Kinsey report. *J. Am. statist. Ass.,* **48,** 1953, 673–715; Principles of sampling. *J. Am. statist. Ass.,* **49,** 1954, 13–35.

Drew L. J. (1967) Grid-drilling exploration and its application to the search for petroleum. *Econ. Geol.,* **62,** 698–710.

Drew L. J. and Griffiths J. C. (1965) Size, shape and arrangement of some oil fields in the U.S.A. Coll. of Mines, University of Arizona, AIME meeting, March, 1965, vol. 3, 31 p.

Duncan A. J. (1971) Comments on 'A general theory of sampling'. *Mater. Res. Stand.,* **11,** Jan., p. 25.

Engels J. C. and Ingamells C. O. (1970) Effect of sample inhomogeneity in K–Ar dating. *Geochim. cosmochim. Acta,* **34,** 1007–17.

Fisher R. A. (1946) *Statistical methods for research workers, 10th edn* (Edinburgh: Oliver and Boyd, 1946), 354 p.

Gaddum J. H. (1945) Lognormal distributions. *Nature, Lond.,* **156,** 463–6.

Griffiths J. C. (1958) Petrography and porosity of the Cow Run Sand, St. Mary's West Virginia. *J. sedim. Petrol.,* **28,** 15–30.

Griffiths J. C. (1959) Size and shape of rock-fragments in Tuscarora Scree, Fishing Creek, Lamar, Central Pennsylvania. *J. sedim. Petrol.,* **29,** 391–401.

Griffiths J. C. (1960*a*) Relationships between reservoir petrography and reservoir behavior in some Appalachian oil sands. *Tulsa geol. Soc. Digest,* **28,** 43–58.

Griffiths J. C. (1960*b*) Frequency distributions in accessory mineral analysis. *J. Geol.,* **68,** 353–65.

Griffiths J. C. (1966*a*) A genetic model for the interpretive petrology of detrital sediments. *J. Geol.,* **74,** 655–72.

Griffiths J. C. (1966*b*) Exploration for natural resources. *J. Ops Res.,* **14,** 189–209.

Griffiths J. C. (1967*a*) *Scientific method in analysis of sediments* (New York: McGraw-Hill), 508 p.

Griffiths J. C. (1967*b*) Mathematical exploration strategy and decision-making. In *Origin of oil, geology and geophysics* (Amsterdam, etc.: Elsevier), 599–604. (*Proc. 7th World Petrol. Congr., Mexico, 1967,* vol. 2)

Griffiths J. C. and Drew L. J. (1966) Grid spacing and success ratios in exploration for natural resources. In *Proc. 6th ann. Symp. and short course on computers . . . Pennsylvania State Univ.,* vol. 1, Q1–24 (*Miner. Inds Exp. Stn Penn. St. Univ. Spec. Publ.* 2–65)

Griffiths J. C. and Ondrick C. W. (1968) Sampling a geological population. *Kansas geol. Surv. Computer Contr.* 30, 53 p.

Griffiths J. C. and Ondrick C. W. (1970) Structure by sampling in the geosciences. In *Random counts in physical science* Patil G. P. ed. (University Park, Pa.: The Pennsylvania State University Press), 31–55.

Gy P. (1967) L'échantillonnage des minerais en vrac.

Tome 1: théorie générale. *Mém. Bur. Recher. géol. min.* no. 56, 186 p.

Gy P. (1971) Modèle général de l'échantillonage des minerais. *Revue Ind. minér.,* **53,** 77–99.

Hazen W. S. Jr. (1958) A comparative study of statistical analysis and other methods of computing ore reserves . . . *Rep. Invest. U.S. Bur. Mines* 5375, 188 p.

Kendall M. G. (1941–42) A theory of randomness. *Biometrika,* **32,** 1–15.

Kendall M. G. (1948, 1951) *The advanced theory of statistics* (London: Griffin), vol. 1, *4th edn,* 1948, 457 p.; vol. 2, *3rd edn,* 1951, 521 p.

Kendall M. G. (1958) *A course in multivariate analysis* (London: Griffin), 185 p.

Kendall M. G. and Babington-Smith B. (1938) Randomness and random sampling numbers. *J. R. statist. Soc.,* **101,** 147–66.

Kish L. (1967) *Survey sampling* (New York: Wiley), 643 p.

Kleeman A. W. (1967) Sampling error in the chemical analysis of rocks. *J. geol. Soc. Austr.,* **14,** 43–8.

Koch G. S. Jr. and Link R. F. (1971*a*) *Statistical analysis of geological data* (New York: Wiley), vol. 1, 1970, 375 p., vol. 2, 1971, 438 p.

Koch G. S. Jr. and Link R. F. (1971*b*) The coefficient of variation—a guide to the sampling of ore deposits. *Econ. Geol.,* **66,** 293–301.

Krumbein W. C. (1938) Size frequency distributions of sediments and the normal phi curve. *J. sedim. Petrol.,* **8,** Dec., 84–90.

Krumbein W. C. (1960) The 'geological population' as a framework for analysing numerical data in geology. *Liv. Manchr Geol. J.,* **2,** 341–68.

Krumbein W. C. and Graybill F. A. (1965) *An introduction to statistical models in geology* (New York: McGraw-Hill), 475 p.

Krumbein W. C. and Rasmussen W. C. (1941) The probable error of sampling beach sand for heavy mineral analysis. *J. sedim. Petrol.,* **11,** April, 10–20.

Landy R. A. (1961) Variation in chemical composition of rock bodies: metabasalts in the Iron Springs quadrangle, South Mountain, Pennsylvania. Ph.D. dissertation, The Pennsylvania State University, University Park.

Link R. F. Koch G. S. Jr. and Schuenemeyer J. H. (1971) Statistical analysis of gold assay and other trace-element data. *Rep. Invest. U.S. Bur. Mines* 7495, 127 p.

Mann C. J. (1970) Randomness in nature. *Bull. geol. Soc. Am.,* **81,** 95–104; 3195–6.

Milner H. B. (1962) *Sedimentary petrography, 4th edn* (London: Allen and Unwin), vol. 1, 643 p.; vol. 2, 715 p.

Munro A. H. (1966) A review of the use of statistical techniques in ore valuation. In *Symposium: mathematical statistics and computer applications in ore valuation* (Johannesburg: South African Institute of Mining and Metallurgy), 3–12.

Ondrick C. W. (1968) Petrography and geochemistry of the Rensselaer Graywacke, Troy, New York. Ph.D.

thesis, The Pennsylvania State University, University Park.

Ottley D. J. (1966) Pierre Gy's sampling slide rule. *Can. Min. J.*, **87**, July, 58–62.

Otto G. H. (1938) The sedimentation unit and its use in field sampling. *J. Geol.*, **46**, 569–82.

Pearson E. S. and Hartley H. O. eds. (1954) *Biometrika tables for statisticians* (London: Cambridge University Press), 238 p.

Popper K. R. (1961) *The logic of scientific discovery* (New York: Science Editions), 479 p.

Rodionov D. A. (1965) *Distribution functions of the element and mineral contents of igneous rocks* (New York: Consultants Bureau), 80 p.

Rowland R. St. J. and Sichel H. S. (1960) Statistical quality control of routine underground sampling. *J. S. Afr. Inst. Min. Metall.*, **60**, Jan., 251–84.

Savinskii I. D. (1965) *Probability tables for locating elliptical underground masses with a rectangular grid* (New York: Consultants Bureau), 110 p.

Scheffe H. (1959) *Analysis of variance* (New York: Wiley), 476 p.

Simpson G. G. (1970) Discussion of **Mann C. J.** (1970). *Bull. geol. Soc. Am.*, **81**, 3185–6.

Singer D. A. and Wickman F. E. (1969) *Probability tables for locating elliptical targets with square, rectangular, and hexagonal point-nets* (University Park: The Pennsylvania State University), 100 p. (*Miner. Sci. Exp. Stn Spec. Publ.* 1–69)

Smalley I. J. (1970) Discussion of **Mann C. J.** (1970). *Bull. geol. Soc. Am.*, **81**, 3191–4.

Stephan F. F. and McCarthy P. J. (1963) *Sampling opinions; an analysis of survey procedure* (New York: Wiley), 451 p.

'Student' (1919) An explanation of deviations from Poisson's law in practice. *Biometrika*, **12**, 211–5.

'Student' (1931) The Lanarkshire milk experiment. *Biometrika*, **23**, 398–406.

Symposium: mathematical statistics and computer applications in ore valuation (Johannesburg: South African Institute of Mining and Metallurgy, 1966), 382 p.

The Rand Corporation (1955) *1 million random digits and 100,000 normal deviates* (Toronto: Burns and MacEachern).

Tocher K. D. (1963) *The art of simulation* (Princeton, N.J.: Van Nostrand), 184 p.

Visman J. (1969) A general sampling theory. *Mater. Res. Standards*, **9**, 8–13; 51–6; 62; 66.

Vistelius A. B. (1960) The skew frequency distributions and fundamental law of the geochemical processes. *J. Geol.*, **68**, 1–22.

von Mises R. (1957) *Probability, statistics and truth* (London: Allen and Unwin), 244 p.

Wilson A. D. (1964) The sampling of silicate rock powders for chemical analysis. *Analyst, Lond.*, **89**, 18–30.

Wood G. V. and Griffiths J. C. (1963) Modal analyses of three quartzites and their textural implications. *J. Geol.*, **71**, 405–21.

Yates F. (1949) *Sampling methods for censuses and surveys* (London: Griffin), 318 p.; *2nd edn*, 1953, 401 p.

Part III

NEW STATISTICAL METHODS

Editors' Comments
on Papers 19 Through 23

The title of this section is to some extent a misnomer for not all of the methods to be mentioned were discovered by geologists. What is new, however, is their application to geology.

The necessity for a new approach to the estimation of ore grades in South African gold mines became apparent during the work of Watermeyer (1919) and later Krige (1951, 1960). As a result, an empirical technique was developed based on the correlogram, a plot of correlation between sample values and intersample distances. It was discovered that Krige's technique yielded more reliable estimates than did classical statistics. Georges Matheron of the Mathematical Morphology Center at Fontainbleu extended the scope of the statistical theory by replacing the correlogram with the variogram relating variance with distance (Paper 19). The principal difference between the new geostatistics theory and classical statistics is that classical statistics assumes a random distribution of errors, with variance independent of distance between samples, whereas geostatistics allows variance to be a function of the intersample distance, and statistical parameters are computed by taking successive differences instead of deviation from an universal mean.

Krige and Matheron exemplify the two lines of development within geostatistics: practical and theoretical. Krige, pioneer of the practical empirical approach, believed he was operating within classical statistical methodology and that his new approach was a logical extension of this method to meet his own particular requirements of providing accurate estimates of ore grades; Matheron manipulated these techniques and built them into a new branch of theoretical statistics. However, to date there has been little objective investigation of the claims to statistical validity made by proponents of the technique. Nevertheless with publicity in languages other than French (Matheron, 1967; Blais and Carlier, 1968; Huijbregts and Matheron, 1971; Journel, 1973, 1975; Royle and Hosgit, 1974; Haas and Viallix, 1975; Olea, 1975; Guarascio, David, and Huijbregts, 1976; David, 1977), it will become accessible for analysis by the majority of the world's statistical and geological communities. The daunting mathematics required to explain the principles of geostatistics become relatively simple to master once the underlying assumptions (and their difference from those of classical statistics) are understood. Matheron's paper reprinted here was the first to clearly present the assumptions and principles of geostatistics in English and remains the benchmark exposition in the field.

Another area where classical statistics is problematic is the study of stratigraphic or lithological sequences. The difficulty in its simplest form is to produce a comparative expression of lithological characteristics within successions—or a process model for the purpose of stratigraphic correlation—or to gain geological insight into processes of sedimentation. Michael Dacey and William Krumbein of Northwestern University present one solution (Paper 20). Markov chain models, first suggested by Vistelius (1949), can be thought of either as a statistical technique using the concept of transition probability or as a simulation method (Harbaugh and Bonham-Carter, 1970). However, Dacey and Krumbein also show that the statistical development of Markov chain analysis has advanced further than the observational techniques required to test the Markovian model. Nevertheless, simple Markov chain experiments have proved useful in lithostratigraphic analysis (Krumbein and Dacey, 1969; Schwarzacher, 1975).

Since Dacey and Krumbein only briefly introduce the transition matrix that forms the core of Markov chain analysis, it is explained here in more detail:

If we have a squence such as AABBACBBCCBBAB . . ., where each letter represents a unit interval of one particular lithology, then we may obtain a set of transitions AA, AB, BB, BA, AC, and so forth. By counting all the transitions AB observed and dividing by the total number of transitions from A, the probability that A will pass upward into litho-

logy B can be estimated. A compilation of all possible transition probabilities form the transition matrix. Then by disqualifying all AA, BB,
CC transitions, the embedded Markov chain described by Dacey and
Krumbein is derived—that is, the transition matrix obtained by considering only the transitions between different lithologies.

John Davis of the Kansas Geological Survey and James Cocke of
Central Missouri State University also employ a transition probability
matrix as the basis of their solution (Paper 21). The problem studied by
Davis and Cocke is the simplification of a large number of different
lithological units in a complex cyclic sequence in order to extract useful
information. An iterative approach to classifying the transitions is
adopted. Lithologic states are grouped together if high mutual substitutability is determined—that is, their upward and downward transitions are similar (e.g., ADB and ACB are transition subsequences for C
and D; therefore C and D are mutually substitutable). Performance of
substitutability analysis, tested by autoassociation, highlights the simplification of a Kansas seventeen-state complex cyclothem to a system
with oscillation between two states and extreme fluctuations in the
autoassociation. Although the technique allows insight into the geology
of cyclic sequences, it also leaves room for subjectivity, in particular, in
the number of lithological units accepted as significant. However, as
Davis (1973) emphasizes, few papers on applications of the method
have appeared, because of the relative obscurity of the technique rather
than its potential usefulness.

Richard Howarth of Imperial College, London, applied an existing,
rigorously established statistical method—Bayesian decision theory—for
the first time to a geological problem (Paper 22). In this situation, the
objective is to classify "unknown" geological samples into one of a
number of predefined categories. Howarth adopts a flexible Bayesian
approach to produce a particularly effective form of discriminant
analysis (see also Howarth, 1971b, 1973). His technique is established
in mainstream probability theory and used in pattern-recognition
studies, and its superiority over the classical, linear or polynomial, least-
squares approach is well illustrated in the paper reprinted herein.

Finally, a method of factor analysis that combines advantages of
both R-mode (variable) and Q-mode (sample) techniques has been
developed by Benzecri (1973) under the title correspondence analysis.
It has recently attracted considerable attention in geology (Melguen,
1973, 1974; David, Campiglio, and Darling, 1974; Teil, 1975, 1976;
Teil and Cheminee, 1975) in spite of the warnings offered by Hill
(1975) and Joreskog, Klovan, and Reyment (1976) as to the application to continuous variables of a technque originally designed for
scored observations of a contingency table.

"Briefly stated, the aim of the method is to obtain simultaneously, and on equivalent scales, what we have termed R-mode factor loadings and Q-mode factor loadings that represent the same principal components of the data matrix" (Joreskog, Klovan, and Reyment, 1976). A lucid description of this process and a successful application of the stated principles combine to make the contribution by Michel David and his colleagues, C. Campiglio and R. Darling of the Ecole Polytechnique in Montreal, a benchmark in the development of a new statistical tool in geology (Paper 23).

In conclusion, at the present stage of development of the science, a number of statistical techniques are accepted by the geological community as established methods whatever their statistical credentials. Other approaches, requiring either a comprehensive mathematical knowledge or unfamiliar modes of interpretation, remain largely ignored—again, irrespective of their statistical value. Furthermore, procedures such as those advocated by Howarth or Davis and Cocke remain untouched for no apparent reason, other than perhaps a lack of sufficient publicity. The editors hope they have in part alleviated this deficiency.

19

Reprinted from *Econ. Geol.* **58**:1246–1266 (1963)

PRINCIPLES OF GEOSTATISTICS

G. MATHERON

ABSTRACT

Knowledge of ore grades and ore reserves as well as error estimation of these values, is fundamental for mining engineers and mining geologists. Until now no appropriate scientific approach to those estimation problems has existed: geostatistics, the principles of which are summarized in this paper, constitutes a new science leading to such an approach. The author criticizes classical statistical methods still in use, and shows some of the main results given by geostatistics. Any ore deposit evaluation as well as proper decision of starting mining operations should be preceded by a geostatistical investigation which may avoid economic failures.

RESUME

Pour tout mineur et géologue minier, la connaissance des teneurs et du tonnage et l'appréciation des erreurs sur ces grandeurs est fondamentale. Or, jusqu'à présent, il n'existait pas d'approche scientifique correcte de ces problèmes.

La *géostatistique*, dont les principes sont résumés dans cet article, est la nouvelle science qui permet cette approche. L'auteur indique les méthodes statistiques antérieures et encore courantes et donne quelquesuns des résultats principaux de la géostatistique.

Toute évaluation de gisement et toute décision de mise en exploitation devrait être précédée d'une étude géostatistique permettant de limiter le risque d'une déconvenue ultérieure.

INTRODUCTION AND SHORT HISTORICAL STATEMENT

GEOSTATISTICS, in their most general acceptation, are concerned with the study of the distribution in space of useful values for mining engineers and geologists, such as grade, thickness, or accumulation, including a most important practical application to the problems arising in ore-deposit evaluation.

Historically geostatistics are as old as mining itself. As soon as mining men concerned themselves with foreseeing results of future works and, in particular as soon as they started to pick and to analyze samples, and compute mean grade values, weighted by corresponding thicknesses and influence-zones, one may consider that geostatistics were born. In so far as they take into account the space characteristics of mineralization, these traditional methods still keep all their merit. Far from disproving them, modern developments of the theory have adopted them as their starting point and have brought them up to a higher level of scientific expression.

However, assuming they could provide a correct evaluation of mean values, the traditional methods failed to express in any way an important

224

character of mineralizations, which is their variability or their dispersion. Some scores of years ago, classical probability calculus techniques began to be used in order to take into account this characteristic. If an unskillful application of those techniques has sometimes led to absurdities, it remains certain that, on the whole, results have been profitable. In a way this is a paradox, for classical statistical methods, in so far as they are not concerned with the spatial aspect of the studied distributions, actually cannot be applied. As a matter of fact, the South-African school, which has recorded the most remarkable results with Krige, Sichel, used to say, and believed that they were applying classical statistics. But the methods they were developing differed more and more from classical statistics, and adjusted themselves spontaneously to their object.

The second decisive change appeared when the insufficiency of classical probability calculus was clearly understood as well as the necessity of re-introducing the spatial characters of the distributions. It consisted in realizing on a higher level the synthesis between traditional and statistical methods. Hence, geostatistics started elaborating their own methods and their own mathematical formalism, which is nothing else than an abstract formulation and a systematization of secular mining experience. This formalism has inherited from its statistical origin a language in which one still speaks of variance and covariance, including however in those notions a new content. This similarity in vocabulary must not deceive. At the end of a protracted evolution, the geostatistical theory had to admit that it was facing, instead of random occurrences, natural phenomena distributed in space. And, therefore, its methods are approximately these of mathematical physics and more specially those of harmonic analysis.

INSUFFICIENCY OF CLASSICAL STATISTICAL CONCEPTS

To be brief, we shall limit ourselves, in what follows, to the distribution of ore-grades in a deposit. The results that will be obtained will however have a general range and will be applicable to any character owned by a spatial distribution. In an usual statistical approach, the grades of samples picked in a deposit are classified on a histogram. Such a procedure does not take into account the location of samples in the deposit. But it is not enough to know the frequency of a given ore-grade in a deposit. It is also necessary to know in what way the different grades follow each other on the field, and specially what is the size and the position of economic orebodies. At the starting point of the theory we have to face one fact: the inability of common statistics to take into account the spatial aspect of the phenomenon, which is precisely its most important feature.

More precisely, the aim of the classical probability calculus is the study of aleatory variables. The mere example of the heads or tails game shows clearly what is going on. Let us record $+1$ each time the coin falls on tails and -1 in the opposite case. Before throwing the coin, there is no way of forecasting whether $+1$ or -1 will be recorded; we only know that there is one chance out of two for one or the other of these two opportunities. An aleatory variable has classically two essential properties: 1) The possibility, theorically at least, of repeating indefinitely the test that assigns to the variable a numerical value; we can for example, throw the coin as often as we want. 2) The independence of each test from the previous and the next ones; if all the 100 first attempts have given tails, there remains however one chance out of two for the 101st attempt to give heads.

It appears clearly that a given ore-grade within a deposit cannot have those two properties. The content of a block of ore is first of all unique. This block is mined only once and there is no possibility of repeating the test indefinitely. When the grade of a sample is concerned, which may be a groove sample of a given size for example, the result is exactly the same, because the grade of a groove located in a point with coordinates (x, y) is unique and well determined. However it is possible to pick a second sample close to the first, then a third one, etc. . . . which shows an apparent possibility of repeating the test. Actually, it is not exactly the same test but a slightly different one. But even assuming this possibility of repetition, the second property will surely not be respected. Two neighboring samples are certainly not independent. They tend, in average, to be both high-grade if they originate from a high-grade block of ore, and vice-versa. This tendency, more or less stressed, expresses the degree of more or less strong continuity in the variation of grades within the mineralized space.

The misunderstanding of this fact and the rough transposition of classical statistics has sometimes led to surprising misjudgments. Around the fifties, in mining exploration, it was advised to draw lots to locate each drilling (i.e., to locate them exactly anywhere). Miners of course went on still using traditional regular grid pattern sampling, and geostatistics could later prove they were right. Or else again, it was urged that the accuracy of ore evaluation of a deposit depended only on the number of samples (and not on their location) and varied as the square root of this number. This unskillful transposition of the theory of errors led to absurdities. For example, if a given deposit is explored by drilling, it would suffice to cut the cores in 5 mm pieces instead of 50 cm pieces to obtain 100 times more samples, and therefore 10 times higher accuracy. This, of course, is wrong. The multiplicity of samples thus obtained is a fallacy, and does nothing more than repeat indefinitely the same information, without yielding anything else. Geostatistics actually show that accuracy is the same with pieces of 5 mm and 50 cms, as every miner understands instinctively.

NOTION OF REGIONALIZED VARIABLE

Thus a grade cannot in any way be assimilated to an aleatory variable. We speak of regionalized variables precisely in order to stress the spatial aspect of the phenomena. A regionalized variable is, *sensu stricto*, an actual function, taking a definite value in each point of space.

In general such a function has properties too complex to be studied easily through common methods of mathematical analysis. From the point of view of physics or geology, a given number of qualitative characteristics are linked to the notion of regionalized variable.

a) In the first place, a regionalized variable is *localized*. Its variations occur in the mineralized space (volume of the deposit or of the strata), which is called *geometrical field* of the regionalization. Moreover such a variable is in general defined on a *geometrical support (holder)*. In the case of an ore-grade, this support is nothing but the volume of the sample, with its geometrical shape, its size and orientation. If, in the same deposit, the geometrical support is changed, a new regionalized variable is obtained, which shows analogies with the first one, but does not coincide with it.

For instance, samples of 10 Kg corresponding to drill cores are not distributed in the same way as samples of 10 tons corresponding to blasts. Often the case of a punctual support will be considered. A punctual grade, for example, will take value 0 or value $+1$ according to whether its support will fall into a barren or mineralized grain.

b) Secondly, the variable may show a more or less steady continuity in its spatial variation, which may be expressed through a more or less important deviation between the grades of the two neighboring samples mentioned above. Some variables with a geometrical character (thickness or dip of a geological formation) are endowed with the strict continuity of mathematicians. Fairly often (for grades or accumulations) only a more lax continuity will exist or, in other words, a continuity "in average." In some circumstances, even this "in average" continuity will not be confirmed, and then we shall speak of a *nugget effect*.

c) Lastly the variable may show different kinds of *anisotropies*. There may exist a preferential direction along which grades do not vary significantly, while they vary rapidly along a cross-direction. Those phenomena are well known under the names of runs, or zonalities.

To those general characters, common to any regionalized variable, specific features can be superimposed. For example, in the case of a sedimentary deposit, a *stratification effect*, will be noted. Large-scale stratification provides individualizable and separately minable strata. Inside each strata it may appear by the existence of beds following one another vertically, and separated by discontinuity surfaces. The grade, almost constant or barely varying inside a given bed, will vary abruptly from one bed to another; however common and familiar this phenomenon appears to be, it is still fundamental, and a theoretical formulation of the problem that would not take it into account would miss the point. It will happen as well that to those vertical discontinuities, stressed by jointing, will be added lateral discontinuities, owing to the lenticular endings of beds. This *bed-relaying phenomenon*, when it does exist, shows up at each stratigraphic level a partitioning of the sedimentation area into micro-basins with almost autonomous evolution, and may appear during operation through *grade-limit effect*.

In the same way in stockwerk types of deposits, high-grade veinlets or granules individualized in a more or less impregnated mass will be observed. This *stockwerk effect*, just as the stratification and bed relaying effects, expresses the appearance of a discontinuity net-work within a homogeneous geometrical field. On a very different scale, that of granularity, the nugget effect appears as a phenomenon of the same nature, the net-work of discontinuities being here that one separating barren from mineralized grains.

Those different specific aspects of spatial distribution of regionalized variables—far apart from classical probability calculus—must compulsorily be taken into account by geostatistics. This is made possible owing to a simple mathematical tool: the *variogram*.

THE VARIOGRAM

The variogram is a curve representing the degree of continuity of mineralization. Experimentally, one plots a distance d in abcissa and, in ordinate, the mean value of the square of the difference between the grades of samples picked at a distance d one from the other. Theoretically, let $f(M)$ be the value taken in a point M of the geometrical field V by a regionalized variable defined on a given geometrical support v (in general support v will be small and the limit may be considered as punctual). The *semi-variogram* $\gamma(h)$, or law of dispersion, is defined, for a vectorial argument h, by the expression:

$$\gamma(h) = \frac{1}{2V} \int \int \int_V [f(M+h) - f(M)]^2 dV. \tag{1}$$

In general, the variogram is an increasing function of distance h, since, in average, the farther both samples are one from the other, the more their grades are different. It gives a precise content to the traditional concept of the *influence zone* of a sample. The more or less rapid increase of the variogram represents, indeed, the more or less rapid deterioration of the influence of a given sample over more and more remote zones of the deposit. The qualitative characteristics of regionalization are very well expressed through the variogram:

a) The greater or lesser regularity of mineralization is represented by the more or less regular behavior of $\gamma(h)$, near the origin. It is possible to distinguish roughly four types (Fig. 1). In the first type the variogram has

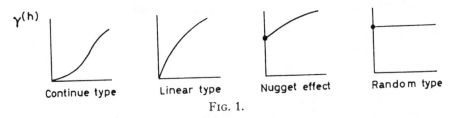

$\gamma(h)$

Continue type   Linear type   Nugget effect   Random type

Fig. 1.

a parabolic trend at the origin, and represents a regionalized variable with high continuity, such as a bed-thickness.

The second type, or linear type, is characterized by an oblique tangent at the origin, and represents a variable which has an "in average" continuity. This type is the most common for grades in metalliferous deposits.

The third type reveals a discontinuity at the origin and corresponds to a variable presenting not even an "in average" continuity, but a nugget effect.

The fourth type is a limit case corresponding to the classical notion of random variable. Between type I (continuous functional) and type 4 (purely random) appears a range of intermediates, the study of which is the proper object of geostatistics.

b) The variogram is not the same along different directions of the space. Function $\gamma(h)$ defined in (1) does not only depend upon the length, but also upon the direction of vector h. Preferential trends, runs, and shoots are revealed through the study of the distortion of variogram when this direction is altered. Geological interpretation of such anisotropies is often instructive.

c) Structural characters are also reflected in the variogram. For instance, the bed-relaying phenomenon appears in the experimental curve as a level stretch of the variogram beyond a distance, i.e., a range equal to the mean diameter of the autonomous micro-basins of sedimentation. And the

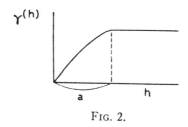

FIG. 2.

fact that these ranges are not the same along different directions makes it possible to determine the directions of elongation, and the average shape of the micro-basins.

This tool, the variogram, does not represent the totality nor the local details of the mineralizing phenomenon, but it expresses in a synthetic form their essential characters. The harmonic analysis of a vibratory phenomenon assigns for each harmonic a phase and an amplitude. The local outline of the phenomenon depends mostly upon phases, but energy depends only upon the square of amplitudes. The spectral curve giving the squares of the amplitudes does not describe the whole phenomenon but gives an account of the essential, i.e., the energetic characteristics. The variogram (or more precisely its Fourier's transformed curve) plays exactly the part of such a spectral curve.

In the following paragraphs, a few of the possible applications of variograms will be run over. It is obviously out of the question to give here a systematic study. I will merely mention some examples and several characteristic formulae. For more details I kindly ask the reader to refer to my "Treatise of Applied Geostatistics." [1]

[1] Editions *Technip,* Paris, Tome I (1962)—Tome II (Le Krigeage) (in press). Tome III (l'effet de pépite et les phénomènes de transition) to be published.

ABSOLUTE DISPERSION (OR INTRINSIC) LAW

The semi-variogram defined in (1) is bound to the geometrical field V of the regionalized variable. If, instead of the total field V, only a portion V' of it would have been considered, a function $\gamma'(h)$ possibly different from $\gamma(h)$ would have been obtained. However, we have the intuitive notion that in a geologically homogeneous geometrical field there might be something intrinsic, independent from location, in the characteristics representing the variabilities of regionalized variable.

Formulated in an accurate way, that intuition leads to the hypothesis of an absolute dispersion or intrinsic law expressed through the equation:

$$\gamma'(h) = \gamma(h)$$

which means that the variogram is independent from the portion V' of the deposit V selected for its calculation. It may be said at once, that this hypothesis is not really essential to the development of the theory and it is possible to eliminate it because of some mathematical complications.[2]

Nevertheless it makes the statement of the theory much easier, and for that reason it will be followed here. A slow deviation of the variogram in space is generally ascertained through experience and if this drift does not take too much importance, the results yielded by the hypothesis of an absolute dispersion law provide an excellent approximation of reality (on the condition that $\gamma(h)$ actually employed has been calculated from the actual portions of the considered deposit).

When this hypothesis is verified, the semivariogram $\gamma(h)$ itself acquires an intrinsic significance. It is often designated under the name of intrinsic (or absolute) *dispersion law* or, more shortly, intrinsic function of regionalized variable.

VARIANCES AND COVARIANCES

Let us consider in the first place, a regionalized variable (which will be called grade in order to simplify) defined in a field V, on a punctual support and submitted to an intrinsic dispersion law $\gamma(h)$. Let $f(M)$ be the value taken by the grade in a point M of the field V. Instead of the punctual grade $f(M)$ we usually are concerned with the grade $y(M)$ of a sample v, of a given size, shape and orientation, picked at point (M).[3] This new variable is deducted from the previous one through an integration performed within the volume v centered in M.

$$y(M) = \frac{1}{v} \int_v f(M + h)dv. \tag{2}$$

To this variable will be bound a parameter measuring its dispersion inside V, called variance, as in classical probability calculus. The mean

[2] Loc. cit., Tome III.
[3] This means that the center of gravity of v is located at point M.

value of the punctual variable inside V being m,

$$m = \frac{1}{V} \int_V f(M) dV$$

the variance of $y(M)$ inside V is defined as the average value within V of the square of the expression $[y(M) - m]$, let:

$$\sigma^2 = \frac{1}{V} \int_V [y(M) - m]^2 dv. \tag{3}$$

It will be noted that this notion has, at the outset, a geometrical and not a probabilistic meaning. It will not deter us from calculating these variances, in the applications, from experimental data with common statistical methods. Should they be taken according to their spatial order, as in integral (3), or previously rearranged in histograms, the same expressions $(y - m)$ are appearing, with the same weights in both the calculation procedures. But, on a conceptual ground, definition (3) has a physical content that the statistical motion has not. From expression (1) of the variogram, of (2), and the definition (3) of the variance, one may deduce, reversing the order of the integrations.

$$\sigma^2 = \frac{1}{V^2} \int_V dV \int_V \gamma(h) dV' - \frac{1}{v^2} \int_v dv \int_v \gamma(h) dv'. \tag{4}$$

Each one of these sextuple integrals has a very clear meaning: it represents the average value of the $\gamma(h)$ inside V (or v) when both the extremities of vector h sweep, each one for its own account, the volume V (or v).

If we write:

$$F(V) = \frac{1}{V^2} \int_V dV \int_V \gamma(h) dV,$$

i.e., $F(V) = $ average value of $\gamma(h)$ inside V.

One gets:

$$\sigma^2 = F(V) - F(v). \tag{5}$$

Thus knowledge of the variogram of punctual grades allows the "a priori" calculation of the variance of any sample v within any portion V of a deposit. It will be noted that this variance does not depend only upon the sizes of volumes v and V, but also upon their shapes and orientation.

Physical meaning of relation (4) is highly instructive. The variance of a macroscopic sample v, considered as the juxtaposition of a great number of microsamples dv, does not depend in any way on the number of those micro-samples nor on their variances, but only on the average value of intrinsic function $\gamma(h)$ inside the geometrical volume v. Classical statistics, considering these micro-samples as independent, should lead to a variance in terms of $1/v$. There does not actually exist any deposit in which 10 ton blasts would have a variance a thousand times lower than that of 10 kg

cores. Formula (4) shows why. The grades of micro-samples are not independent at all. They are inserted into a spatial correlation lattice, the
nature of which is bound to the more or less steady continuity of mineralization, and which is expressed precisely through the intrinsic dispersion law
$\gamma(h)$. The grades of the micro-samples are much less different, on the average, than classical statistics would indicate, and in consequence 10 ton blasts
have a much higher variance than the thousandth of the variance of 10
kg cores.

The expression of the variance in form (5) shows a law of additivity.
If we consider panel V' and samples v within a field V and if $\sigma^2(V', V)$,
$\sigma^2(v, V)$ and $\sigma^2(v, V')$ designate the variances of V' inside V, of v inside V
and v inside V', we get:

$$\sigma^2(v, V) = \sigma^2(v, V') + \sigma^2(V', V).$$

This formula is known as *Krige's Formula*. It has been established by
D. G. Krige in the case when the grades are distributed according to a
(statistical) lognormal law. Its validity is actually not linked to a special
statistical distribution law, but only to the existence of an intrinsic dispersion law.

Besides the variance, geostatistics introduce the notion of *covariance*.
If $y(M)$ and $z(M + h)$ are the grades of two samples v and v' centered in
two points M and $M + h$, covariance (inside V) of y and z is the function
of h defined by:

$$\sigma_{yz} = \frac{1}{V^2} \int_V [y(M) - m][z(M + h) - m]dv.$$

It can be expressed through the variogram with a relation similar to (4):

$$\sigma_{yz} = F(V) - \frac{1}{vv'} \int_v dv \int_v \gamma(k)dv'. \tag{6}$$

The second integral represents the average value of $\gamma(k)$, when both extremities of vector k sweep, respectively, volume v and volume v', at a
distance h one from the other.

Let us consider, as a particular case, the isotropic de Wijs's [4] scheme.
It is defined by an intrinsic isotropic function of the form:

$$\gamma(r) = 3\alpha \ln r \tag{7}$$

in which $r = |h|$ represents the modulus of the vectorial argument h, or
otherwise the distance between the two points M and $M + h$. When
symbol ln represents the natural logarithm, parameter α is called *absolute
dispersion*. It characterizes indeed the dispersion of grades independently
from the shape and the volume of the samples and of the deposit. In the

---

[4] The starting point of development of the present theory is the original De Wijs's reasoning
which is a remarkable example of transition from classical statistics to geostatistics. Reference
to "Traite de Geostatistique Appliquee," where bibliographical references will be found.

particular case where the volume of the samples is geometrically similar to the volume V of the deposit, formulae (4) and (7) give:

$$\sigma^2 = \alpha \ln \frac{V}{v}. \qquad (8)$$

This formula, which is the *Wijs's formula*, does express a principle of similitude. It ceases to be appliable generally as soon as the deposit is not geometrically similar to the samples. It is however possible to associate to any geometrical volume v its *linear equivalent d* devined by relation:

$$\ln d - \frac{3}{2} = \frac{1}{v^2} \int_v dv \int_v \ln r dv'. \qquad (9)$$

Formula (4) entails that sample v has the same variance in any deposit as the linear sample of length d. If D and d are the linear equivalents of the deposit and of the samples respectively, the variance may be set into the form:

$$\sigma^2 = 3\alpha \ln \frac{D}{d}.$$

The linear equivalents have been calculated and tabulated for a certain amount of geometrical figures, and, in addition, we have at our disposal some simple approximation formulae. For example for a rectangle with sides a and b we have:

$$d = a + b.$$

For a parallelogram, with sides a, b, and surface S:

$$d = \sqrt{a^2 + b^2 + 2S}.$$

For a triangle with sides a, b, c, and Surface S:

$$d = \sqrt{\frac{a^2 + b^2 + c^2}{3} + 2S}.$$

For a trapezium with basis $\quad a = \dfrac{L + l}{2},$

$$b = \frac{L - l}{2}$$

Median: m

Surface: S

$$d = \sqrt{L^2 + l^2 + m^2 - \frac{l^2 m^2}{3L^2} + 2S}$$

For a rectangular parallelepiped with sides $a > b > c$,

$$d = a + b + \frac{c}{2}.$$

For an oblique parallelepiped with edges r_1, r_2, r_3, faces S_1, S_2, S_3 and volume V, we put up:

$$\begin{cases} R^2 = r_1^2 + r_2^2 + r_3^2 \\ S^2 = S_1^2 + S_2^2 + S_3^2, \end{cases}$$

and we obtain the following approximate equivalent:

$$d = \sqrt{R^2 + 2S + \frac{V^2 R^2}{S^3}}.$$

This notion of linear equivalent allows an easy comparison between samplings of different natures, at least in the case, common in metalliferous deposits, where the law of dispersion has the form (7).

ESTIMATION VARIANCE AND EXTENSION VARIANCE

One of the most practical problems geostatistics are supposed to resolve is the size of the possible error in the evaluation of a deposit. The general characteristics of regionalized variables indicate that this error does not only depend upon the amount of picked samples, but first of all upon their shapes, their sizes and their respective locations, in other words, on the whole, upon *the geometry of achieved mining workings*. These indications get a precise meaning through the geostatistical notion of the estimation variance. Let us suppose that, in order to estimate the real unknown grade z of a deposit or of a panel V, we know the grade x of a given net-work of mining workings Mw. The estimation error $(z - x)$ has a simple, well determined value, although unknown, for a given panel V as for the net-work Mw located preferentially. In order to make out of this error a regionalized variable, geostatistics consider the panel or the deposit to be estimated as a panel extracted from a very large fictive deposit K. This deposit is supposed to be ruled by the intrinsic dispersion law $\gamma(h)$ defined by the experimental variogram controlled in mining works Mw. We shall see that the shape and the sizes assigned to K do not actually intervene. Let us imagine that panel V which is being estimated travels across the large deposit K, drawing with its attached mining works, the error $(z - x)$ then appears as a regionalized variable with an average value equal to zero and a variance:

$$\sigma^2 = \sigma_z^2 + \sigma_x^2 - 2\sigma_{zx}. \tag{10}$$

This variance called *estimation variance* is calculated after variances σ_z^2, σ_x^2 and covariance σ_{zx} of the variables z and x inside the field K, which are themselves given by formulae of type (4) or (6). Field K interferes in the expression of $\sigma_z{}^2$; $\sigma_x{}^2$ and σ_{zx} by the simple constant $F(K)$ which is eliminated in equation (10), so that the estimation variance σ^2 is independent from the choice of K, and is calculated after the formula:

$$\sigma^2 = \frac{2}{VV'} \int_V dV \int_{V'} \gamma(h) dV'$$

$$- \frac{1}{V^2} \int_V dV \int_V \gamma(h) dV' - \frac{1}{V'^2} \int_{V'} dv \int_{V'} \gamma(h) dv'. \quad (11)$$

In (11) V is the volume of the deposit being estimated and V' that of mining works Mw. The estimation variance σ^2 is calculated after integration of the intrinsic function $\gamma(h)$ inside the geometrical volumes of the deposit and of the samples. In the same way, as the variogram could give to the concept of the influence zone of a sample a precise content, one may say that the estimation variance (11) can give a precise meaning to the "influence" of mining works over the whole deposit.

In practical calculations, formula (11) should be difficult to use. Mining works usually frame a discontinuous net-work in which the samples themselves may be picked discontinuously (for instance, groove samples cut off according to a regular grid pattern in the drifts on a vein developed at different levels). Volume V' interfering in (11) is the discontinuous volume set up by the lattice of samples actually cut off and analyzed. An influence zone is traditionally assigned to each individual sample, in the center of which it is located and supposed to represent the grade. The error usually performed in extending the grade of such an individual sample to its influence zone can be represented by a type (11) variance, where V is the volume of influence zone and V' that of the sample.

Such a variance is called *elementary extension variance* and can be calculated for a given $\gamma(h)$ in terms of geometrical parameters of the sample and its influence-zone. On condition of certain approximation hypothesis, it is possible to prove that an estimation variance of type (11) can be calculated by composing the elementary extension variances.

In practice, two cases are to be distinguished essentially. The elementary samples network, for an isotropic function $\gamma(h)$ may be isotropic[5] or not. Let us mention, as an easy example of isotropic network, the square grid pattern drilling. The errors made for an isotropic network by extending to each influence zone the grade of its central sample may be considered as independent (in other words having a geostatistical covariance equal to zero). In this case *estimation variance is obtained by dividing the extension variance $\sigma_E{}^2$ of each sample within its influence zone by the number N of these influence zones.*

$$\sigma^2 = \frac{1}{n} \sigma_E{}^2. \quad (12)$$

[5] More generally, for any given function $\gamma(h)$, the lattice may or not be adjusted to the anisotropy of function $\gamma(h)$. For questions concerning the different types of anisotropy, one should refer to the *Treatise of Applied Geostatistics*.

If, on the contrary the network is not isotropic, we are led to rearrange the samples along lines or planes of maximum density, and to compose extension variances of different natures. For example let us suppose a vein-type deposit developed by drifts and channel sampled. In the first place, we have to consider the extension variance $\sigma_{E_1}^2$ of a channel within the length of a drift from which it has been cut off. If N is the total number of channels, one can see that the estimation variance $(1/N)\sigma_{E_1}^2$ represents the error obtained by extending the grade deduced from channel samples over the mining works themselves. We consider afterwards the extension variance $\sigma_{E_2}^2$ of the grade (supposed to be perfectly well known) of a drift inside its influence zone. The influence zone is here the panel composed by joining both the half-levels located above and below the drift. If n is the number of developed levels, the estimation variance $(1/n)\sigma_{E_2}^2$ represents the error obtained by extending the average grade supposed to be perfectly well known of the mining works to the whole deposit. The resulting estimation variance becomes:

$$\sigma^2 = \frac{1}{N}\,\sigma_{E_1}^2 + \frac{1}{n}\,\sigma_{E_2}^2. \tag{13}$$

It is usually necessary to add an additional variance to this expression, representing the sampling and analyses errors. The second term in such an expression is usually broadly predominating. The greater part of the error proceeds from the extension of data from the mining works to the deposit. In particular, it would be no use to increase indefinitely the number N of samples without carrying out supplementary mining works. In fact, the estimation variance coincides very soon with the $(1/n)\sigma_{E_2}^2$ limit below which it cannot decrease.

Tables and graphs giving the numerical values of elementary extension variances have been established [6] for a given number of intrinsic functions (especially for type (7) of de Wijs's function). They allow a fast computation of estimation variances assigned to different drilling and underground exploration schemes.

We offer for example a vein deposit conformable to a type (7) isotropic de Wijs's scheme and developed by drifts. Let us also assume that drifts have been sufficiently well sampled as to reduce the first term of equation (13) to zero.

Let h be the raise between two consecutive levels (measured inside the plane of the vein). The extension variance of a drift of length l within an influence panel lh is proved to be:

$$\sigma_E^2 = \alpha\,\frac{\pi}{2}\frac{h}{l}.$$

This formula is valid only if h is small compared to l, but it may be used until $h = l$. When $h > l$, it must be replaced by a different formula. Let

[6] *Treatise of Applied Geostatistics*, Vol. I, for the de Wijs's functions. Vol. III for the case of a nugget effect.

us assume that lengths $l_1, l_2 \cdots l_n$ are all superior to h. The estimation variance is obtained by weighting the extension variance of each drift within its influence panel by the square of the surface of this panel:

$$\sigma^2 = \frac{l_1^2 \sigma_{L_1}^2 + l_2^2 \sigma_{E_2}^2 + \cdots}{(l_1 + l_2 + \cdots)} = \alpha \frac{\pi}{2} h \frac{l_1 + l_2 + \cdots}{(l_1 + l_2 + \cdots)^2}.$$

The explored mineralized surface being $S = h(l_1 + l_2 \cdots + l_n)$ and the total developed length being $L = l_1 + l_2 + \cdots l_n$, we obtain the following remarkable formula:

$$\sigma^2 = \alpha \frac{\pi}{2} \frac{S}{L^2}.$$

Once the estimation variance has been calculated, one has still to interpret it for practical uses under the form of *conventional error spread*. This aim is reached by allocating to this variance a probabilistic meaning. By implicit reference to a gaussian model, we shall take it that the actual average grade of the deposit is included within a 95% probability in the range $m \pm 2\sigma$, m being the estimated grade. In other cases, particularly if 2σ is not small towards m, we shall take the spread $m \exp(\pm 2\sigma/m)$, by reference to a lognormal model.

These implicit references to probabilistic models are mainly arbitrary. Actually, the notion itself of statistical distribution of an estimation error is doubtlessly meaningless. The only thing which has an objective physical meaning is the variance. This is why we speak about conventional spreads. Their practical interest resides in the fact that they draw a more intuitive picture of the possible errors than variances themselves.

KRIGING

A second application of major importance is provided by a geostatistical procedure called "kriging." It consists in estimating the grade of a panel by computing the weighted average of available samples, some being located inside others outside the panel. The grads of these samples being x_1, $x_2, \cdots x_n$, we attempt to evaluate the unknown grade z of the panel with a linear estimator z^* of the form:

$$z^* = \sum a_i x_i.$$

The suitable weights a_i assigned to each sample are determined by two conditions. The first one expresses that z^* and z must have the same average value within the whole large field V and is written as:

$$\sum a_i = 1.$$

The second condition expresses that the a_i have such values that estimation variance of z by z^*, in other words the kriging variance, should take the smallest possible value.

This is formulated with a linear equation system related to a_i, the coefficients of which are expressed with the help of the variances and covariances of the samples and of the panel. It is thus possible to tabulate, for each intrinsic function, the coefficients and the kriging variance in terms of geometrical parameters, appropriately for different configurations. Numerous drilling and underground work configurations have thus been tabulated in

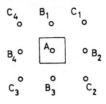

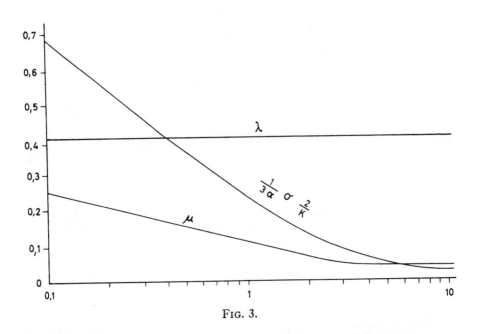

FIG. 3.

the case of an isotropic scheme of de Wijs. For information we show an example in Figure 3. The studied configuration is useful for the appraisal of a deposit explored by drilling, or for open cast selective mining. It consists, in the case of a square grid pattern drilling, in the kriging of the influence blocks of a drilling A with help of the grade of this central drilling A, and those of the 8 nearest drillings rearranged into two "aureolae" $B_1B_2B_3B_4$ and $C_1C_2C_3C_4$. Let u be the grade of A, v and w the average grades of drillings

B and C, the estimator to be used is:

$$z^* = (1 - \lambda - \mu)u + \lambda v + \mu w.$$

In Figure 3, is plotted in abscissa the ratio h/a between the width h of the formation and the size a of the mesh of the drilling grid and the numerical values of λ and μ are read on the curves as well as these of the expression $(1/3a)\sigma_K^2$. The multiplication of this last expression by three times the value of absolute dispersion, $3a$, yields the kriging variance.

Theoretically it is advantageous to "krige" each panel by all the samples located in the deposit, inside and outside this panel. In addition to the great complexity of computation which grows very fast, it appears in numerical examples that it is usually unnecessary to take into account remote samples. In general the one or two proximate aureolae of external samples are enough to remove practically the whole effect of remaining external samples. This is, in particular, the case of the configuration studied in Figure 3 where both aureole B and C form an almost perfect screen towards all other external drillings.

One can even notice that for high values of the h/a ratio, the weight μ of the second aureole becomes slight, so that the aureole made out of the four B drillings constitutes a screen by itself alone. This *screen effect* is a general phenomenon and plays an important part in the kriging theory.

From a practical point of view, the advantage of kriging is double. First of all, as a result of the definition itself of this procedure, it leads to achieve the best possible estimation for a given panel, that is to say the estimation with minimal variance. It can pay most appreciable services by improving, for example, the monthly output forecast for different mine-sections, and especially in the case where the mine operator is compelled to supply ores with characteristics as constant as possible.

However appreciable they are, the improvements of accuracy provided by the kriging would not always justify the amount of calculations it requires. In most cases, the major interest of the procedure does not come for the reduction of estimation variances but from its being able to eliminate the cause of systematical error. A deposit seldom happens indeed to be payable in the whole. Only some panels chosen as payable according to the grades of the samples cut off within them, are considered as payable. D. G. Krige [7] has proved that the results based only on inside samples inevitably led to over-estimating rich panels and underestimating poor ones. The geostatistical notion of kriging allows to expound this phenomenon easily and to rectify its effects. The selected panel being a rich one, the aureola of outside samples has, in general, a lower grade than that of inside samples. Yet its influence on the panel to be estimated, is not negligible, since it is allocated a weight different from zero by the kriging. Not to take in account this external aureole inevitably introduce therefore a *cause of systematical error by over-estimation* which can be eliminated by kriging.

---

[7] D. G. Krige's original reasoning constitutes a second example of an implicit passage from classical statistics to geostatistics. It is essentially based on the fact that the variance of a panel is always lower than that of its inside sampling. For references, see "Treatise of Applied Statistics."

THE NUGGET EFFECT

In the presence of a strong nugget effect the general rules outlined in the above paragraphs may suffer some apparent objections. The nugget effect has been defined in Figure 1 by a variogram characterized by a discontinuity at the origin, and corresponding to a regionalized variable that does not have the "in average" continuity. Its nature may be purely granulo-metrical, as in gold or diamond deposits, or, more generally, it may reveal the existence of discontinuous micro-structures. The presence of veinlets or microfractures with high-grade fillings in a stockwerk may promote such an effect. In gold deposits, the grades of two very close or even adjoining samples may be different if, by chance, one of them contains a large nugget. The smaller the samples, the more important this effect is, and it may reach a considerable magnitude for samples of several liters in volume. A translation of some millimeters only of the geometrical support of a sample is enough for it to contain or not a large nugget able to modify its grade in a proportion of 1 to 10 or 1 to 100. The possibility for a marginal nugget to be embodied, or not, inside a sample appears as an entirely random event. Actually, however, the behavior of the grade can be considered as random locally only. If it were not so, the panels of several thousand tons, on which marginal nuggets have no more detectable effect, would present almost constant grades (their variance being then a million times lower than that of samples of several kilograms). It is well known that actually, even in gold deposits, there are rich panels and poor panels. But this random effect may locally be so strong that it entirely hides the underlaying regionalization. The frequency of some expressions such as "erratic," "monstrous," or "mammoth grades" etc. . . . alluding to an hypothetical anomalous behavior of mineralization in the literature devoted to these deposits is striking. Certainly the classical statisticians were right when they noted that there was no actual anomaly, and that those monster grades, actually existing in the deposit, appeared from time to time in the sampling, with frequency determined by random laws. Historically, a clear distinction between the notions of regionalized and aleatory variables was doubtlessly hampered for a long while by the fascination aroused by this nugget effect. It appears, from the geostatistical point of view, that in fact, the ingenious terminology was not wrong while suggesting the existence of some anomaly; but the aberrant fact is not the presence of some "anomalously" high grades, but rather the locally aleatory behavior of all the grades, high or low, as well as in the deterioration of the spatial correlations grid. Those mammoth grades of the ingenious terminology are not aberrant by themselves, but the fact that they are not assorted with influence zones is so. And, on the other hand, classical statisticians were right stressing the fact that the apparitions of these aberrant grades are ruled by random laws. But they failed to note that the phenomenon can be considered as aleatory locally only.

Without trying to make a systematical statement,[8] let us show briefly how geostatistics allow us to represent a nugget effect. Let us examine the

[8] See *Treatise of Applied Geostatistics*, Vol. III.

$\gamma(r)$ semi-variogram representing the third type of Figure 1. We shall stick here to the case where $\gamma(r)$ is an isotropic function (in other words depending only upon the r modulus of the h vectorial argument). The C discontinuity, or jump, noticed at the origin on the $\gamma(r)$ of a variable with punctual support, is called *nugget constant*. $H(r)$ being Heaviside's function, thus defined:

$$\begin{cases} H(r) = 1 & r > 0 \\ H(r) = 0 & r = 0 \end{cases}$$

the semi-variogram may be divided into 2 components:

$$\gamma(r) = CH(r) + \gamma_1(r). \tag{15}$$

The first component $CH(r)$ represents the pure nugget effect. The second one $\gamma_1(r)$, continuous at the origin, represents the underlaying regionalization. All the variances and the covariances that have to be introduced, may then be calculated as if the variable $x(M)$ with punctual support was the sum:

$$x = x_0 + \epsilon \tag{16}$$

of a theoretical regionalized variable x_0 following the $\gamma(r)$ dispersion law continuous at the origin, and of an aleatory ϵ variable with a zero average and C variance.

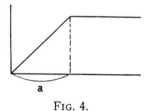

FIG. 4.

The x_0 and the ϵ are independent, and the ϵ assigned to two distinct points even very close, are independent as well. If we limit our study to the variation of the punctual grade x in the proximity of a given point, or, in other words, we consider only the small values of the distance r, $\gamma_1(r)$ will vary so slightly that it might be taken for a constant equal to C. The locally detectable variations are to be assigned almost solely to ϵ. That is what we mean when we say that the regionalized variable behaves locally as an aleatory variable. But, on a larger scale, i.e., for higher values of r, the increase of the continuous component $\gamma_1(r)$ can no longer be neglected and the regionalization of x_0 becomes perceptibly apparent.

As a matter of fact, the Heaviside function does not represent with entire satisfaction the random aspect of the behavior of a punctual variable. Unless we suppose the constant C to be infinite, the term $CH(r)$ will lose all influence over the variance of a sample of a size different from zero. It is automatically eliminated in formula (4). It means that the mean value of the ϵ independent aleatory variables, located in infinite number inside an unpunctual support, has compulsorily a zero variance.

The notion of a random variable ϵ with a punctual support has actually no physical meaning. The actual physical phenomenon will never involve a true discontinuity at the origin but a narrow transition zone in the proximity of $r = 0$. $II(r)$ must be replaced by the transition function $T(r, a)$ defined by:

$$\begin{cases} T(r, a) = \dfrac{r}{a} & \text{if} \quad r \leq a \\ T(r, a) = 1 & \text{if} \quad r > a. \end{cases}$$

The a constant, or *range*, gives the scale of the transition zone, that is to say the size of the nuggets. In the case of homogranular nuggets of same volume u it is shown that:

$$u = \frac{\pi}{3} a^3.$$

The intrinsic function $\gamma(r)$ of a punctual grade is decomposed in the following way:

$$\gamma(r) = CT(r, a) + \gamma_1(r).$$

C is still the nugget constant, and $\gamma_1(r)$ the continuous component.

The punctual grade x can be given by a sum similar to (16) in which ϵ is a regionalized variable admitting $CT(r, a)$ as its intrinsic function. Now the ϵ are only independent for distances superior to the range a. For smaller distances they are bound by a linear variogram. The nugget effect will therefore reflect itself on samples of size v different from zero. If v is large in regard to the grain size a^3 the transition zone will be diluted in the integration volume v, and the nugget effect will yield an additional variance of the type a^3/v. Indeed, let σ_P^2 (nugget variance) be the share of $CT(r, a)$ for the variance of sample v. According to (4) we have to compute integrals of the type:

$$\frac{C}{v^2} \int\int\int_v dv_1 \int\int\int_v T(r, a) dv_2.$$

If all sizes of v are supposed to be large in regard to a, each point inside v brings to the sextuple integral the following part:

$$C(v - \tfrac{4}{3}\pi a^3) + \frac{C}{a} \int_0^a 4\pi r^3 dr = C\left(v - \frac{\pi}{3} a^3\right).$$

This is valid only for points located at a distance superior to (a) from the boundary of v; but, when v is large, the boundary points only interfere with superior order terms. With such an approximation, the sextuple integral is equal to $C(1 - (\pi/3)(a^3/v))$.

As the integral inside V is computed in the same way, we finally have

$$\sigma_P^2 = C\frac{\pi}{3}\left[\frac{a^3}{v} - \frac{a^3}{V}\right]. \tag{17}$$

Practically a^3/V is negligible and the nugget variance is in terms of a^3/v, i.e., in an inverse ratio to the number of grains contained inside the sample.

Any time a nugget effect does exist, i.e., anytime a regionalized variable shows a locally aleatory behavior, an additional variance is assigned to macroscopic samples, called nugget variance, inversely proportional to their size.

The variance of those samples appears as the sum:

$$\sigma^2 = \sigma_P{}^2 + \sigma_\theta{}^2$$

of the nugget variance and of the theoretical variance $\sigma_\theta{}^2$ calculated with the continuous component $\gamma_1(r)$ of the intrinsic function.

When v is increasing, the theoretical variance is decreasing much slower than the nugget variance. In the presence of a very strong nugget effect $\sigma_p{}^2$ may happen to be widely predominating for samples of several kilograms. The underlying regionalization is almost completely hidden at the scale of these samples. If we limit the variation of the volume v in the interval of a few liters up to tens of liters, the experimentally observable variations of the variance will be those of the nugget variance effect only, and we may take the risk to conclude that the variance varies in inverse ratio of the volume.

Whereas if we consider samples of several tens of tons, the term $\sigma_p{}^2$ decreases and disappears, and the theoretical variance $\sigma_\theta{}^2$ becomes prominent. The effect of the underlying regionalization appears again and the variance is steadily decreasing as v is increasing, but much slower than $1/v$.

We have somewhat insisted upon the nugget effect in order to show, through a crucial example, how geostatistical concepts allow us to rediscover the local results that are fluently obtained from common statistical reasoning (nugget variance inversely proportional to volume) but inserting them in the general prospect of an underlying regionalization. As for the practical use of this theory, let us succcinctly mention the two following points:

In the presence of a nugget effect, the extension and the estimation variances are both increased by a term $C(\pi/3)(a^3/v)$ inversely proportional to the total volume of available samples and, therefore, in particular to the number n of those samples. In this regard, the additional estimation variance due to the nugget effect behaves itself as the sampling and analyses variances, and may be rearranged with them.

As for the kriging, the nugget effect results in partly removing all the screens. Practically, we are led to use the special forms of kriging called "aleatory kriging" which are not different from those proposed formerly by D. G. Krige himself, in connection with the gold deposit of the Rand, in which the nugget effect is probably very strong.

SEARCH FOR OPTIMUM IN MINING EXPLORATION

Geostatistics are able, through estimation variances, to provide an accurate measurement of the information yielded, by a given amount of underground workings on a deposit. Generally, these workings are expensive, and their cost must be weighed against the economic value of the provided information. Thus appears the possibility to determine the optimum amount of credits to be allocated for the exploration of a deposit, and particularly the possibility to choose the suitable moment for stopping the exploration, as well as for taking a positive or negative decision towards starting the exploitation of the deposit. These methods, permit one to solve, at least partly, one of the main problems raised by mining exploration, will be published in another connection, and cannot be treated here. Let us only, as a conclusion, stress the fact that they appear as the natural extension of geostatistics. The possibility of their adjustment was bound to the preliminary elucidation and to the thorough scientific study of the different ideas which have been summarized in this paper.

20

Reprinted from *Math. Geol.* 2(2):175–191 (1970)

Markovian Models in Stratigraphic Analysis[1]

Michael F. Dacey[2] and W. C. Krumbein[2]

A stratigraphic section may be divided into lithologic units which in turn may be divided into beds. This paper gives a mathematical formulation of stratigraphic sections that takes these two levels into account and uses bed properties to yield the thickness and number of beds in lithologic units. The model is a semi-Markov chain in which the succession of lithologic bed types forms a Markov chain and is an independent random variable. The model is tested against stratigraphic data obtained from micrologs. There is close agreement between the observed and calculated thicknesses of lithologic units. Tests for the degree of agreement between observed and calculated numbers of beds in lithologic units are hampered by inability to observe thin beds on micrologs. Some implications of this limitation to stratigraphic analysis are noted.

INTRODUCTION

Although Markov models have been applied to a large variety of geologic phenomena, their most widespread use has been in the study of stratigraphic sequences that show evidence of cyclical or recurrent events. A stratigraphic section composed of lithologic units of sandstone, shale, and limestone commonly shows repetitions of particular sequences of these lithologic types. Thus, if we let sandstone represent state A, shale state B, and limestone C, a typical stratigraphic section observed upward from the bottom (to maintain a sense of relative time) could show a sequence ABABCBCAB . . . , if each successive lithologic unit is recorded without regard for its thickness.

Thickness can be taken into account in either of two ways. In the first, the thickness of each lithologic unit is measured and recorded separately for each state. The alternative is to select some uniform vertical interval (which may range from 6 in. to 10 ft depending on the nature of the section), and to "probe" the lithologic section successively upward from some initial point t_0. In this method the same lithology may be encountered in successive probes, depending upon the thickness of each lithologic unit. Thus, if the vertical interval is 2 ft, the translation into thickness of the sequence AAAABBABBBBBCBBBBAABBB . . . would be a sandstone unit 8 ft thick, followed by a 4-ft unit of shale, followed in turn by a 2-ft unit of sandstone, a 10-ft unit of shale, and so on.

If the succession of lithologies is described by a transition matrix that gives the probability that lithologic state i is followed by lithologic state j, the two methods of

---

[1] Manuscript received 26 January 1970.
[2] Northwestern University (USA).

structuring data result in different properties for the matrix of transition probabilities. As shown by Krumbein and Dacey (1969), the method structured on equal-intercept observation leads to a matrix in which the diagonal elements are greater than zero, whereas the other leads to a matrix in which the diagonal elements are identical to zero. If the succession of lithologies displays a Markov property, we have referred to these two types of models as, respectively, a Markov chain and an embedded Markov chain.

It was pointed out that successively higher levels of structure in the Markov models and in the observational data become important as our insight into geologic phenomena increases. In early exploratory stages of using Markov models, two informal tests of the model were to see whether simulations "looked like" natural sections and whether the simulated proportions of each lithology agreed with those seen in nature. The next level makes use of the fact that if vertically spaced probes of a stratigraphic section are described by a Markov chain the number of successive occurrences of a lithologic type within a lithologic unit has the geometric distribution, whereas for a section described by an embedded Markov chain there are, of course, no successive occurrences of the same lithologic unit. The next higher level of structure in stratigraphic sections, considered in this paper, concerns the internal makeup of the lithologic units themselves in terms of the thickness distributions of the beds that make up the unit.

LITHOLOGIC UNITS AND SEDIMENTARY BEDS

The concepts of lithologic unit and sedimentary bed, as we shall use the terms here, require some clarification. Both the lithologic unit and the bed are elements of a stratigraphic section that are homogeneous in some geologic sense (usually in terms of texture, composition, or internal structure), at different scales of observation. Lithologic units are homogeneous at the coarsest lithologic scale of observation, the scale at which a stratigraphic section is partitioned into sandstone, shale, or some other broadly defined lithologic type. Most lithologic units are measured on a thickness scale of feet or meters.

A bed (or stratum) is a subunit within a lithologic unit, homogeneous at a scale of thickness measured in inches or centimeters, although thicker massive beds also occur. The beds themselves may be composed of even thinner units called laminae, usually measured in fractions of an inch or in millimeters; various authors, in fact, limit laminae to layers a centimeter or less in thickness. If thin sections are cut through laminated rocks normal to the stratification, even thinner microscopic layering may be visible.

There is thus a hierarchy of layering in most sedimentary rocks, each having homogeneity at its scale of observation. Beds may occur without laminae, however, and the laminae are not always arranged into beds. In practice one can perhaps think in terms of at least three categories of layer thickness, but in this paper we shall limit ourselves to two major categories, the lithologic unit and its first scale of subdivision. Normally this will be bedding, though for some rocks the next finer scale may be in

the thickness range of laminae. Our operational definition for distinguishing a lithologic unit from its next finer subunit is based on textural, compositional, or structural
homogeneity, rather than on absolute thickness. Thus, if one or more massive "beds"
of sandstone, say, occur in a body of shale, we are inclined to refer to these as lithologic units of sandstone in their own right, rather than as minor parts of a shale lithologic unit. This choice can of course lead to difficulties as we move down the thickness
scale, but the problem of making a decision is a real one in terms of some of the alternative ways of measuring stratigraphic sections.

In conventional practice a stratigraphic section is divided into lithologic units
by identifying the upper and lower contacts of differing rocks types and then measuring
the thickness between contacts. A different procedure, which probably will become
more common as instrumental devices (including remote sensors) are developed, is a
"probing" of the section at fixed intervals normal to the stratification. These probes,
which are essentially observations at a point or over a limited area of rock face, yield
a discretized version of the stratigraphic section. Other devices, such as borehole
sondes, produce a continuous analog record of the rocks penetrated by the drill,
although most of those readily available (electric logs) record a combination of rock
properties plus the fluids contained in the rocks. Discretized versions of the sonde
data also are available, however.

Our major interest in the bedding problem is in structuring data for Markov
models. We already have examined the situation of equally spaced probes of a stratigraphic section to obtain a Markov chain at the scale level of the lithologic unit
(Krumbein and Dacey, 1969), and in this paper we extend the scope of our study to
examine two levels of homogeneity in a stratigraphic section—the lithologic unit and
its major subdivision, the bed. Two levels are used because it is possible to construct a
mathematical formulation of properties of beds that yields interesting properties of
lithologic units. In order to reach this next stage of analysis, workable operational
definitions for measuring beds are required, and some discussion of this problem is
given in a later section.

For our present purposes, it is accepted that lithologic unit and bed are useful
objects of geologic investigation. The objective of this paper is to show how the
concept of bed may be integrated into a mathematical structure that yields properties
of lithologic units. This mathematical structure combines the number of beds comprising a lithologic unit with the thickness of these beds to obtain statements concerning lithologic unit thickness. Whereas the relationship between number of beds and
thickness of beds may be modeled in various manners, the model considered in this
paper is a formulation that takes into account only the succession of individual beds
and the thicknesses of these beds. The beds are classified according to their lithology,
and the sequence of beds forms a Markov chain. Furthermore, the thicknesses of beds
are treated as independent random variables. A mathematical model that incorporates
the succession of lithologic bed types and bed thicknesses is called a semiMarkov
chain. The mathematical properties of this type of Markov chain are identified before
considering applications to stratigraphic analysis.

The geologic interpretation attached to parameters of this model is critical, because varying the geologic meaning of elements of the semiMarkov chain yields markedly different formulations for lithologic unit thickness. This paper identifies several ways of using semiMarkov chains for the analysis of lithologic structure, and the differences in these formulations may suggest factors that need to be taken into account when constructing operational definitions for lithologic unit and bed.

SEMIMARKOV CHAINS

A semiMarkov chain combines a Markov chain with transition matrix P and starting vector Π with random variables W_{ij}, defined for every nonzero element p_{ij} for the transition matrix, that represent the waiting time in state i that is followed by state j. The P and Π may be any transition matrix and starting vector that define a Markov chain. The random variable W_{ij} may be discrete or continuous, but to simplify analysis it is commonly assumed to be positive and finite valued with probability 1. The distribution function of W_{ij} is denoted by Φ_{ij}. When W_i depends only upon the initial state i, it is appropriate to consider the waiting time W_i with distribution function Φ_i.

In the following formulations waiting time W_i is treated as a discrete random variable that depends only upon the initial state. So,

$$P\{W_i \leqslant k\} = \Phi_i(k)$$

and

$$P\{W_i = k\} = \Phi_i(k)$$

The geologic interpretation of waiting time will depend upon the geologic interpretation of states of the transition matrix. Two examples are given.

First, suppose each state is a type of lithology (such as sandstone, shale, or limestone), and p_{ij} is the probability of transition from a lithology of the ith type to a lithology of the jth type. If all $p_{ii} = 0$ so that each transition is to a different type of lithology, a geologic interpretation of the semiMarkov-chain model is that the waiting time W_i corresponds to thickness of a lithologic unit of the ith type. Alternatively, suppose all elements of P are positive so that a transition may be to the same lithology. A geologic interpretation of this example may be that the lithology is observed at points $t_0, t_1, \ldots, t_n, \ldots$. Instead of specifying equal spacing between these points as in Krumbein and Dacey (1969), suppose that the locations of these points are determined by a sampling design so that the distance between successive points t_n and t_{n+1} is the realization of a random variable that depends upon the lithology observed at t_n. For this model, the random variable W_i may represent distance between successive points if the ith lithology is observed at the earlier (or lower) point, but, in general, W_i can not be used to represent thickness of a lithologic unit.

Second, suppose each state is a bed of lithologic type i, and p_{ij} is the probability of transition from a bed of the ith type of lithology to a bed of the jth type of lithology. The random variable W_i may be interpreted as representing the thickness of a bed of the ith type of lithology. In this model the thickness of a lithologic unit is equal to the sum of thicknesses of the beds comprising the unit. Accordingly, W_i represents the thickness of a lithologic unit only if a unit has one bed with probability 1, i.e., $p_{ii} = 0$.

This paper examines several methods of using semiMarkov chains in stratigraphic analysis. It provides a transition between our earlier paper, that identifies properties of two types of regular Markov chains that seem particularly suited to stratigraphic analysis, and a forthcoming paper that considers geologic applications of a broadly defined class of semiMarkov chains. Because the present paper is transitional, it uses only a few basic properties of semiMarkov chains. Though the forthcoming paper gives a more complete description of semiMarkov chains, the relevant mathematical literature is noted now. SemiMarkov chains were introduced simultaneously by Lévy (1954) and Smith (1955), and the latter paper provides a particularly clear introduction to the subject. Many properties of semiMarkov chains are identified by Pyke (1961a, 1961b), and Cinlar (1970) provides a rather complete and mathematically sophisticated account of their properties.

DEFINITIONS AND NOTATION

The terminology used in the remainder of this paper corresponds to that of Krumbein and Dacey (1969) and is briefly reviewed here.

Let $P = [p_{ij}]$ and $\Pi = [\pi_i]$ denote the transition matrix and starting distribution vector of a Markov chain. This chain has the five following properties:

(A) The number of states J is finite and, in practice, seldom larger than 5 or 6.
(B) There is no absorbing state so that $p_{ij} < 1$ for all i and j.
(C) All $p_{ii} > 0$. This convention is not necessary. It is used in order to avoid notation problems in expressions, such as eq (1), where probabilities are divided by p_{ii} only when $p_{ii} > 0$.
(D) The Markov chain is regular in the sense that there is a positive integer N such that P^N has no zero entries, which implies that each state may be reached from every other state in N transitions.
(E) The lithologic state X_0 at the initial observation point t_0 is specified so that $\pi_i = 0$ or $\pi_i = 1$. In practice, $X_0 = i$ implies $\pi_i = 1$ and all other starting probabilities are 0.

In stratigraphic analysis a Markov chain with these properties may be used for the study of lithologic sequences where the observed data give the lithologic type occurring at discrete, equally spaced points $t_0, t_1, \ldots, t_n, \ldots$.

Let $R = [r_{ij}]$ and $\Pi = [\pi_i]$ denote the transition matrix and starting distribution vector of an embedded Markov chain in which the diagonal elements are exactly zero. R is obtained from P by putting

$$r_{ii} = 0$$
$$r_{ij} = p_{ij}/(1 - p_{ii}), \quad i \neq j \tag{1}$$

This embedded Markov chain has the properties (A), (B), (D), and (E) listed above for all Markov chains, whereas the property corresponding to (C) is

(C') All $r_{ii} = 0$.

In stratigraphic analysis an embedded Markov chain with these properties is used in the study of lithologic successions when the observed data are arranged into a transition matrix by noting only the sequence of lithologic types. In addition, a separate record may be kept of the thickness of each type of lithologic unit.

MARKOV CHAIN WITH TRANSITION MATRIX P

Lithologic unit thickness was obtained in our previous paper for a Markov-chain formulation of lithologic succession. The results are summarized here and provide a basis of comparison with lithologic unit thickness if lithologic structure is modelled by a semiMarkov chain.

The succession of lithologies at equally spaced points $t_0, t_1, \ldots, t_n, \ldots$ is described by a Markov chain with transition matrix P, with all $p_{ii} > 0$, and starting vector Π. The random variable T_i represents the number of successive points having the ith lithologic type, and T_i has the geometric distribution with parameter $(1 - p_{ii})$ or

$$P\{T_i = k\} = (p_{ii})^{k-1}(1 - p_{ii}), \quad k = 1, 2, \ldots \tag{2}$$

Let L_i represent the thickness of a lithologic unit of the ith type. If the interval between successive points t_n and t_{n+1} is h units of distance, it is assumed that lithologic thickness is

$$L_i = hT_i$$

So,

$$P\{L_i = hk\} = (p_{ii})^{k-1}(1 - p_{ii}) \tag{3}$$

The generating function $E\,t^{L_i}$ of L_i is required for subsequent use. To obtain this generating function observe that if X has the geometric distribution with parameter θ, then the generating function of X is

$$f(t) = E\,t^X = \sum_{k=1}^{\infty} t^k P\{X = k\} \tag{4}$$

$$= \theta t \sum_{k=0}^{\infty} [t(1-\theta)]^k$$

$$= \theta t / [1 - (1-\theta)t]$$

This is a common form of the generating function of the geometric distribution, with parameter θ, which has the probability mass of X distributed on the positive integers. Put $Y = hX$, and the generating function Et^Y is

$$g(t) = \sum_{k=1}^{\infty} t^{hk} P\{Y = hk\} = \sum_{k=1}^{\infty} t^{hk} P\{X = k\} = f(t^h) = t^h / [1 - (1-\theta)t^h] \tag{5}$$

This is the generating function of the geometric distribution, with parameter θ, which has the probability mass of Y distributed on the lattice points $\{hk\}$ where k is a positive integer.

Comparing (3) with (5) establishes that L_i has the generating function $g(t)$ with θ

replaced by $(1 - p_{ii})$, which implies that L_i has the geometric distribution on the lattice points $\{hk\}$. In contrast, T_i has the geometric distribution on the positive integers.

SEMIMARKOV CHAIN WITH TRANSITION MATRIX R

This semiMarkov chain has transition matrix R, with all $r_{ii} = 0$, starting distribution vector Π, and associated with each element in the ith row of R is the waiting time W_i. These random variables are mutually independent and W_i is defined by the probability mass function ϕ_i.

The states of this semiMarkov chain are lithologic types so that r_{ij} is the probability that a lithologic unit of the ith type is followed by one of the jth lithologic type. Because $r_{ii} = 0$, so that successive occurrences of a lithologic type are precluded, the waiting time W_i associated with the ith state is interpreted as the thickness of the ith lithologic type. Let L_i represent the thickness of a lithologic unit of the ith type. Because $r_{ii} = 0$, successive occurrences of a lithologic type are precluded so that each lithologic unit consists of a single lithologic state. Because this lithologic state has thickness W_i, it follows that $L_i = W_i$.

Whereas this type of semiMarkov chain may describe lithologic unit thickness, it contributes little additional information for the geologic analysis of stratigraphic processes. The model could be considered as trivial in the sense that the geologically interesting results are the succession of lithologic units and the thickness of lithologic units, but these correspond to the transition matrix R and the distribution ϕ_i of waiting times that are used initially to construct the semiMarkov chain.

The semiMarkov chain with transition matrix R and waiting times W_i, where each W_i has a lognormal distribution, was used by Potter and Blakely (1968) to structure a simulation procedure for studying lithologic transitions in stratigraphic analysis. The results of the simulation are that, except for variations attributable to sampling, the simulated stratigraphic sections have lithologic transitions described by R and lithologic unit thickness described by lognormal distributions.

SEMIMARKOV CHAIN WITH TRANSITION MATRIX P

This semiMarkov chain has transition matrix P, with all $p_{ii} > 0$, starting distribution vector Π, and the waiting time W_i is associated with each element in the ith row of P. These random variables are mutually independent and W_i is defined by the probability mass function ϕ_i.

The states of this semiMarkov chain are the beds within lithologic units so that p_{ij} is the probability that a bed of the ith type of lithology is followed by a bed of the jth type of lithology. The waiting time W_i associated with the ith state is interpreted as the thickness of a bed of the ith lithologic type. A lithologic unit consists of consecutive beds of the same lithologic type, and the thickness of a unit is the sum of thicknesses of these beds. Both bed thickness W_i and the number of beds in a lithologic unit are random variables. Let $N(i)$ represent the number of beds comprising a lithologic unit of the ith type, and let the random variable $L_{N(i)}$ represent the thickness of the ith kind of lithologic unit that has $N(i)$ beds. If $W_{iN(n)}$ represents the thickness of the nth bed in a sequence of beds forming a lithologic unit of the ith type, then the thickness of the lithologic unit is

$$L_{N(i)} = W_{i1} + W_{i2} + \cdots + W_{iN(i)}$$

where the W_{in} are independent random variables identically distributed as the random variable W_i with mass function ϕ_i.

This semiMarkov chain now is analyzed in order to derive the probability distribution of the thickness $L_{N(k)}$ of a lithologic unit of the ith type of lithology. Because equivalent events have the same probability of occurrence,

$$P\{L_{n(i)} = k\} = P\{W_{i1} + W_{i2} + \cdots + W_{iN(i)} = k\}$$

The conditional probability distribution of $L_{N(i)}$ given $N(i)$ is

$$P\{L_{N(i)} = k \mid N(i) = n\} = P\{W_{i1} + W_{i2} + \cdots + W_{in} = k$$

From the fundamental formula for conditional probabilities (Feller, 1968, p. 114), it follows that the probability distribution of $L_{N(i)}$ is

$$P\{L_{N(i)} = k\} = \sum_{n=1}^{\infty} P\{L_{N(i)} = k \mid N(i) = n\} P\{N(i) = n\}$$
$$= \sum_{n=1}^{\infty} P\{W_{i1} + W_{i2} + \cdots + W_{in} = k\} P\{N(i) = n\}$$

So, the generating function of $L_{N(i)}$ is

$$h_i(t) = E \, t^{L_{N(i)}} = \sum_{k=1}^{\infty} t^k P\{L_{N(i)} = k\}$$
$$= \sum_{k=1}^{\infty} \sum_{n=1}^{\infty} t^k P\{W_{i1} + W_{i2} + \cdots + W_{in} = k\} P\{N(i) = n\}$$
$$= \sum_{n=1}^{\infty} \left(\sum_{k=1}^{\infty} t^k P\{W_{i1} + W_{i2} + \cdots + W_{in} = k\} \right) P\{N(i) = n\}$$
$$= \sum_{n=1}^{\infty} g_i(t, n) P\{N(i) = n\} \tag{6}$$

where $g_i(t, n)$ is the generating function of $W_{i1} + W_{i2} + \ldots + W_{in}$. It remains to identify the generating function of this sum and to identify the probability distribution of $N(i)$.

The generating function of the sum is obtained first. Let $g_i(t) = E t^{W_i}$ be the generating function of W_i. Because the W_{in} are independent random variables, the generating function $g_i(t, n)$ of the sum $W_{i1} + W_{i2} + \ldots + W_{in}$ is equal to the product of the generating functions of $W_{i1}, W_{i2}, \ldots, W_{in}$. The W_{in} are identically distributed as the random variable W_i with generating function $g_i(t)$ so that

$$g_i(t, n) = g_i(t)^n, \quad n = 1, 2, \ldots \tag{7}$$

The probability distribution of $N(i)$ is obtained next. This random variable represents the number of beds in a lithologic unit of the ith type of lithology. The sequence of beds forms a Markov chain with transition matrix P in which p_{ij} is the probability that a bed belonging to the ith type of lithology is followed by a bed belonging to the jth type of lithology. Accordingly, $N(i)$ corresponds to the number of successive observations in a Markov chain with transition matrix P. Hence, $N(i) = T_i$, as given by eq (2), so that $N(i)$ has the geometric distribution with parameter $(1 - p_{ii})$, or

$$P\{N(i) = n\} = (p_{ii})^{n-1}(1 - p_{ii}), \quad n = 1, 2, \dots \tag{8}$$

Substituting (7) and (8) into (6) gives

$$h_i(t) = \sum_{n=1}^{\infty} [g_i(t)]^n (p_{ii})^{n-1}(1 - p_{ii})$$

$$= (1 - p_{ii})g_i(t) \sum_{n=0}^{\infty} [g_i(t)p_{ii}]^n$$

$$= \frac{(1 - p_{ii})g_i(t)}{1 - p_{ii}g_i(t)} \tag{9}$$

This equation yields the generating function, and hence the probability distribution, of thickness $L_{N(i)}$ of the ith type of lithologic unit if the generating function $g_i(t)$ of bed thickness is specified.

EXAMPLES

Two examples are given of lithologic structures described by a semiMarkov chain with transition matrix P. These particular examples are chosen because they describe lithologic structures for which lithologic unit thickness has the geometric distribution. The lithologic structure described by a Markov chain in the section on a Markov chain with a transition matrix P also has a geometric distribution of lithologic unit thickness. These two models, along with the model of the section on a semiMarkov chain with a transition matrix R, illustrate how the results of modeling are affected by the interaction between the selection of a mathematical model and the geologic meaning of the components of the model.

The first example shows that a lithologic structure described by a semiMarkov chain may have properties identical with those of the lithologic structure represented by a Markov chain in the section on a Markov chain with a transition matrix P. In this model lithologic type was observed at intervals of h units and lithologic unit thickness was h times the number of successive occurrences of a lithologic type. A semi-Markov chain model with similar properties is obtained by specifying that every bed has a thickness of h units. This implies that each W_i has the degenerate distribution defined by

$$P\{W_i = hk\} = \phi_i(hk) = 1, \quad k = 1$$
$$= 0, \quad \text{elsewhere}$$

253

This degenerate distribution has the generating function

$$\sum_{k=0}^{\alpha} t^{hk}\phi_i(hk) = t^h$$

When W_i has the generating function $g_i(t) = t^h$, eq (9) becomes

$$h_i(t) = \frac{(1-p_{ii})t^h}{1-p_{ii}t^h} = \frac{0t^h}{1-(1-0)t^h}$$

where $0 = 1 - p_{ii}$. Comparing this result with (5) establishes that $L_{N(i)}$ has the same generating function as the random variable L_i identified in the section on a Markov chain with a transition matrix P.

The second, less trivial example of a semiMarkov chain that also yields a geometric distribution for lithologic unit thickness follows from the specification that the bed thickness W_i is related to the geometric distribution. Specifically, W_i/h has the geometric distribution with parameter u_i, so that

$$0_i(hk) = u_i(1-u_i)^{k-1}, \quad k = 1, 2, \ldots$$

Then, using (5), W_i has the generating function

$$g_i(t) = u_i t^h / [1 - (1-u_i)t^h]$$

Substituting this generating function into (9) gives

$$h_i(t) = \left[(1-p_{ii})\frac{u_i t^h}{1-(1-u_i)t^h} \right]\left[1 - \frac{p_{ii}u_i t^h}{1-(1-u_i)t^h} \right]^{-1}$$

$$= \frac{u_i(1-p_{ii})t^h}{1-(1-u_i+p_{ii}i)t^h}$$

$$= \frac{0_i t^h}{1-(1-0_i)t^h}$$

where $0_i = u_i(1-p_{ii})$. Comparing this result with (5) establishes that $L_{N(i)}$ has the same generating function as the random variable L_i identified in the section on a Markov chain with a transition matrix P. This means that $L_{N(i)}$ has the geometric distribution on the lattice $\{hk\}$, where k is a positive integer. It is pertinent that the three random variables W_i, $N(i)$, and $L_{N(i)}$ are defined in terms of the two parameters p_{ii} and u_i.

MULTISTORY LITHOLOGIES

A "multistory" transition matrix was introduced by Carr and others (1966) and later used by Potter and Blakely (1968). Their description of lithologic structure may be formulated as a semiMarkov chain with transition matrix P and waiting time W_i. Evidently, all $p_{ii} > 0$ but it is anticipated that diagonal elements commonly are small. Also, in the simulations by Carr and others and by Potter and Blakely, the W_i are treated as random variables of the continuous type that have lognormal distributions.

The geologic interpretation of the model is that each state of the transition matrix represents a lithologic type, whereas W_i is the thickness of the ith lithologic type. Because $p_{ii} > 0$, there is a positive, though possibly small, probability for the occurrence of two or more lithologies of the same type. This property leads to difficulties in both the mathematical formulation and the geologic interpretation of the model.

As the model is described by Carr and others, if a succession of the same broadly defined lithologic type occurs the lithology is given a multistory interpretation by treating each occurrence as a variant of the basic lithologic type. For example, shale followed by shale may be interpreted as blocky gray layers at the base and thin-bedded greenish layers at the top. If, instead, shale is followed by clay, there are several possible interpretations. The shale may be taken as a blocky gray layer which may be followed by, but never preceded by, thin-bedded greenish layers. This interpretation suggests that the classification of lithologic types should be enlarged to include two or more types of shale and, hence, implies an increase in the number of states in the transition matrix. Moreover, this transition matrix would be characterized by zeros in the main diagonal so that, for example, a blocky gray layer is not followed by another blocky gray layer, and zeros in some off-diagonal elements so that, for example, a blocky gray layer is not preceded by a thin-bedded greenish layer. If this interpretation is accepted, a multistory lithology may be formulated as a semiMarkov chain for which the transition matrix is an embedded Markov chain. The resulting model is of the type considered in the section on a semiMarkov chain with a transition matrix R.

CONTINUOUS-THICKNESS MEASURES

The preceding formulations have treated bed thickness as a discrete random variable. However, most experimental situations treat measurements of bed thickness as a continuous variable and, for such situations, it is preferable to represent bed thickness by a random variable of the continuous type. If bed thickness is a continuous variate, then lithologic unit thickness is, of course, also a continuous variate. The derivation in the section on a semiMarkov chain with a transition matrix P is extended readily to the continuous example.

If W^*_{jn} is the thickness of the nth bed of a lithologic unit of the jth type, $L^*_{N(j)}$ is the thickness of a lithologic unit of the jth type, and the random variable $N(j)$ is the number of beds in a lithologic unit of the jth type, then

$$P\{L^*_{N(j)} \leqslant t\} = P\{W^*_{j1} + W^*_{j2} + \cdots + W_{jN(j)} \leqslant t\}$$

$$= \sum_{n=1}^{\infty} P\{W^*_{j1} + W^*_{j2} + \cdots + W^*_{jn} \leqslant t\} P\{N(j) = n\}$$

If the W^*_{jn} are independent random variables identically distributed as the random variable W^*_j with characteristic function $g^*_j(e^{it}) = E\, e^{it W^*_{jn}}$, where $i = \sqrt{-1}$, then an argument similar to that leading to eq (9) shows that lithologic unit thickness $L^*_{N(j)}$ has the characteristic function

$$E\,e^{itL^*_{N(j)}} = h^*_j(e^{it}) = \frac{(1-p_{jj})g^*_j(e^{it})}{1-p_{jj}g^*_j(e^{it})}$$

The specification of the probability distribution of W^*_j may be based on theoretical, empirical, or other considerations. The exponential distribution is a reasonable selection in that for continuous random variables it takes over the role of complete lack of memory that the geometric distribution provides for discrete random variables.

If W^* has the exponential distribution with parameter λ_j, $g^*_j(e^{it}) = \lambda_j/(\lambda_j-it)$ and, hence,

$$h_j(e^{it}) = (1-p_{jj})\frac{\lambda_j}{\lambda_j-it}\left(1-\frac{p_{jj}\lambda_j}{\lambda_j-it}\right)^{-1}$$

$$= \theta_j/(\theta_j-it)$$

where $\theta_j = \lambda_j(1-p_{jj})$, or $L^*_{N(j)}$ has the exponential distribution with parameter θ_j.

As with the discrete formulation, the three random variables W^*_j, $N(j)$, and $L^*_{N(j)}$ are defined in terms of the two parameters p_{jj} and λ_j. This property is relevant to application or testing of the model because parameter values are obtained by study of any two of the three identified properties of stratigraphic sections.

OBSERVATIONAL DATA FOR A SEMIMARKOV PROCESS

This section considers the problems of obtaining observational data for structuring and evaluating the semiMarkov process described in the preceding section. Observations are required on the number and thickness of individual beds and thickness of lithologic units in a stratigraphic section.

Some difficulties in developing an adequate operational definition of bed have been noted. A data source allowing a first approximation of the internal structure of lithologic units is provided by mechanically recorded resistivity curves on micrologs of boreholes. These curves record the electrical responses of penetrated rocks and reflect variations in the chemical and physical properties of the rocks and their contained fluids.

Figure 1 shows an enlarged portion of the resistivity curves on a microlog originally recorded at a scale of 1/200, with the lightweight divisions on the microlog equivalent to a 2-ft spacing. The solid curve on the right is the 1×1-in. microinverse curve, and the dashed curve is the 2×2-in. micronormal curve. These curves are obtained by measuring the electrical resistivity of the rock formation between electrodes spaced, respectively, 1 and 2 in. from a common pole. The microlog sonde consists of three electrodes mounted on a rubber pad pressed firmly against the borehole wall. As the sonde is raised at a constant rate, analog curves are recorded in a monitoring device at the wellhead which traces the galvanometer readings with light beams focussed on a slowly revolving film. The limit of resolution is controlled by the width of the light beam, the rate of raising the sonde in the borehole and the spacing of the electrodes. If the log is scaled up to full size, the "equivalent width" of the light beam is about 4 in., which sets the minimum vertical and horizontal distances that can

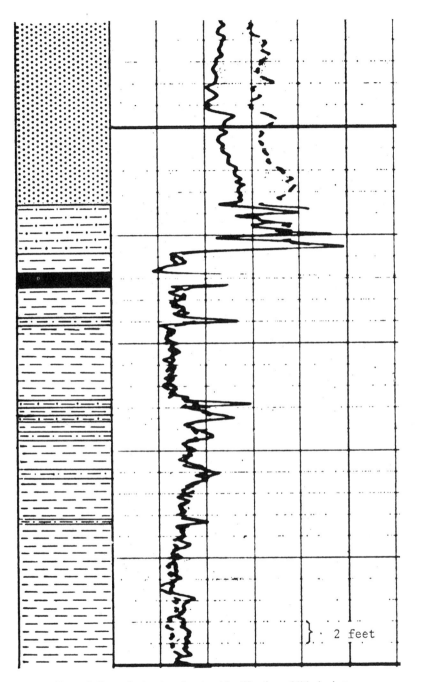

Figure 1. Part of microlog showing identification of lithologic types.

be measured. Although this is a rather coarse scale for bedding, it is sufficiently smaller than the thickness scale of lithologic units to give some idea of internal structure.

Figure 1 has two resistivity curves on the right of the graphic log that show the types of lithologic units penetrated. The 1×1-in. microinverse curve (solid line) is commonly used for measurement inasmuch as it reflects rock composition at the borehole wall, a zone commonly penetrated in part by the drilling fluid, purposely charged with electrolytes. The 2×2-in. micronormal curve (dashed line) reflects compositional features away from the borehole wall and involves the normal fluids in the rocks. Both curves in Figure 1 show numerous minor fluctuations and a smaller number of high peaks.

The resistivity curves show changes in the electrical properties of the rocks and their contained fluids. Because the interpretation of the recorded properties may or may not be equivalent to what is commonly understood as "bedding," the term "electrobed" refers to the resistivity defined smaller units within the lithologic units. By placing the original log under a binocular microscope with micrometer ocular at magnification 24, the smallest unit of measurement on the log is 0.05 mm. This limit is critical because it means that, with the available instrumentation, electrobeds have a minimum thickness of about 10 cm (4 in.).

The well log used in this experiment is the same as that reported in Krumbein and Dacey (1969), from Scherer (1968). Scherer's data identify lithologic type at probe points spaced 2 ft apart. The first step in the use of electrologs is identification of lithologic unit boundaries. The identification of lithologic units (as shown on the left in Fig. 1) uses additional information, such as borehole diameter and self-potential (not shown) that are commonly recorded along with the resistivity curves. The next step is to measure the vertical distance between successive fluctuations in the micro-inverse curve that define electrobeds; the lengths (thicknesses) of the electrobeds are measured to the nearest 0.05 mm. At this initial stage of developing appropriate operational definitions, numerous subjective elements are involved, especially difficulties induced by the sharp and abrupt resistivity peaks in siltstone and lignite.

The more detailed examination of the log under the microscope established that a number of thin lithologic units had not been intercepted in the earlier 2-ft probe experiment. Scherer accordingly reexamined the microlog item by item and found a total of 152 lithologic units of shale. Figure 2 shows the thickness distribution of the shale units; there is good correspondence with the exponential probability law, which is evidence that lithologic unit thickness corresponds to the exponential form specified by the model in the preceding section. The shale units have a mean thickness of $\bar{X} = 4.789$ ft so that the estimated value of the parameter of the exponential distribution is $\lambda = 1/\bar{X} = 0.2088$.

A thorough examination of the model also requires evaluation of the assumptions that the number of beds in a lithologic unit has the geometric distribution and that bed thickness has the exponential distribution. However, the available data are not adequate for this evaluation. This deficiency occurs because the observations are on properties of electrobeds and the instrumentation limits the identification of electrobeds to those with a thickness of at least 4 in. This means that if electrobeds

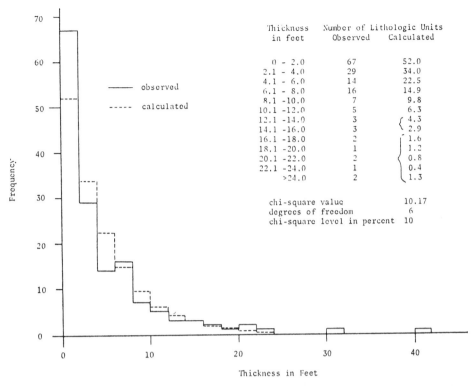

Thickness in feet	Number of Lithologic Units Observed	Calculated
0 – 2.0	67	52.0
2.1 – 4.0	29	34.0
4.1 – 6.0	14	22.5
6.1 – 8.0	16	14.9
8.1 –10.0	7	9.8
10.1 –12.0	5	6.3
12.1 –14.0	3	4.3
14.1 –16.0	3	2.9
16.1 –18.0	2	1.6
18.1 –20.0	1	1.2
20.1 –22.0	2	0.8
22.1 –24.0	1	0.4
>24.0	2	1.3

chi-square value   10.17
degrees of freedom   6
chi-square level in percent   10

Figure 2. Thickness of shale units in stratigraphic section. Observed data are from Scherer (1968), revised by him in 1970. Calculated values are from exponential probability law with parameter 0.2088.

correspond to conventionally defined beds, only beds of 4 in. or more are identifiable while smaller beds are lumped together. As a consequence there are fewer, but thicker, electrobeds than conventionally defined beds. Because of this bias, the domain of the model is presently restricted to stratigraphic sections composed of beds that are all 4 in. or more thick.

Further application of the semiMarkov-process model for thin-bedded sediments requires one of several developments. One is the development and application of instrumentation allowing the identification of beds that are 1 in. or less thick. Another is the formulation of a semiMarkov model for electrobeds that takes into account the limitations of presently available instrumentation.

CONCLUDING REMARKS

As indicated in the previous section, the experiment in counting and measuring electrobeds failed to supply critical data for testing the models of the sections on a semiMarkov chain with a transition matrix P and on continuous-thickness measures. One essential element in testing for geometric or exponential distributions is the presence or absence of a mode in the smallest class interval that can be measured. A procedure that fails to identify the thinner beds or combines them with thicker beds may shift the mode out of the smallest class and thus alter the form of the distribution in this critical range.

Despite this limitation of micrologs, the experiment did succeed in part of its conceptual aim—that of looking into the internal structure of lithologic units. The following comments are appropriate.

(1) Micrologs provide one basis for developing operational definitions to measure subunits within lithologic units, but considerably more study is needed in order to relate thickness and number of electrobeds to their more conventional geologic counterparts. Normally, definitions of beds are independent of the fluids contained in the rocks, but the fluid is a factor in electrobeds. Major difficulties were encountered in identifying electrobeds in siltstone and lignite, and the experiment included neither carbonates nor evaporites. Despite these dificulties, micrologs do supply objectively recorded variations within lithologic units, and it is possible to develop mathematical models for incorporating electrobed data into special kinds of semiMarkov chains. These will not, however, be the same as the models in the sections on a semiMarkov chain with a transition matrix P and on continuous-thickness measures.

(2) The measurement scale in commercially available micrologs is too coarse for realistic counts of conventionally defined sedimentary beds or measurements of their thicknesses. The electrode spacing even in the 1×1-in. microinverse curves undoubtedly averages lithologic changes over this interval, and the equivalent width of the light beam on the microlog trace obscures features less than about 4 in. thick. It is possible that a careful comparison of micrologs and cores from the same depth may lead to some sort of conversion factor that could be useful for scaling down the electrobeds.

(3) The microlog, especially if studied under the microscope, affords a satisfactory method for identifying and measuring lithologic units for conventional Markov chains of the type described in the Section on a Markov chain with a transition matrix P. Analysis of stratigraphic sections by equally spaced "probes" can be conducted on a scale of about 6 in. on micrologs. We are experimenting now with logs digitized at this interval, in which the borehole diameter, self-potential, and the inverse and normal resistivity are recorded as input to a computer program that identifies the lithology and produces the transition matrix P for a Markov chain at the lithologic unit scale level.

In summary, the experimentation is now at a juncture that is becoming more common as new types of models are introduced into geology. Observational data of the types necessary to test mathematical models are not available in the published literature, nor have adequate operational definitions been developed to obtain the needed information. This is particularly true for probabilistic models, where the critical data commonly relate to the presence of frequency distributions that have modes at or near zero on the measurement scale.

ACKNOWLEDGMENTS

This work was supported by the Geography Branch, Office of Naval Research, Contract Nonr 1228 (36), ONR Task Nos. 389–150 and 389–153. This support is gratefully acknowledged. We also are indebted to Wolfgang Scherer, a graduate student at Northwestern University, for the microlog measurements.

REFERENCES

Carr, D. D., and others, 1966, Stratigraphic sections, bedding sequences, and random processes: Science, v. 154, no. 3753, p. 1162–1164.

Cinlar, E., 1970, Markov renewal theory: Jour. Appl. Prob., in press.

Feller, W., 1968, An introduction to probability theory and its applications (3rd ed.): John Wiley & Sons, Inc., New York, 509 p.

Krumbein, W. C., and Dacey, M. F., 1969, Markov chains and embedded Markov chains in geology: Jour. Intern. Assoc. Math. Geology, v. 1, no. 1, p. 79–96.

Lévy, P., 1954, Processus semi-Markoviens: Proc. Int. Congress Math., v. 3, p. 416–426.

Potter, P. E., and Blakely, R. F., 1968, Random processes and lithologic transitions: Jour. Geology, v. 76, no. 2, p. 154–170.

Pyke, R., 1961a, Markov renewal processes: definitions and preliminary properties: Ann. Math. Stat., v. 32, p. 1231–1242.

Pyke, R., 1961b, Markov renewal processes with finitely many states: Ann. Math. Stat., v. 32, p. 1243–1259.

Scherer, W., 1968, Applications of Markov chains to cyclical sedimentation in the Oficina Formation, eastern Venezuela: Unpubl. master's thesis, Northwestern Univ., 93 p.

Smith, W. L., 1955, Regenerative stochastic processes: Proc. Roy. Soc. London, Ser. A, v. 232, p. 6–31.

21

INTERPRETATION OF COMPLEX LITHOLOGIC SUCCESSIONS BY SUBSTITUTA-

BILITY ANALYSIS

John C. Davis and J. M. Cocke

Kansas Geological Survey and

East Tennessee State University

ABSTRACT

Many stratigraphic successions are characterized by repetitive patterns of lithologies. These patterns are most apparent if lithologies are grouped into relatively few categories, and become increasingly obscure as rock types are classified into finer subdivisions. Most cyclothems and megacyclothems, for example, are patterns composed of only four or five distinctive lithologies. Unfortunately, the gross classification necessary to reveal a cyclic pattern results in lithologic categories which yield meager environmental information.

A section through supposedly cyclic lower Pennsylvanian rocks in eastern Kansas was examined and the lithologies classified into 17 states. Although this degree of subclassification is typical of lithofacies studies, the variety of rock types conceals any cyclicity that might be present. Seemingly different lithologies appear at common positions within cyclothems, obscuring the repetitive pattern in the sequence. These lithologies "substitute" for one another in successive cycles, but may be identified by substitutability analysis, a classification procedure that groups states on the basis of their context in a sequence. States with common high conditional probabilities on subjacent and superjacent states are considered equivalent. Results suggest that lithologies must be combined into fewer than eight states before a cyclic pattern emerges. Analyses also suggest that the lower Pennsylvanian cyclothems studied represent interaction of two depositional processes rather than a single megacyclic process.

INTRODUCTION

Sedimentary deposits exhibit differing degrees of perfection of a cyclic pattern. Varved deposits, for example, consist of a precise alternation between two seasonally controlled states. Certain flysch deposits exhibit an almost perfect repetitive succession of upward fining units. Some lithologies are so intimately coupled in their origins that they rarely occur out of sequence. Examples include the couplet formed by seaearth and coal. Formation of one implies the formation of the other; whereas seaearths are found without overlying coals, and coals found without underlying seaearths, the two rarely occur in reverse order. Commonly, the sequence of evaporite lithologies, limestone → dolomite → anhydrite → salt is less well developed. Chemical considerations dictate a fixed order to the succession; observed variations from the predicted order require explanations involving other geologic mechanisms.

Beyond these relatively simple examples are more complex types of sedimentary alternations. Included are the cyclothems and megacyclothems described by many authors (e.g., Merriam, 1964; Duff, Hallam, and Walton, 1967; Weller, 1960). In general, these terms refer to sequences which consist of a more-or-less regular recurrence of a variety of lithic types vertically within the stratigraphic section. Because lengthy sections rarely repeat in a precise manner, workers may refer to an "idealized cyclothem" (Weller, 1930; Weller and Wanless, 1939; Moore, 1936) which is expressed as a series of naturally occurring approximations called "typical cyclothems." The ideal cyclothem is what would develop, presumably, if the cyclothem-causing mechanism functioned without interference from other depositional controls. Its influence is impressed onto actual sections and appears as the common elements of typical cyclothems; the degree of their correspondence to the ideal reflects the relative influences of the cyclothem mechanism and other controls.

The example presented in this paper will be concerned with cyclothems and megacyclothems developed in a segment of Upper Carboniferous strata in the American Midcontinent. These deposits consist of complex alternations of a wide variety of lithic types, almost all marine in origin. No entirely satisfactory explanation has been advanced for the origin of the cyclic aspect of these deposits, although some authors (Moore and Merriam, 1965) strongly imply that the cycles resulted from widespread marine transgression and regression.

The cyclothem mechanism is a physical process, which leaves its imprint on the lithologies and faunal content of the sediment. A detailed examination of the lithologic succession should provide information which could be analyzed quantitatively to estimate the relative influence of the cyclic process. Quantitative studies of the nature of the cyclic pattern in these rocks have been made (Pearn, 1964; Schwarzacher, 1967, 1969) but these deal with idealized representations of the succession and involve simplistic definitions of lithologies. In this paper, we will attempt to measure quantitatively the cyclic aspect of these rocks, using actual data typical of that gathered in a modern stratigraphic study. If a cyclic pattern does not emerge from the analysis, the lithologic data will be iteratively generalized until a pattern becomes apparent. The degree to which lithologies must be generalized should provide information not only on the nature of the cyclothem process, but also on the relative equivalencies of rock types. Although consideration is confined to one stratigraphic interval, the methods employed should be applicable to other successions as well.

DETECTION OF CYCLICITY

A stratigraphic succession may be considered as having two principal attributes: lithology (including biotic composition) and thickness. The range of possible lithologies in sedimentary rocks is finite, but an infinite number of subdivisions are possible within this range. Meaningful, consistent identifications of rock lithologies must be made in order to recognize repetitions. On a sufficiently fine scale of classification, no repetitions are possible because every unit will be placed in an unique lithologic category. At the other extreme, an entire stratigraphic section may be placed into only one or two gross lithologic groups. Stratigraphic units in a cyclothem, or any system of repetitive lithologies, must be classified into states which are rigorously and consistently defined and which are mutually exclusive. Stratigraphic position and lithologic classification must be completely independent if any assignment of units to states is to have meaning. Assignment of a unit to a lithology on the basis of its position within a cyclothem introduces an element of circular reasoning into the definition of the cyclothem (cf. Schwarzacher, 1969, p. 29).

Cycles within a succession of lithologic states may be obscure for a variety of reasons. They may be incompletely developed, in which situation the cyclic pattern must be recognized from a succession of fragments of cycles. In the extreme, it may be impossible to deduce the nature of the ideal cycle or even to determine if cycles are present at all.

Commonly, comparatively few cycles of the magnitude of cyclothems or megacyclothems are present in any stratigraphic interval. If the cycles are incomplete, there may be insufficient repetition to verify statistically their existence. This is a severe problem when working with actual, as opposed to idealized, stratigraphic sections. Nonstationarity of the cyclic pattern presents similar problems. Examination of the Upper Paleozoic succession in Kansas, for example, shows pronounced variation in the nature of lithic repetitions from the Middle Pennsylvanian into the Lower Permian.

Inconsistency in the assignment of rock units to lithologic classes can obscure or bias the pattern of a succession. This may result from assignment of units of a single lithic type to several classes, from the inclusion of units of dissimilar lithologies into a single class, and from the generalization of a heterogeneous interval into a single unit.

Every practical precaution was taken in this experiment to attain consistency of lithologic identification and to maintain objectivity in measurement and analysis. The data were gathered on a stratigraphic section representing most of the Missourian Stage of the Pennsylvanian System, which is well exposed in quarries near Kansas City, Missouri, and along adjacent highways. The Missourian section was selected because it contains "typical" cyclothems, less perfectly developed than those in the overlying Virgilian Stage but far better than many that have been recognized elsewhere. A composite section 342 feet long was obtained, with observations taken at 6-inch intervals. The section does not contain any covered or missing intervals, nor are there any discernible lateral changes in facies between segments of the column.

All segments of the composite section, along with alternate segments, were examined prior to measurement; at that time, decisions were made as to the lithologic states present. These states are based primarily on lithology, although paleontologic content and fabric are important in the definition of some. Seventeen distinct lithologies were recognized; most samples can easily be catagorized into a single state although some ambiguity is present with certain intermediate samples. Limestone lithologies were defined by a carbonate petrographer; clastic rock lithologies were defined by a petrologist specializing in shales. During measurement of the sections, the status of ambiguous samples was decided mutually. Final data consist of 684 ordered descriptions and classifications. Because the sampling interval is constant, thickness of units may be estimated to the nearest one-half foot. However, in this study only the succession of changes in lithology are considered.

During field measurement, no attempt was made to recognize either cyclic patterns or the formal boundaries of rock units. In this manner it was hoped to avoid any tendency to assign units to "expected" classes in the cyclothem model.

Techniques for quantitative assessment of cyclicity in sedimentary successions are based on time-series analysis (Schwarzacher, 1964) or on determination of the conditional probabilities between successive states (Vistelius and Feigel'son, 1965; Krumbein, 1967; Krumbein and Dacey, 1969). No exact tests for the presence of a cyclic component are available, because stratigraphic sections almost invariably violate one or more of the fundamental assumptions required for these tests. A serious constraint is imposed by the nature of most stratigraphic data; the states are nominal classes, so many of the powerful techniques of parametric statistics are not available. Some attempts have been made to apply Fourier analysis or autocorrelation methods, but these necessitate an arbitrary scaling of states (Mann, 1967; Carss, 1967) which has an unassessed effect on the analysis. Some workers have used time-series methods to investigate variations in thicknesses of beds (Anderson and Koopmans, 1963), electrical properties (Preston and Henderson, 1964), or composition (Anderson, 1967) with somewhat more assurance.

Autoassociation (Sackin and Merriam, 1969) is a technique similar to autocorrelation, but based on the counting of common nominal states in two matched sequences. Assumptions are much less restrictive than those of autocorrelation, and the method seems appropriate for the examination of sedimentary successions. If cycles exist in a series of states, they will appear as significant high autoassociations at repeating intervals whose length is equal to the period of the cycles.

The transition probability matrix of a sequence is a matrix of conditional probabilities, and will exhibit no cyclic structure if the sequence is random. The hypothetical section shown in Figure 1 was generated from a random number table and produces a transition probability matrix in which all rows tend to be identical and the columns reflect relative abundances of the various states (in the example, all states are equally abundant). In constrast, Figure 2 shows a purely deterministic succession. The transition probability matrix is completely structured and cyclic. Intermediate between these two extremes are sequences which show partial dependency or conditional probability of one state or another. Such sequences exhibit Markovian behavior, and although pronounced conditional behavior does not necessarily imply cyclicity, cyclic successions must be Markovian.

The structure of transition probability matrices may be shown by plotting them as directed networks or flow graphs (Berge and Ghouila-Houri, 1965), considering the transition probabilities from one state to another to be vectors whose magnitude reflects the probability of one state succeeding another. A purely deterministic succession such as Figure 2 will produce a flow graph in the form of a ring. In contrast, the random succession of Figure 1 will result in a complex flow pattern in which a preferred path cannot

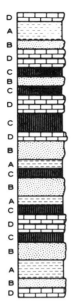
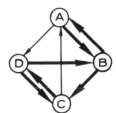

Figure 1. - Hypothetical stratigraphic section containing four
            lithologies, generated from random number table. Flow
            graph for succession does not contain preferred path.

be found.

    Anderson and Goodman (1957) describe a test of the Markov
property, and this has been programmed by Krumbein (1967). Unfortunately, the necessary assumptions severely restrict the applicability of this test. Because no single definitive test for the
presence of cyclicity is available, several of the methods described
were used concurrently in the search for cyclothems in the test
section. These include examining the autoassociation of the sequences, testing of transition probability matrices for the Markov
property, and examination of flow graphs for the presence of pronounced rings.

SELECTION OF LITHOLOGIC STATES

    The stratigraphic interval studied has been classified by
Moore (1949) into fourteen cyclothems and six megacyclothems. He
recognized ten units in an ideal cyclothem, of which seven are
lithologically distinct (Moore, 1936). In a later analysis, Pearn
(1964) reduced these to five distinct states, and Schwarzacher
(1967) further reduced the number of states to four in an attempt
to measure the period of a cycle. From the experiences of other

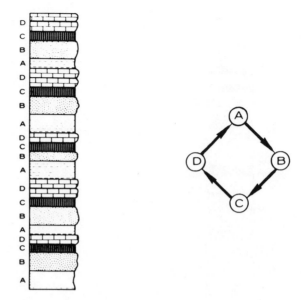

Figure 2. – Hypothetical stratigraphic section which is entirely
          deterministic. Resulting flow graph consists of
          single ring.

workers, it seems likely that some of the lithologic classes which
we have recognized in the stratigraphic succession must be combined
and generalized before cyclicity will become apparent.

     A lithologic succession may be simplified in a number of
ways. Lithologic states, judged on some basis to be similar, may
be combined into single composite states. These composite states
may be combined further, resulting in a reduced number of states
having gross characteristics. The process of combination follows
a tree-like path, consisting of successive estimates of relative
similarity between states, as in cluster analysis. Estimates of
similarity may be based on intuition, measurements of lithologic
variables, assessment of similarities of placement within the
succession, or on some other criterion.

     Alternatively, stratigraphic units which contain more than
one lithology may be considered to consist only of the dominant
lithology. For example, thick shale units may contain sandstone
or limestone beds; these would be ignored using this approach.
Similarly, the complex alternation of carbonate rock types common
in thick limestones would be reduced to a single lithology. Although not explicitly stated, this method of simplification seems
to be used in most quantitative studies of sedimentary successions

(e.g., Carr and others, 1966; Krumbein, 1967). A disadvantage of
this technique is that two complex stratigraphic units containing
the same collection of lithic types may be assigned to different
states if the relative proportions of the constituent lithologies
are different. This may actually defeat the attempt to reduce the
number of states being considered, unless the classification scheme
is extremely rudimentary.

Simplification of the lithologic succession results in a reduction in rank of the transition probability matrix and makes
incipient patterns in the flow graph more apparent. Examinations
of the flow graph or the lithologic succession itself might suggest
several combinations of states that could be made. However, we
will consider a method of assessing the similarity of two states
by their tendency to occur in equivalent positions within the sequence. The hypothetical stratigraphic section shown in Figure 3
is characterized by unusual numbers of successions A → B → C   and
A → D → C. This suggests that states B and D are somehow alike,
as they occur in a common context; i.e., between states A and C.
States B and D are said to be mutually substitutable, as either may
take the place of the other without altering the succession in any
other way. It should be possible to measure the degree of mutual
substitutability between all states in a sequence, and identify
any which form natural groupings. We would expect that groups of

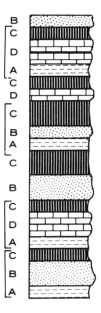

Figure 3. - Hypothetical stratigraphic section containing four
          lithologies arranged so unusual numbers of successions
          A → B → C and A → D → C appear.

mutually substitutable lithologies would correspond to those selected on the basis of geologic considerations, if the lithologic
succession reflects a cyclic mechanism.

MEASURES OF SUBSTITUTABILITY

Following the development of Rosenfeld, Huang, and Schneider
(1968) we can regard the sedimentation mechanism as being capable
of producing any of the states a_1, a_2, ... a_n. Occurrence of any
state is conditional upon the occurrence of all other states. The
sequence may be denoted ΣS, whose elements are

$$\{a_{i1}, \ldots a_{ik} \mid \text{each } a_{ij} \in \{a_1, \ldots a_n\}; \text{ k an integer} \geq 0\}.$$

For all states α, β, γ in the sequence ΣS, let $f_{\alpha\beta}(\gamma)$ be the probability that, given any segment $\alpha \to \xi \to \beta$, we have $\xi = \gamma$. We may
define a function g such that

$$g(\xi, n) = \Sigma f_{\alpha,\beta}(\xi) \, f_{\alpha,\beta}(n) \, / \, \sqrt{\Sigma f_{\alpha,\beta}(\xi)^2 \, \Sigma f_{\alpha,\beta}(n)^2}$$

Because of the Schwarz inequality, g will assume values only in the
range 0, 1. The function $g(\xi, n)$ is the <u>mutual substitutability</u> of
ξ and n is simply the normalized cross-correlation of $f_{\alpha,\beta}(\xi)$ and
$f_{\alpha,\beta}(n)$, regarded as functions of the parameters α and β. The
cross-correlation will be high only if $f_{\alpha,\beta}(\xi)$ and $f_{\alpha,\beta}(n)$ may be
simultaneously high. This occurs only if both states ξ and n occur
with high probability in many common contexts $\alpha \to \square \to \beta$.

Direct application of this concept requires knowledge of the
infinite set of conditional probabilities between all elements in
the sequence. For practical purposes, with finite sequence, higher
order conditional probabilities can be assumed zero and only the
first-order conditional probabilities considered. We are concerned
then only with the set of conditional probabilities $\Pr_{\alpha \to \gamma \to \beta}$ for
all states γ and all combinations of α and β. This requires specification of n^3 conditional probabilities, where n is the number
of states. Rosenfeld, Huang, and Schneider (1968) suggest that
prior and posterior first-order conditional probabilities be considered separately, reducing the number of necessary conditional
probabilities to n^2. Let $\Pr_{i \to j}$ be the conditional probability that
state a_j follows a_i. The first-order left substitutability of
states a_r and a_s is defined as

$$L_{rs} = \Sigma \Pr_{i \to r} \Pr_{i \to s} \, / \, \sqrt{\Sigma \Pr_{i \to r}^2 \, \Sigma \Pr_{i \to s}^2}$$

The first-order right substitutability may similarly be defined

$$R_{rs} = Pr_{r \to j} \; Pr_{s \to j} \; / \sqrt{\Sigma Pr^2_{r \to j} \; \Sigma Pr^2_{s \to j}}$$

Because the probability $Pr_{i \to r \to j} = Pr_{i \to r} \cdot Pr_{r \to j}$, we can create a measure of mutual substitutability between states a_r and a_s by

$$M_{rs} = L_{rs} \cdot R_{rs}$$

This procedure may be used to classify lithologic states according to their tendency to occur between common pairs of states. It is necessary to compute upward and downward transition probability matrices for the sequence of rocks and then compute two symmetrical matrices of right and left substitutability. A product matrix which contains mutual substitutabilities can then be formed by element-by-element multiplication of the two substitutability matrices. The matrices can be clustered by one of many techniques designed to express measures of relative similarity. In this study, an unweighted pair-group method suggested by Sokal and Sneath (1963) was used. This algorithm is incorporated in the NTSYS computer package implemented at The University of Kansas.

ITERATIVE COMBINATION OF STATES

An iterative approach was used to analyze the test section for cycles as the number of lithologic states is reduced. First, the plot of autoassociation versus lag was examined for significant matches, the matrix of transition probabilities tested for the first-order Markov property, and the flow diagram examined for patterns. Next, right-, left-, and mutual substitutabilities were calculated for all pairs of states, and clusters of higher mutual substitutability determined. Clusters of states then were merged to create new, composite states and the examination process repeated. Eventually, the process results in a reduced set of composite states whose succession exhibits a cyclic property. The composition of the composite states and the degree of reduction necessary to make cyclicity evident reflects the importance of the cyclothem mechanism in the stratigraphic section. If cyclothems are expressed as relatively regular patterns of lithologic successions, the pattern will emerge rapidly from the analysis. A comparatively large number of composite states will be retained, most of which will contain a limited number of lithologies which are combined in a geologically meaningful manner. A few composite states may consist of a heterogeneous collection of lithologies that occur infrequently in the original sequence. These may be regarded as lithologic "accidents" in an otherwise orderly cyclic succession.

However, if a cyclic pattern does not actually exist in the original succession, many iterations of the reduction process will be necessary before a repetitive pattern emerges. That a cyclic pattern eventually must emerge should be apparent, because the last possible iteration will result in two composite states which must of necessity oscillate. Measures of mutual substitutability will be low and the resulting composite states will be heterogeneous collections of lithologies that make no geologic sense.

APPLICATION TO MISSOURIAN CYCLOTHEMS

The stratigraphic section examined is shown in Figure 4, adapted from Zeller (1968). This section has been classified by

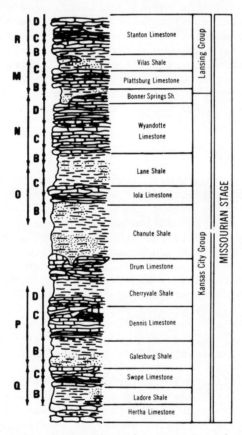

Figure 4. - Generalized stratigraphic section showing major lithologic units in interval studied. Symbols B through D represent cyclothems and M through R represent megacyclothems within interval as defined by Moore (1949) (after Zeller, 1968).

Moore (1949) into a series of cyclothems and megacyclothems which
are indicated on the figure. Moore recognized three types of ·cy-
clothems which comprise the larger system of megacyclothems. His
idealized "B"-type cyclothem consists of a basal sandstone unit
overlain by shale which may contain coal, overlain by a thin,
dense and massive limestone. An idealized "C"-type cyclothem contains, in upward succession, a basal black shale, a thin light-
colored shale, capped by a thick, even-textured limestone which
contains fusulinids near the bottom and algae or oolites in the
top. The limestone is overlain by unfossiliferous shale. Moore's
"D"-type cyclothem is not recognized on the basis of distinctive
lithologies. Moore (1949, p. 82) states "Wherever recognized,
("D"-type cyclothems) are clearly separable from subjacent and
superjacent cyclothems by deposits which denote retreat of marine
waters, whereas the central part of such cyclothems records more
or less off-shore sedimentation during marine submergence."

The measured section contains units which have been assigned
to 6 B-type cyclothems, 6 C-type cyclothems, and 2 D-types. These
are grouped into 6 megacyclothems which consist of either a B cyclothem followed by a C-type cyclothem, or by a succession of B-,
C-, and D-type cyclothems.

Moore (1936) recognized ten major units in a ideal cyclothem,
comprised of seven distinct lithologies. The cycle which he recognized is

        9.   Shale (and coal)
        8.   Shale, typically with molluscan fauna
        7.   Limestone, algal, molluscan fauna
        6.   Shale, molluscoid fauna
        5.   Limestone, with fusulinids and molluscoids
        4.   Shale, molluscoid fauna
        3.   Limestone, molluscan fauna
        2.   Shale, molluscan fauna
        1c. Coal
        1b. Underclay
        1a. Shale with land-plant fossils
        0.   Sandstone

Some of these lithologies are not present in the particular
section measured, although they may occur in equivalent strata at
other locations. For example, no coals or underclays were encountered. Sandstones of the type referred to by Moore are likewise absent. Although Moore (1949) gives descriptions of the
lithology of units in this interval, there is no exact correspondence between Moore's states and those chosen for this study. Moore
describes the general characteristics of relatively large stratigraphic units over a broad area. Ours are detailed lithologic

classifications at small intervals along one specific line of
section. The states recognized in this study are:

B. Siltstone and very fine sandstone, thin bedded,
   micaceous and carbonaceous
C. Silty shales with burrows, micaceous and carbonaceous
D. Shale, brown, slightly silty, unfossiliferous
E. Shale, brown to grey, burrowed, with marine fossils
F. Shale, mottled red and green
G. Shale, green, with calcilutite cobbles
H. Shale, black, blocky and fossiliferous
I. Shale, black, papery and phosphatic
J. "Marl," fine-grained and earthy
K. Calcilutite, mottled, unfossiliferous
L. Calcilutite, laminated, with chert
M. Calcilutite, algal and sparry algal
N. Calcilutite, spar blebbed, poorly fossiliferous
O. Calcilutite, highly fossiliferous
P. Calcarenite, skeletal
Q. Calcarenite, pelletal
R. Calcarenite, oolitic skeletal

It is obvious that these states are subdivisions of a wide
spectrum of lithic types. Further subdivision is possible, and
probably would be made if a detailed petrographic study were undertaken of these sediments. The nine clastic lithologies contain
some states which are distinguished by relatively subtle differences. States B, C, and D, for example, are differentiated by their
relative amounts of sand and silt. Similar gradations exist between some carbonate states such as O and P, which are classified
on the basis of the relative amount of skeletal material. Although
samples with intermediate characteristics lead to ambiguous classifications, most can be consistently assigned to these seventeen
categories.

The initial sequence of seventeen recurring lithic states
shows little tendency toward cyclic behavior. The flow graph
(Fig. 5) contains several oscillatory pairs but only one ring,
involving a calcilutite, a calcarenite, and a shale. Shale and
limestone are distinctly separate in the diagram. If a succession
were generated from this flow graph, it would quickly evolve into
a series of alternations between fossiliferous calcilutite and
about three other limestone lithologies, plus an occasional fossiliferous marine shale. Alternatively, it might evolve into a series
of oscillations between fine sandstones and silty shales. (These
lithologies, which commonly form thick sequences in the Pennsylvanian succession of Kansas, are called "outside shales" because
they separate limestone formations. Other shales, which occur as
thin beds within limestone formations, are referred to as "inside

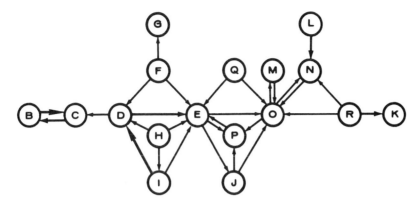

Figure 5. - Flow graph representing transitions between seventeen
           lithic states in interval studied. Note general ab-
           sence of rings.

shales" (Schwarzacher, 1969).) The probability of a transition
between these two end states is small.

     The autoassociation plot (Fig. 6) reflects the pronounced
tendency for oscillation between pairs of states which is apparent
in the flow diagram. At least three oscillations tend to occur
in succession; positive autoassociations at lags of 25 to 32 also
reflect oscillatory sets of lithologies. Only three states are
involved to a significant extent in producing high matches. These
are algal calcilutite, fossiliferous calcilutite, and skeletal
calcarenite.

     The original sequence was reduced by successively combining
states which have high mutual substitutabilities. During initial
iterations, four pairs of lithologies, all carbonates, were combined. In the final iterations, all of these combined to form a
single category containing all limestones except algal calcilutite.
By the ninth iteration, the system was reduced to ten lithologies
and all remaining mutual substitutabilities were below 0.3. The
flow diagram is shown in Figure 7. Note that shales were relatively
unaffected by the reduction process. Limestone lithic types are
similar in that they have a high probability of alternating with
algal calcilutite. Shales, in contrast, show low preferred transitions into other states and hence have low mutual substitutabilities.

     The autoassociation plot (Fig. 8) shows a pronounced tendency
to oscillate. This is caused by interbedding of various shale
lithologies and limestone, and by limestone-algal calcilutite
alternations. Up to six successive oscillations tend to occur

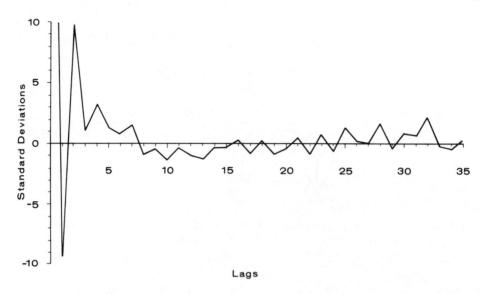

Figure 6. – Autoassociation plot of stratigraphic succession coded
into seventeen lithic states. Abscissa is given in
lags, ordinate in units of standard deviation from
binomial mean. Binomial mean is calculated as expected number of matches of random sequence with itself.

together. High autoassociations at about lag 30 result almost
entirely from matches between limestone units. High autoassociations also occur at about lag 70, and also result from matches
between limestones. The occurrence of beds of each state in the
sequence is shown diagrammatically in Figure 9.

Although the mutual substitutability criterion succeeded in
reducing the number of lithic states, the resulting succession

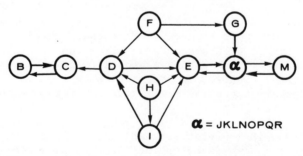

Figure 7. – Flow graph of stratigraphic section after combining
states having high mutual substitutabilities.

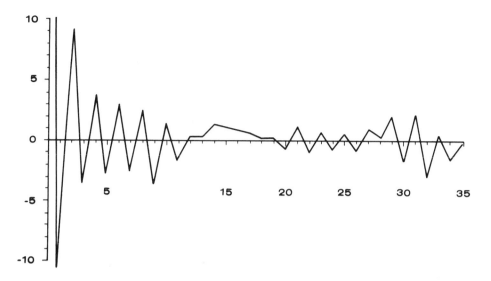

Figure 8. – Autoassociation plot of stratigraphic section after
           combination of states by mutual substitutability.
           Note pronounced oscillation caused by interbedded
           shales and limestones.

does not display the cyclic pattern that might be expected. However, right substitutabilities between lithic states tend to be
much higher than left substitutabilities. As right substitutability is a function of upward transition probabilities, it seems
reasonable that it might provide a more meaningful measure of similarity than mutual substitutability, which is also dependent upon
downward transitions.

     From the initial sequence, a series of six iterations reduced
the maximum right substitutability below 0.5 and resulted in eight
lithologic states. The composite states include an "outside shale",
an "inside shale", a black shale, a calcilutite, and a calcarenite.
Two other shales remain uncombined, as does oolitic calcarenite.
The flow diagram is shown in Figure 10. A pronounced ring exists,
"inside shale" → calcilutite → calcarenite → "inside shale". All
other lithologies tend to feed into this ring, with very diffuse
return.   The tendency toward two-state oscillation is reduced
(Fig. 11), probably as a result of the relative importance of
the three-component ring and the combination of states B and C
into a single category. High autoassociations at about 25 lags
result from repetitious occurrences of calcilutite → calcarenite
oscillations separated by intervals of calcilutite → calcarenite →
shale successions. Using the symbols on Figure 10, the sequence
first oscillates α → Ψ → α → Ψ ..., then   γ → α → Ψ → γ → α → Ψ...

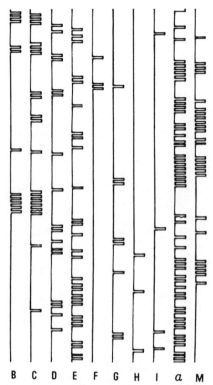

B C D E F G H I α M

Figure 9. – Diagrammatic section showing occurrence of lithic
         types within interval measured. Deflection to right
         in line represents occurrence of that lithology. See
         text and Figure 9 for key to lithologies.

Within about 15 lags, the sequence again becomes $\alpha \to \Psi \to \alpha \to \Psi...$,
giving the longer period phenomenon seen in Figure 11. The other
lithologies occur in the sequence, but not in an apparent pattern.
The sequence of states is shown in Figure 12.

The iteration procedure was continued for six additional cycles, reducing the sequence four states, an "outside shale", an
"inside shale", a calcarenite, and a calcilutite. The flow graph
(Fig. 13) shows two rings, "inside shale" → calcilutite → calcarenite, and the reverse. "Outside shales" may enter the pattern at
any point, but have low return probabilities. The maximum right
substitutability after twelve iterations is less than 0.4, between
inside and outside shales.

Figure 14 is a plot of autoassociation and shows the influence

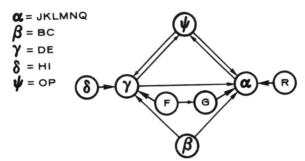

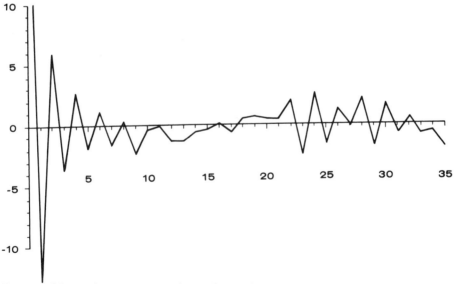

Figure 10. – Flow graph for stratigraphic section after states
            have been reduced by right substitutability. Diagram
            shows system after six iterations.

of both oscillations between pairs of states and transitions between three states. High autoassociations at initial lags result
from alternations between calcilutite and calcarenite and between
calcarenite and "inside shale". High autoassociations at lags 14
and 18 result from repetitions of the two limestone types. High
autoassociations at about lag 30 reflect changes in the repetitive
pattern from calcilutite → calcarenite → calcilutite to sequences
of calcarenite → "inside shale" → calcarenite alternations and
calcilutite → calcarenite → "inside shale". The sequences are the
same as those apparent in the system at six iterations, $\alpha \to \Psi \to \alpha \to$

Figure 11. – Autoassociation plot of stratigraphic section after
            six iterations of combining states by right substitu-
            tability.

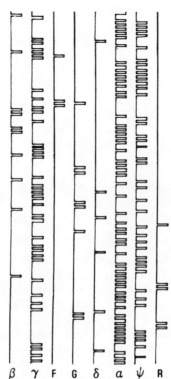

β γ F G δ a ψ R

Figure 12. – Diagrammatic section showing occurrences of lithologies after six iterations of combination of states by right substitutability. See Figure 10 for key to states.

Ψ... and γ → α → Ψ → γ → α → Ψ... Although the order γ → α → Ψ ... is dominant, these three states also occur in other combinations. "Outside shales" sporadically appear within this system. Figure 15 shows the succession of four lithic states within the stratigraphic interval.

As a comparison, two runs were made in which lithic types were

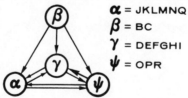

α = JKLMNQ
β = BC
γ = DEFGHI
ψ = OPR

Figure 13. – Flow graph for stratigraphic section after twelve iterations of combining states by right substitutability. Diagram shows pronounced flow involving inside shale, calcilutite, and calcarenite.

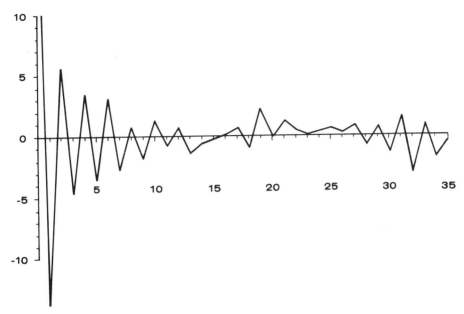

Figure 14. - Autoassociation plot for stratigraphic section after
reduction to four lithologic states by right substitutability.

combined into states that correspond closely to those in the classic cyclothem model, and to lithologies used in other mathematical
studies of cyclicity. The first of these uses a five state system of an "inside shale", an "outside shale", a black shale, a
calcarenite, and a calcilutite. It differs from the model found
by left substitutability only in the retention of black shale as
a separate lithic category, and in the substitution of pelletal
calcarenite for skeletal calcilutite in the calcarenite lithology.
The flow diagram (Fig. 16) is similar to Figure 10. Note that the
black shale lithology does not contribute to the cycle, but occurs
without apparent pattern in the succession (Fig. 17). The plot
of autoassociation resembles Figure 14, except for slight shifts
in peak positions. These suggest that the distinction of black
shales from other "inside shales" does not enhance the appearance
of cyclicity.

Next, all limestones were combined into a single lithic category, a procedure used in many studies of idealized stratigraphic
successions. "Inside" and "outside shales" were retained as distinct categories, as was black shale. The section oscillates between limestone and "inside shale", with minimal contributions by
the other shale lithologies (Fig. 18). The autoassociation dia-

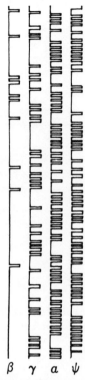

β γ a ψ

Figure 15. – Diagrammatic section showing occurrence of lithologies
            after reduction to four states by right substituta-
            bility. See Figure 13 for key to states.

gram becomes strongly oscillatory, with a beat frequency of about
16 lags caused by appearance of occasional black shales and "outside shale" (Fig. 19).

        The extreme fluctuations apparent in Figure 19 suggest that

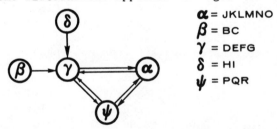

α = JKLMNO
β = BC
γ = DEFG
δ = HI
ψ = PQR

Figure 16. – Flow graph for stratigraphic section with lithologies
            classified into five arbitrary groups. These closely
            correspond to Figure 13, except that black shale (δ =
            HI) is retained as separate category.

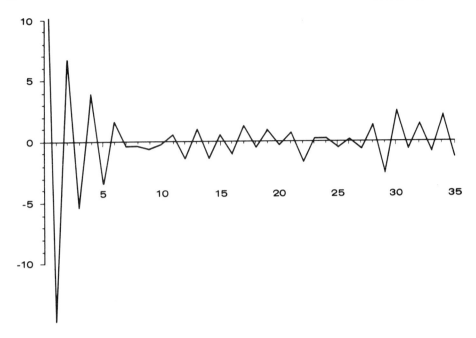

Figure 17. - Autoassociation plot for stratigraphic section containing five lithologic states. Note similarity to
Figure 14.

the succession has deteriorated into a disturbed two-state system.
Although such a system will of necessity "cycle" as it oscillates
between the two states, it is trivial as a model for cyclothems or
megacyclothems. The flow diagram and autoassociation plot suggest
that the only distinctive roles played by "outside" or black shales
are as occasional perturbations to an otherwise stable, alternating
system of limestone and marine shale.

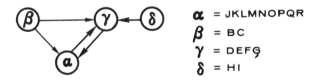

Figure 18. - Flow graph for stratigraphic section after combination
of all lithologies into four states: inside shale, outside shale, black shale, and limestone. Oscillation between limestone and inside shale is only pronounced
flow pattern.

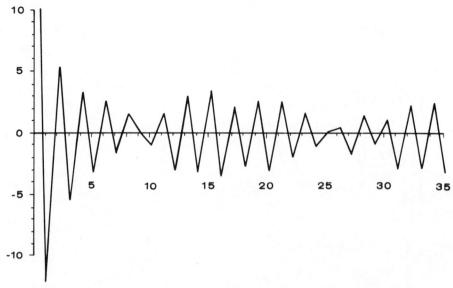

Figure 19. – Autoassociation plot for stratigraphic section after combination of lithologies into four states. Note pronounced beat frequency resulting from perturbation of two-state oscillation between inside shale and limestone.

CONCLUSIONS

The flow graph developed from seventeen-state data for the Missourian section shows no evidence of large cyclic patterns, nor does the plot of autoassociation. Mutual substitutability analysis is capable of reducing the number of lithic states to nine by combining all limestones except algal calcilutite into a single category. However, the shales remain separate, reflecting in part their low probabilities of occurrence and their sporadic pattern of succession. Right substitutability, which is based on upward transition probabilities, provides a more effective criterion for reducing the number of states. After twelve iterations, four lithic categories remain: a calcilutite, a calcarenite, an "inside shale", and an "outside shale". Using these states, the succession reveals a strong tendency for three-state cycles. "Outside shales" occur as sporadic interruptions in this cyclic pattern. The set of lithic states obtained by right substitutability seems at least as adequate as empirical models in which black shales are retained as a separate category.

At the four-state level, the stratigraphic interval measured consists of an alternating series of calcilutites, calcarenites, and marine shales. Sporadic incursions of coarser sediments may interrupt this pattern at any point. We may infer from this that the various limestone and "inside shale" lithologies are intimately interrelated. The occurrence of a specific lithologic type at a certain point within the sequence seems to have no great significance. In contrast, the coarser "outside" shales and siltstones seem to represent a clastic influx independent of the basic depositional pattern.

Although the section examined supposedly contains many examples of megacyclothems, these are not expressed in a detailed examination of the rocks, nor do they become apparent if lithologies are combined into increasingly gross categories. Cyclothems and megacyclothems may be characteristics apparent only in idealized successions, and they may not occur in the actual stratigraphic intervals from which these idealized successions are derived. Alternatively, the recognition of cyclothems may depend upon characteristics not expressed in the lithology of the rocks themselves.

ACKNOWLEDGMENTS

Several persons contributed to the lively debate which accompanied this investigation. These include D.F. Merriam, W. Schwarzacher, P.H. Heckel, and C.D. Conley. Special thanks are extended to Schwarzacher and Conley who re-examined the stratigraphic interval to aid in the confirmation of lithologic assignments. R.C. Moore graciously discussed his concepts of cyclic sedimentation in sessions with members of the Geologic Research Section of the Kansas Geological Survey. A. Rosenfeld provided unpublished documents containing algorithms from which the substitutability measures were programmed. R.J. Sampson wrote the FORTRAN programs and graphic display routines used in the study.

REFERENCES

Anderson, R. Y., 1967, Sedimentary laminations in time-series study, in Colloquium on time-series analysis: Kansas Geol. Survey Computer Contr. 18, p. 68-72.

Anderson, R. Y., and Koopmans, L. H., 1963, Harmonic analysis of varve time series: Jour. Geophysical Res., v. 68, no. 3, p. 877-893.

Anderson, T. W., and Goodman, L. A., 1957, Statistical inference about Markov chains: Am. Math. Stat., v. 28, p. 89-110.

Berge, C., and Ghouli-Houri, A., 1965, Programming, games and
    transportation networks: Methuen and Co., Ltd., London, 260 p.

Carr, D. D., and others, 1966, Stratigraphic sections, bedding se-
    quences, and random processes: Science, v. 154, no. 3753,
    p. 1162-1164.

Carss, B. W., 1967, In search of geological cycles using a tech-
    nique from communications theory, in Colloquium on time-series
    analysis: Kansas Geol. Survey Computer Contr. 18, p. 51-56.

Duff, P. M. D., Hallam, A., and Walton, E. K., 1967, Cyclic sedi-
    mentation: Elsevier Publ. Co., Amsterdam, 280 p.

Krumbein, W. C., 1967, FORTRAN IV computer programs for Markov
    chain experiments in geology: Kansas Geol. Survey Computer
    Contr. 13, 38 p.

Krumbein, W. C., and Dacey, M. F., 1969, Markov chains and embedded
    Markov chains in geology: Jour. Intern. Assoc. Math. Geol.,
    v. 1, no. 1, p. 79-96.

Mann, C. J., 1967, Spectral-density analysis of stratigraphic data,
    in Colloquium on time-series analysis: Kansas Geol. Survey
    Computer Contr. 18, p. 41-45.

Merriam, D. F., ed., 1964, Symposium on cyclic sedimentation:
    Kansas Geol. Survey Bull. 169, 636 p.

Moore, R. C., 1936, Stratigraphic classification of the Pennsyl-
    vanian rocks of Kansas: Kansas Geol. Survey Bull. 22, 256 p.

Moore, R. C., 1949, Division of the Pennsylvanian System in Kansas:
    Kansas Geol. Survey Bull. 83, 203 p.

Moore, R. C., and Merriam, D. F., 1965, Upper Pennsylvanian cyclo-
    thems in the Kansas River Valley: Field Conf. Guidebook,
    Kansas Geol. Survey, 22 p.

Pearn, W. C., 1964, Finding the ideal cyclothem, in Symposium on
    cyclic sedimentation: Kansas Geol. Survey Bull. 169, p. 399-
    413.

Preston, F. W., and Henderson, J. H., 1964, Fourier series charac-
    terization of cyclic sediments for stratigraphic correlation,
    in Symposium on cyclic sedimentation: Kansas Geol. Survey
    Bull. 169, p. 415-425.

Rosenfeld, A., Huang, H. K., and Schneider, V. H., 1968, An application of cluster detection to text and picture processing: Univ. Maryland Computer Science Center, College Park, Md., Office Naval Res. Grant Nonr 5144(00), Tech. Rept. 68-68, 64 p.

Sackin, M. J., and Merriam, D. F., 1969, Autoassociation, a new geological tool: Jour. Intern. Assoc. Math. Geol., v. 1, no. 1, p. 7-16.

Schwarzacher, W., 1964, An application of statistical time-series analysis of a limestone-shale sequence: Jour. Geology, v. 72, no. 2, p. 195-213.

Schwarzacher, W., 1967, Some experiments to simulate the Pennsylvanian rock sequence of Kansas, in Colloquium on time-series analysis: Kansas Geol. Survey Computer Contr. 18, p. 5-14.

Schwarzacher, W., 1969, The use of Markov chains in the study of sedimentary cycles: Jour. Intern. Assoc. Math. Geol., v. 1, no. 1, p. 17-39.

Sokal, R. R., and Sneath, P. H. A., 1963, Principles of numerical taxonomy: W. H. Freeman and Co., San Francisco, 353 p.

Vistelius, A. B., and Feigel'son, T., 1965, The theory of formation of sedimentary beds: Doklady Akad. Nauk SSSR, v. 164, no. 1, p. 158-160.

Weller, J. M., 1930, Cyclical sedimentation of the Pennsylvanian Period and its significance: Jour. Geology, v. 38, p. 97-135.

Weller, J. M., 1960, Stratigraphic principles and practices: Harper & Bros., New York, 725 p.

Weller, J. M., and Wanless, H. R., 1939, Correlation of minable coals of Illinois, Indiana, and western Kentucky: Am. Assoc. Petroleum Geologists Bull., v. 23, p. 1374-1392.

Zeller, D. E., ed., 1968, The stratigraphic succession in Kansas: Kansas Geol. Survey Bull. 189, 81 p.

22

Copyright © 1971 by the Plenum Publishing Corporation

Reprinted from *Math. Geol.* 3(1):51–60 (1971)

An Empirical Discriminant Method Applied to Sedimentary-Rock Classification from Major-Element Geochemistry[1]

R. J. Howarth[2]

Classification of sandstones, greywackes, pelites, limestones, dolomites, and acid-igneous and basic-igneous rocks, using a literature sample of 183 post-1920 analyses for the 11 major oxides has achieved an 80-percent success rate. The method is based on nonparametric estimation of a probability density function for each category to be classified, using the Bayes decision rule. The method is suitable for use with small training sets and gives much improved results over a linear discriminant function. Classification following data compression using principal components also has given satisfactory recognition rates. KEY WORDS: entropy, principal components analysis, geochemistry, mineralogy.

INTRODUCTION

This work was undertaken as part of a preliminary investigation into the use of statistical decision theory for the automatic classification of reconnaissance geochemical data.

The data set chosen is a relatively small one; 183 "good" post-1920 wet-chemical analyses, for which determinations of all the 11 major-element oxides were available, were obtained from the literature. Interest in sedimentary-rock analyses and the stipulation that all 11 oxides should be determined, combined to restrict the number of admissible analyses. The sample was composed of: 24 quartzose sandstones, 5 arkoses, 4 protogreywackes, 30 greywackes, 83 pelites and semipelites, 7 limestones, 5 dolomites, and a selection of 10 acid-igneous and 15 basic-igneous rocks. These analyses were grouped into seven classes: (1) quartzose sandstones and arkoses; (2) protogreywackes and greywackes; (3) pelites and semipelites; (4) limestones; (5) dolomites; (6) basic igneous; and (7) acid igneous.

The purpose of discriminant analysis is to determine from the observed characteristics of all the samples grouped into classes an optimum method for the classification of any unknown object presented to the classifier. The nature of the categories is defined by the investigator. While it is hoped that the boundaries of each category may

---

[1] Manuscript received 25 August 1970.
[2] Geology Department, Imperial College (UK).

be exclusive, in a real-world situation partial overlap of the classes is relatively common.

A statistically representative "training" set of samples is used to design the classi-fier, in the sense that the criteria upon which the decision is based are derived by examination of the training sample characteristics for each category. In order to evaluate the performance of the recognition system a second "testing" set of data is required. This consists of a series of samples not included in the training set, which are presented to the classifier following the training phase. From a comparison of the true categories of the testing set samples with those assigned by the classifier one arrives at a measure of the overall recognition (success) rate of the classification system.

Although it can be argued that it is sufficient to design the classifier in order to perform well on the training set (if used as a testing set), in practice the training set is always too small and extrapolation to new data hazardous (Nagy, 1968). For this reason the analytical data were split into two groups. Approximately one-third of the total number of samples were selected at random, within each of the seven classes, to form the training set and the rest of the samples formed the testing set. Some of the smaller classes had only two or three training samples. No transformations were applied to the data.

DISCRIMINANT-FUNCTION METHOD

The technique used here is the exponential form of the polynomial discriminant method of Specht (1967a). It is based on the nonparametric estimation of a probability density function for each category to be classified so that the Bayes decision rule may be implemented. The method seems to have a good performance even when the training sets are small. The explanation given below closely follows that of Specht (1967b).

Let us assume that for each class Ω_j, $j = 1, 2, \ldots, k$, we have observations on a p-feature vector $\mathbf{X} = \{x_1, x_2, \ldots, x_p\}$, and that the a priori probabilities h_j, $j = 1, 2, \ldots, k$, of occurrence of each class are known. Let the multivariate probability density function for the jth category be $f(\mathbf{X})$, that is, the probability that $\mathbf{X}$ belongs to category j. These functions may be of any form provided that they are everywhere nonnegative, integrable, and that their integrals over all space equal unity. The classifier must perform the classification on the basis of this information with a minimum of misrecog-nition.

Defining a decision function $d(\mathbf{X})$, where $d(\mathbf{X}) = d_i$ means that $\mathbf{X}$ is assigned to Ω_i, let l_i be the loss (penalty) incurred if $d(\mathbf{X}) = d_j$ when $\mathbf{X}$ is a member of Ω_i. It is assumed that the loss is zero for a correct decision. The problem is now to choose a decision criterion such that the average loss over all classes is minimized. Fu (1968) has shown that under the optimal decision rule, in the sense of minimizing the average loss, $\sum_i l_i h_i f_i(\mathbf{X})$ is smaller than under any other decision rule. Using a symmetrical loss function

$$d(\mathbf{X}) = d_i; \, l_i = 0$$
$$d(\mathbf{X}) = d_j, \, j \neq i; \, l_i = \text{const}$$

when $\mathbf{X}$ is a member of Ω_i, then the Bayes decision rule is to assign $\mathbf{X}$ to the category for which $h_r l_r f_r(\mathbf{X})$ is a minimum.

While it may be possible to estimate the a priori probabilities and loss-function values to the correct order of magnitude for each category, it may be impossible to know the probability density functions. Again following Specht, we will estimate the probability densities for each category as a sum of exponentials based upon the set of training samples, which all have a positive probability of occurring, and we will assume that samples not in the training set but near a given sample point (in p space) will have about the same probability of occurring as the training samples.

Assuming that the estimated probability density function for a category is smooth and continuous, and that the first partial derivatives are small, Specht proposed that an interpolation function $g(\mathbf{X}, \mathbf{X}_i)$ be found such that

$$f(\mathbf{X}) = \frac{1}{m}\sum_i g(\mathbf{X}, \mathbf{X}_i) \tag{1}$$

where m is the number of training patterns available and $g(\mathbf{X}, \mathbf{X}_i)$ is the contribution of the ith training pattern to the estimated density. If it is assumed that each training pattern contributes independently to the overall density distribution, and that $g(\mathbf{X}, \mathbf{X}_i)$ is a function of the Euclidean distance of $\mathbf{X}$ from the ith training pattern point in p space, following Specht we write

$$g(\mathbf{X},\mathbf{X}_i) = \frac{1}{(2\pi)^{p/2}\sigma^p} \cdot \exp\left[\frac{-(\mathbf{X}-\mathbf{X}_i)'(\mathbf{X}-\mathbf{X}_i)}{2\sigma^2}\right] \tag{2}$$

where σ is a "smoothing parameter." The estimated density function for the ath category is then

$$f_a(\mathbf{X}) = \frac{1}{(2\pi)^{p/2}\sigma^p} \cdot \frac{1}{m}\sum_{i=1}^{m} \exp\left[\frac{-(\mathbf{X}-\mathbf{X}_{ai})'(\mathbf{X}-\mathbf{X}_{ai})}{2\sigma^2}\right] \tag{3}$$

where $\mathbf{X}_{ai}$ is the ith training pattern from category a.

The smoothing effect achieved by increasing σ is shown for a one-dimensional case in Figure 1. As σ increases, the five distinct modes corresponding to the positions of the training samples are progressively smoothed until a symmetrical unimodal density function is obtained. A rigorous analysis of this operation will be found in Specht (1967a, p. 310–311). It might be helpful in certain circumstances to have σ vary for each class, although this has not been implemented in this work. The exponential form used here has been programmed for a CDC 6600 computer and is extremely fast in execution time. Specht (1967b) describes excellent results achieved using this method.

The polynomial discriminant method also developed by Specht (1967a) has been used by the Laboratory for Agricultural Remote Sensing at Purdue University in crop-classification work (LARS, 1968) but has not yet been used in a geologic context so far as the writer is aware.

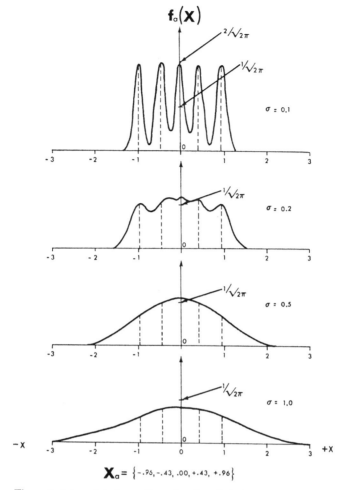

Figure 1. Interpolated one-dimensional probability density function for set of five training samples with increasing values of smoothing parameter σ (redrawn from Specht, 1967a).

ROCK CLASSIFICATION BASED ON MAJOR-ELEMENT DATA

It will be recalled that both the a priori probability for the occurrence of a given class and the loss function are required in the assignment of an unknown sample to a particular class under the Bayes rule. In all situations it was assumed that the a priori probability of occurrence was equal for each class ($h_i = 0.143$, $i = 1, \ldots, 7$). The loss function was varied in an initial experiment, following the observation that the class with the lowest successful recognition rate tended to be that for the acid-igneous rocks (class 7).

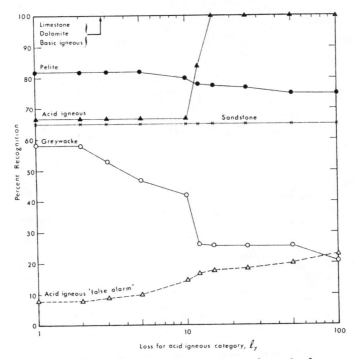

Figure 2. Variation in classification success rate for each of seven categories of rocks as penalty for misrecognition of acid-igneous class is increased.

Accordingly, the loss for misclassification of the acid-igneous category l_7 was varied from 1 to 100 while $l_i = 1, i = 1, \ldots, 6$. The results in Figure 2 show what might be expected, that achievement of 100-percent-correct recognition of this class is obtained at the expense of the recognition rate for other classes, particularly the grey-wacke samples. The overall misclassification rate, the acid-igneous "false alarms," increases with increase in l_7. It was therefore decided to use an equal (0, 1) loss function for all classes during the classification experiments.

The only other variable to be chosen is the smoothing parameter σ. The variation of the overall recognition rate with σ for constant a priori probability and loss-function values for the original data is shown in Figure 3. The behavior might be expected from the appearance of the interpolated probability density functions in Figure 1. A rapid rise in the recognition rate for the testing samples occurs as σ increases from 0.02 to 0.2. A plateau of maximum success exists from σ values of 0.4 to 3.5, after which the recognition rate falls to a lower level, becoming stable at 58-percent correct when σ is greater than 10. This final state must correspond to a constant misclassification rate resulting from overlap between the interpolated density functions. The recognition rate of the training samples (when treated as test data) remains close to 100 percent at values of σ below 1, thereafter falling off for the reasons given above. It is clear that, as Specht

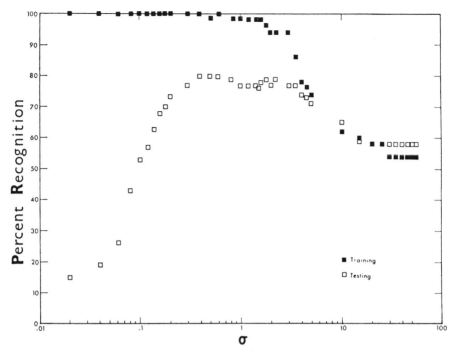

Figure 3. Variation in overall classification success rate for training and testing sets of original data with increase in smoothing parameter σ.

suggested, values of σ within a fairly wide range are satisfactory and it need not be determined a priori with a high degree of accuracy. With values of σ in excess of 3 the acid-igneous false-alarm rate increased rapidly.

The most successful classification achieved an overall 80-percent success rate using a smoothing parameter of 0.6. The misclassification table in which the categories assigned to the testing samples by the classifier are set out (Table 1) allows the results to be compared for each class. Several of these seeming misclassifications are found to be permissible from the geologic context. For example, two pelites which were assigned to the basic-igneous category were volcanic clays from the Pacific, and a greywacke similarly assigned contained 21 percent basic-igneous rock fragments by volume in the mode. Two arkoses and a greywacke assigned to the acid-igneous class contained 44, 53, and 60 percent normative feldspar. Two protogreywackes assigned to the sandstone class contained 82 and 84 percent SiO_2, and several siliceous mudstones with 72–79 percent SiO_2 in the pelite class were assigned to the greywacke category. In the majority of situations the acid-igneous rocks were assigned to either the greywacke or quartzose sandstone class.

The "best" classification rate achieved would probably have been higher if the subjective assessment of the nature of the original rock samples had been more definitive. In many of the literature sample descriptions supporting data to justify the rock

Table 1. Misclassification of Testing Samples for Original Data[a] ($\sigma = 0.6$; 102 correct sample classifications)

		Class assigned						
		1	2	3	4	5	6	7
True class	1	9	3	1	–	–	*	2
	2	2**	13	3	–	–	1*	1
	3	–	8	53	–	–	3**	2
	4	–	–	–	5	–	–	–
	5	–	–	–	–	3	–	–
	6	–	–	–	–	–	13	–
	7	1	3	–	–	–	–	2

[a] Explanation: Equal a priori probabilities and (0, 1) loss function for each class. Classes and number of training samples in each: 1 = sandstones (14); 2 = greywackes (12); 3 = pelites (15); 4 = limestones (2); 5 = dolomites (2); 6 = basic igneous (2); 7 = acid igneous (3); * = assigned classification allowable on geologic evidence (see text), each asterisk corresponding to one sample.

name assigned by the original investigators were lacking. Clearly, this could unwittingly cause erroneous members to be included in the original training sets.

No adjustments for nonnormality or unequal dispersion in the classes were made using the empirical discriminant method and it is of interest that attempted classification using the same training and testing sets with a linear discriminant function (the IBM Scientific Subroutine Package MDISC program) achieved a maximum of only 27 percent success over the seven classes.

CLASSIFICATION RESULTS FOLLOWING DATA COMPRESSION

It will be seen from (3) that for a given number of m training patterns and constant σ, the estimated density-function value will decrease as the number of variables p increases. In the event that $f(X)$ is small, computer truncation errors may produce seeming zero values for $f(X)$, which could result in classification of the pattern as an "unknown." Indeed, the relative probability values may be high although $\max_i [f(X)]$ is itself small. For example, the largest $f(X)$ values obtained in this study have been, in some situations, as small as 10^{-230} although the sample in question has been assigned to the correct class. On the CDC 6600 computer the exponent may be as small as 10^{-294}, but had the problem been calculated on an IBM 7094 machine, incorrect results would have been obtained had the exponents been smaller than 10^{-38}. If these truncation effects can be minimized the correct classification may yet be achieved.

The aim, therefore, would be to reduce the number of original variables to some smaller number such that the estimated density-function values will tend to be maximized, and the information loss incurred in the compression minimized. Although the example given here is a relatively trivial one, multielement geochemical reconnaissance data may involve 30 or more element determinations per sample, and the risk of incurring truncation errors in the determination of the probability density functions would be enhanced.

Workers in the field of pattern recognition (Watanabe, 1965; Chien and Fu, 1968; Favella, Reineri, and Righini, 1969) derived an optimal coordinate transform for data compression. Known as the Karhunen–Loève (KL) coordinate system, it was originally conceived within the framework of square-integrable functions in a continuous domain (Watanabe, 1965). However, its derivation produces a transform identical to that obtained in a principal components analysis.

The eigenvectors derived from the covariance matrix pooled over all k classes for the p features are reordered according to the descending order of their associated eigenvalues. This set of orthogonal vectors forms the coordinate system and is optimal only if based on the covariance matrix (Chien and Fu, 1968). The set of new (ordered) features $\mathbf{V}_j$ then are obtained from the original features $\mathbf{X}_j$ for the jth class by the transformation

$$\mathbf{V}_{jl} = \sum_r \mathbf{X}_{jr}\mu_{lr} \qquad \begin{array}{l} r = 1, \ldots, p \\ l = 1, \ldots, p \end{array}$$

where μ_{lr} is the rth element of the lth coordinate vector. Proofs of the error-minimizing and entropy-minimizing properties of this transform are given by Watanabe (1965) and Chien and Fu (1968).

As it is fairly common in principal components analysis to standardize the variates, replacing the covariance matrix by the correlation matrix (Lawley and Maxwell, 1963, p. 47), classification of the data previously discussed was attempted using the coordinate transform based on: (1) a product-moment correlation matrix and (2) a covariance matrix, as recommended by Watanabe (1965) in order to increase the discriminating power of the coordinate system. As before, σ was varied [from 0.02 to 50.00 with a (0, 1) loss function and equal a priori probabilities for each category].

The optimum successful recognition rate obtained using all 11 coordinates based on the correlation matrix (Table 2) was smaller than that obtained using the original data (Table 1). However, the results based on the covariance matrix (Table 3), were slightly better than those obtained with the original data for the optimum choice of σ.[3]

In order to determine the effect of the data compression on the recognition accuracy of the classifier, the percentage of samples correctly classified on the basis of the first q variables (major-element oxides or coordinate values) was observed sequentially for $q = 1, \ldots, 11$. An improvement in the performance of the classifier of the order of 10 percent more successful recognition was obtained over the first five variables using the compressed data. The small magnitude of this improvement is attributable to the reporting of major-element geochemical analyses in near-optimal order from the point

---

[3] Chien and Fu (1968) describe an example in which the classification success rate obtained using the covariance-based coordinates shows a much greater improvement over the original data.

Table 2. Misclassification of Testing Samples for Co-ordinate System Based on Product-Moment Correlation Matrix[a] ($\sigma = 0.14$; 98 correct sample classifications)

		Class assigned						
		1	2	3	4	5	6	7
	1	9	5	–	–	–	–	1
	2	2**	14	2	–	–	–	1
True class	3	7	5	51	–	–	2**	1
	4	–	1	2	4	–	–	–
	5	–	1	–	1	1	–	–
	6	–	1	2	–	–	10	–
	7	1	–	–	–	–	–	5

[a] Explanation as for Table 1.

Table 3. Misclassification of Testing Samples for Co-ordinate System Based on Covariance Matrix[a] ($\sigma = 0.6$; 103 correct sample classifications)

		Class assigned						
		1	2	3	4	5	6	7
	1	9	3	1	–	–	–	2
	2	2**	13	3	–	–	–	2
True class	3	–	8	54	–	–	2**	2
	4	–	–	–	5	–	–	–
	5	–	–	–	–	3	–	–
	6	–	–	–	–	–	13	–
	7	1	3	–	–	–	–	2

[a] Explanation as for Table 1.

of view of the information conveyed by each variable. Methods of reporting trace-element geochemical data differ widely, however, and it is clear that the coordinate transform based on a covariance matrix would afford a minimum information loss. The risk of incurring truncation errors in the computation of the probability density function for multielement data also would be minimized.

DISCUSSION

The results obtained using this empirical discriminant method are encouraging. Of particular interest is the good performance based on small training sets. Provided the truncation-error problem does not prove serious (which might be the situation in some computers), the method seems to be well suited for the classification of geologic data.

Further experiments are in progress using geochemical data gathered from reconnaissance stream-sediment surveys, and are producing classifications with a similar success rate to that reported here.

ACKNOWLEDGMENTS

The project of which this forms a part is supported by a grant from the Natural Environment Research Council for an investigation under the direction of Prof. J. S. Webb into the applicability of statistical techniques to the interpretation of geochemical data. The writer wishes to thank Dr. J. C. Gower for critically reading the manuscript and for his helpful comments, and Mrs. E. Kwoo for typing it.

REFERENCES

Chien, Y. T., and Fu, K. S., 1968, Selection and ordering of feature observations in a pattern recognition system: Information and Control, v. 12, p. 394–414.

Favella, L. F., Reineri, M. T., and Righini, G. U., 1969, On a mathematical procedure for detecting significant parameters in the classification of a statistical ensemble of phenomena and its applications: Kybernetik, Bd. 5, Heft 5, p. 187–194.

Fu, K. S., 1968, Sequential methods in pattern recognition and machine learning: Academic Press, London, 227 p.

LARS, 1968, Remote multispectral sensing in agriculture: Laboratory for Agricultural Remote Sensing, Purdue University, Lafayette, Ind., Ann. Rept., v. 3, 175 p.

Lawley, D. N., and Maxwell, A. E., 1963, Factor analysis as a statistical method: Butterworths, London, 117 p.

Nagy, G., 1968, State of the art in pattern recognition: Proc. IEEE, v. 56, p. 836–862.

Specht, D. F., 1967a, Generation of polynomial discriminant functions for pattern recognition: IEEE Trans. Electr. Computers, v. 16, p. 308–319.

Specht, D. F., 1967b, Vectorcardiographic diagnosis using the polynomial discriminant method of pattern recognition: IEEE Trans. Bio-med. Eng., v. 14, p. 90–95.

Watanabe, S., 1965, Karhunen–Loève expansion and factor analysis, theoretical remarks and applications: Proc. 4th Conf. Inform. Theory, Prague, p. 635–660.

23

Reprinted from *Can. J. Earth Sci.* 11:131–146 (1974)

Progresses in *R*– and *Q*–Mode Analysis: Correspondence Analysis and its Application to the Study of Geological Processes

M. DAVID, C. CAMPIGLIO, AND R. DARLING

Department of Geological Engineering, Ecole Polytechnique, Montréal, Québec

Received July 5, 1973

Revision accepted for publication September 5, 1973

This paper essentially introduces a new statistical technique of data analysis together with a practical application. The technique is Correspondence Analysis and it is designed to extract the maximum information from a two dimensional array of positive numbers, such as the concentrations of various elements in several samples. It combines the advantages of *R*– and *Q*-mode analysis and also considerably reduces the problems encountered in the practical application of these techniques: Scaling problems and computer costs. The diagrams, which are produced by projecting at the same time samples and variables on the same plane with proper scaling, aid the interpretation of geological processes. The example that is discussed concerns the Bourlamaque batholith, where igneous differentiation and metamorphism have been clearly identified as the major factors controlling its geochemistry. The complete mathematical formulation is given since it has apparently never been published in the geological literature of North America.

Cet article voudrait présenter une nouvelle technique d'analyse des données en même temps qu'un exemple pratique. Cette technique est l'analyse des correspondances. Son but est d'extraire le maximum d'information possible d'un tableau de nombres positifs, tels que les teneurs en différents éléments de plusieurs échantillons. On y retrouve les avantages des analyses en mode *R* et *Q* et plusieurs de leurs inconvénients sont réduits tels que les problèmes d'unités et d'encombrement de mémoire de l'ordinateur. Les diagrammes qui résultent de cette technique, obtenus en projetant simultanément les échantillons (représentés dans l'espace des variables) et les variables (représentées dans l'espace des échantillons) sur le même plan, sont aptes à faciliter l'interprétation de processus géologiques. On discute en exemple le cas du batholite de Bourlamaque où la différenciation magmatique et le métamorphisme sont clairement identifiés comme facteurs principaux. La formulation mathématique complète est donnée puisqu'il semble que ceci n'ait jamais paru dans la littérature géologique nord-américaine.

Introduction

One of the favourite tools of mathematical geologists in the last decade has been factor analysis, which was expected to be of great help to people manipulating large arrays of data and trying to extract a clear picture of a phenomenon from hundreds of analyses. The most commonly used method was Principal Components Analysis, either in *R*– or *Q*–mode. Although this was probably one of the less deceiving techniques to be adopted, a number of problems remained. One of them was the choice of a weighting procedure. The duality between *R*– and *Q*–mode analysis does not seem to have been fully exploited, even if the diagrams of Le Maitre (1968) are a step in that direction and if Klovan and Imbrie (1971) use it in their algorithm for large *Q*–mode analysis.

In the past few years a new technique called Correspondence Analysis has been developed (Benzecri 1970) and applied in geology (Cazes and Solety 1971). Briefly speaking, this technique is designed to extract the maximum of information from an array of positive numbers. It is used for the analysis of two-dimensional frequency distributions and its generality extends far beyond the treatment of geochemical data.

We here present this new technique, because it is relatively unknown to geologists and because we have so far been very impressed by its advantages, such as large discriminating power and easy interpretation of dual properties, compared to other techniques. We will show that if we can perform an *R*–mode analysis on 1000 samples, a *Q*–mode analysis can also be performed with no computer storage problem. In addition, many of the scaling problems are reduced and the additional property of distributional equivalence is of particular interest to geologists, for whom pooling of variables is a common operation.

In the first section of this paper we will describe the method, assuming that our reader is familiar with principal components analysis

(Parks 1969; McCammon 1968). The next section outlines our computer program and describes the output. The final section describes the application of this method to a geochemical problem in which several interesting interpretations appear. A reader not familiar with mathematical formulation may skip the first part and should find enough information in the second part to understand the example and appreciate the power of the method.

Our example concerns the interpretation of chemical data from an Archean batholith affected by a low-grade metamorphism. Its geochemistry is known to result from at least two distinct processes *i.e.* an early magmatic event responsible for its emplacement as igneous rock and a metamorphism responsible for its present state. The example is used to test the above proposed method of data processing and in particular to show its suitability to separate and identify geological processes such as magmatism and metamorphism.

A geological and geochemical investigation of the Bourlamaque batholith by the second and third authors provided the chemical data used in this study. The first author has introduced the method of correspondence analysis. The geological interpretation presented here forms part of a Ph.D. thesis by the second author.

Mathematical Background

General framework of factor analysis

We here define our notations and review some elementary properties of linear algebra (after Lebart and Fenelon 1971).

We will consider a two-dimensional array, which for the sake of simplicity we will call the matrix of observations of p variables for n samples. A value in the array will be x_{ij}, i referring to the sample and j to the variable. Now the values x_{ij} can be considered as the coordinates of n points in a p-dimensional space R^p or conversely as the coordinates of p points in an n-dimensional space R^n.

The most familiar space is R^p. Identifying groups and relationships in a p-dimensional space is essentially a visual problem; how can we visualise in p dimensions? The problem is similar to the problem of selecting that photograph of a person that would give us as much information as possible. In other words we want to project a

3-D person onto a plane. Similarly, in order to better visualise our information we will try to project our n-D space onto a plane for instance, or onto a single line.

Choosing the best line in R^p

We will project the p-dimensional space onto a line passing through the origin. Let U be a unit column vector on this line. On this line, the projection of a point of R^p is given by the scalar product

$$[1] \quad U'x_{iv} = \sum_{j=1}^{p} u_j x_{ij} \quad U' = (u_1, u_2, \ldots u_p)$$

where U' is the transpose of unit vector U and x_{iv} is the p-component column vector representing the i-th sample.

Minimizing the distance between the points and the line is equivalent to maximizing the sum of the squares of projections of the points on the line or obtaining the maximum spread of the points, which after all is a common way to obtain a good 'picture' of the original p-dimensional group of points. Thus we find the usual condition: To maximize

$$[2] \quad S_p{}^2 = \sum_{i=1}^{n} (U'x_{iv})^2$$

under the condition that $U'U = 1$ since U is a unit vector. We can rewrite $S_p{}^2$ as follows

$$[3] \quad S_p{}^2 = \sum_{i=1}^{n} U'x_{iv}x'_{iv}U$$

and if we call X our original data matrix we have:

$$[4] \quad S_p{}^2 = U'X'XU$$

Simple differentiation of $S_p{}^2 - \lambda(UU' - 1)$ shows that the maximization of $S_p{}^2$ under the condition that $U'U = 1$ yields a vector U, which is the eigenvector corresponding to the largest eigenvalue of $X'X$. Similarly if we want a projection on a plane, the unit vectors defining that plane will be the eigenvectors corresponding to the two largest eigenvalues of X'X, and so on until we reach the true dimensionality of the problem.

Choosing the best line in R^n

If we now consider the same problem in the dual space R^n, the projection of a point representing a variable on a line going through the origin and of unit vector V, will be $V'x_{oi}$ where x_{oi} represents the i-th variable (V and x_{oi} are column vectors).

The sum of the squares of projections is

[5] $$S_n{}^2 = \sum_{i=1}^{p} (V'x_{0i})^2$$

which can be rewritten as [6] $S_n{}^2 = V\,XX'V$

Thus we face the same problem as before and have to maximize $S_n{}^2 = V'XX'V$ under the condition that $V'V = 1$. We know that V will be the eigenvector corresponding to the largest eigenvalue of XX', after what we have seen before.

Now we come to a point that seems to have escaped the attention of geologists, and which may be of considerable interest. It allows the characterization of groups of samples by a given variable or set of variables and considerably reduces computing time since instead of working on an (n × n) matrix XX', everything is done on the (p × p) matrix $X'X$, which in usual circumstances is much smaller. For instance, in usual problems we may have to deal with 200 samples and 10 variables, one matrix is 10 × 10 = 100 the other is 200 × 200 = 40 000. The savings are obvious as pointed out by Klovan and Imbrie (1971).

Duality of R^p and R^n

We have obtained as new reference systems on R^p, the eigenvectors of $X'X$ and in R^n the eigenvectors of XX'.

Considering the two matrices $X'X$ and XX', a capital point can easily be proven: They have the same eigenvalues.

Let V_k be the k-th eigenvector of XX', and U_k be the k-th eigenvector of $X'X$, and λ_k be the k-th eigenvalue of XX'. We have by definition

[7] $$XX'V_k = \lambda_k V_k$$

If r is the rank of matrix XX', then there is a maximum of r nonzero eigenvalues. Now let us premultiply equality [7] by X'.

[8] $$X'X(X'V_k) = \lambda_k(X'V_k)$$

This means that for each k ($k \leqslant r$), $X'V_k$ is an eigenvector of $X'X$ and λ_k is the associated eigenvalue.

To sum up, the eigenvalues (nonzero) of $X'X$ and XX' are the same. Now if we remember that these eigenvalues (divided by their sum) are usually interpreted as percentage of variation explained by each factor, we see that the first factor of R^p and the first factor of R^n yield the same percentage of variation, and so on for the second, third ... to the r-th factor.

A remark on the eigenvectors

Since V_k is a unit vector, we cannot have at the same time $U_k = X'V_k$ as a unit vector. Since the vectors are only defined within a multiplicative constant we can obtain two sets of unit vectors, as follows (using equality [7]).

In one space $U_k = (1/\sqrt{\lambda_k})X'V_k$ $k = 1,2 \dots r$ and in the other $V_k = (1/\sqrt{\lambda_k})XU_k$ $k = 1,2 \dots r$

Another interesting point concerns the co-ordinates of the projections of the n points of R^p on each of the k factorial axes of R^p or the coordinates of the p points of R^n on the kth factorial axis of R^n.

The coordinates of the n points on the kth factorial axis are (scalar products)

$$(x_{1v}{}'U_k, \dots\dots\dots x_{nv}{}'U_k)$$

We see that these are in fact the components of vector XU_k, hence, within a multiplicative constant $(1/\sqrt{\lambda_k})$, the components of V_k.

Similarly the coordinates of the p variables on the kth factorial axis of R^n are the components of U_k times $\sqrt{\lambda_k}$.

Conclusion

The dual properties of the two analyses should always be kept in mind and we see no reason to divorce them and perform only one. In doing so, we simply lose an important quantity of 'visual information', which has proved to be one of the major tools in data analysis, as well as wasting 80% of our computing budget!

A visual inspection of a plot of the sample points and variable points in the same plane will yield information as follows:

(1) Groups of sample points will be interpreted as the result of the same process, or belonging to a specific family.

(2) Similarly, nearby variable points will show the correlation between variables.

(3) Finally a group of sample points will be characterized by the variable points close to that group. This is a great help in the identification of groups and factors. It will be proven later.

One should never forget, however, that these points are in fact two-dimensional projections of 2 multi-dimensional spaces, and although it is the best in average, one should remember that two close projections in some particular cases may represent two points very far from each other.

To be fully protected against unfortunate particular cases, we could compute the contribution of each individual to a factor (Benzecri 1970).

The particular case of Correspondence Analysis

The data that we have to study are nonnegative and arranged in a two-dimensional array, let us say $X = \{x_{ij}\}$.

We may have to compare variables that have very different orders of magnitude; this is a well known scaling problem and in Principal Component Analysis we often use a standardization of variables in order to avoid it. This standardization is an operation that is not symmetrical in i and j.

We will consider our array as a two-dimensional frequency distribution, which means we will divide all the entries by

$$\sum_i \sum_j x_{ij}.$$

They can now be regarded as probabilities and we can also define marginal probabilities:

[9]   Let $T = \sum_i \sum_j x_{ij}$

$p_{ij} = x_{ij}/T$

$p_{.j} = \sum_i p_{ij}$

$p_{i.} = \sum_j p_{ij}$

We will represent our n sample points in R^p by n points having as coordinates $(p_{ij}/p_{i.})$. Thus the coordinates are the relative proportions of each variable in each sample.

We should note that the n points will in fact be in a $(p - 1)$ dimensional space since $\sum_j p_{ij} = p_{i.}$. Also note that two points having the same coordinates, represent two samples with the same relative proportion of each element.

Now in order to characterize the proximity of two samples we have to choose a distance. The usual distance gives to each variable the same weight since the formula is, considering two sample points i and l

[10] $D^2(i, l) = \sum_{j=1}^{p} \left\{ \dfrac{p_{ij}}{p_{i.}} - \dfrac{p_{lj}}{p_{l.}} \right\}^2$

If a column j_o has large numbers (silica), the role of this variable in the distance computation

is too large in usual problems where we are interested in the relative role of each variable; we will thus choose a weighted distance

[11] $D_1^2(i, l) = \sum_{j=1}^{p} \dfrac{1}{p_{.j}} \left\{ \dfrac{p_{ij}}{p_{i.}} - \dfrac{p_{lj}}{p_{l.}} \right\}^2$

In other words we have considered each sample as a frequency distribution and the distance $D_1^2(i,l)$ is the X^2 distance associated to the distribution $p_{.j}$, between the two distributions representing the two samples. Altogether, everything is as if we take as coordinates of our n points

$p_{ij}/p_{i.}.\sqrt{p_{.j}}$

Now giving to each sample point a weight proportional to p_i and noticing that the mean point in R^p has coordinates $\sqrt{p_{.j}}$, the axes we are looking for are the eigenvectors of the usual cross products matrix Z defined as follows

[12] $Z = \{z_{jj'}\} = \left\{ \sum_{i=1}^{n} p_{i.} \left(\dfrac{p_{ij}}{p_{i.} \sqrt{p_{.j}}} - \sqrt{p_{.j}} \right) \right.$

$\left. \times \left(\dfrac{p_{ij'}}{p_{i.} \sqrt{p_{.j'}}} - \sqrt{p_{.j'}} \right) \right\}$

Note that Z can be rewritten, splitting p_i into $(\sqrt{p_{i.}}) (\sqrt{p_{i.}})$:

[12A] $Z = \{z_{jk}\} = \left\{ \sum_{i=1}^{n} \dfrac{(p_{ij} - p_{i.}p_{.j})}{\sqrt{p_{i.}p_{.j}}} \right.$

$\left. \times \dfrac{(p_{ik} - p_{i.}p_{.k})}{\sqrt{p_{i.}p_{.k}}} \right\}$

We thus see that we have applied a symmetrical transformation to the original data matrix,

x_{ij} became $\dfrac{p_{ij} - p_{i.}p_{.j}}{\sqrt{p_{i.}p_{.j}}}$

Also note that the trace of matrix Z is the χ^2 quantity with $(n - 1)(p - 1)$ degrees of freedom which has to be calculated to check the hypothesis that the n and p characters which are put in correspondence through the $\{x_{ij}\}$ are independent.

Various simplifications show that the eigenvectors of z are the same as for matrix Y defined as follows

[13] $Y = \{y_{jk}\} = \left\{ \sum_{i=1}^{n} \dfrac{p_{ij}p_{ik}}{p_{i.} \sqrt{p_{.j}p_{.k}}} \right\}$

Here we are back to the general problem and the conclusions we have already made regarding the coordinates of sample points and variable points apply.

Graphical representation

Let U_q be the qth eigenvector of Y, then the projection of the ith sample point on the qth axis is simply

$$[14] \qquad f_{iq} = \sum_{k=1}^{p} u_{kq} \frac{p_{ik}}{p_{i.} \sqrt{P_{.k}}}$$

If we want the factors to apply to the $p_{ij}/p_{i.}$ data we have to take as component of the factor t_q

$$[15] \qquad t_{kq} = u_{kq}/\sqrt{P_{.k}}$$

The dual property now gives us the components of the w_q factors of R^n

$$[16] \qquad w_{lq} = \frac{1}{\sqrt{\lambda_q}} \sum_{j=1}^{p} \frac{p_{lj}}{p_{l.}} t_{jq}$$

where λ_q is the qth eigenvalue of Y.

This relationship shows that the w_i are, within a scale factor $1/\sqrt{\lambda_q}$, the gravity centers of the t_j, each one of the t_j having a weight $p_{ij}/p_{i.}$, which can be seen as the conditional probability of j once i has happened. In other words each variable point is the center of gravity of the sample points weighted by their relative importance for that variable.

To summarize all these mathematical considerations, we can now consider the output of the program that we used for our study.

The Program and Interpretation of the Output

The only results we need besides a listing of the original data and identification are the two sets of factors in R^p and R^n, the percentage of variation explained by each of them, and a graphical plot of both sample points and variable points in either the same plane of the first two factors, or considering other factors, in whichever other plane is preferred.

First we print the percentage of variation explained by the factors in decreasing order, and the eigenvalues. They give us as usual, a measure of the quantity of information that is retained when we decide to keep only one, two or more factors. For instance, in the example shown in Table 1, the first factor accounts for 46% of the total variation. One should remember that the technique used is distribution free, consequently there is no test of confidence and no rule to accept or reject

the significance of a factor. In a different problem, in which we tried to discriminate between different rock types for instance, we had factors explaining less than 10% of the variance, but which were definitely characteristic of a particular horizon. In the example that we will discuss later, we are dealing with a single rock unit and the different geological processes responsible for factors explaining a few percent of the variance may be highly questionable.

The factors that are printed next are equivalent to the usual factors of Principal Component Analysis and the contribution of each original variable to the factor can also be interpreted within a multiplicative constant as a correlation coefficient between that factor and each variable. Now to find the scores of each sample with respect to these factors one has to remember the weighting procedure and also remember that there is a proportionality coefficient ($\sqrt{\lambda_k}$) to use in order to be able to plot both sample points and variable points on the same plane and at the same scale. As an example we will take a particular sample, and using the first and second factors shown in Table 2 we will compute its scores on axes 1 and 2 and then find the corresponding point on Fig. 3.

Sample 52 has the following trace elements concentrations (ppm): 97 Cu; 86.5 Zn; 18.50 Li; 212.5 Sr; 16.5 Rb; 47 Ni; 37 Co; 45 Cr; 201 V. The sum of all these values is $p_{52.} = 761$ and the first and second eigenvalues are 0.066 and 0.0406.

The score of sample 52 on the first axis is obtained by multiplying the vector of concentrations by the first factor and dividing by 761 and 0.066. We find 0.3069. Similarly the score of sample 52 on the second axis is obtained by multiplying the vector of concentrations by the second factor and dividing by 761 and 0.0406. We find 0.1474. These scores are listed by the program. The columns of scores, which are also computed in standard computer packages, are, within a multiplicative constant, the eigenvectors of the n by n matrix.

Finally the two sets of factors, which happen to be proportional to the coordinates of sample points and variable points, are used to produce a simple plot that is the visual result on which most of the interpretation is made. Every point is identified and because space is required to write this identification many overlaps may

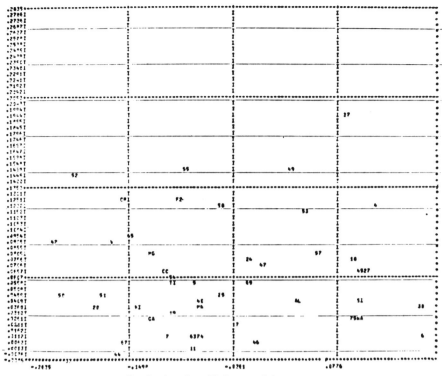

FIG. 1.   Actual graphical output of the program.

occur each of which is listed. In order to avoid this we can produce the graphs at almost any scale. A complete one is shown on Fig. 1; to eliminate definition problems the other graphs have been redrawn without the identifications of the sample points.

The interpretation of the plot should be made remembering that this is nothing more than a simultaneous representation of $R-$ and $Q-$mode analysis with appropriate scaling: the proximity of sample points characterizes members of a similar group while the proximity of variable points denotes a similarity of behaviour. One should always keep in mind however that this is only a projection and that two points that are 'on top' of each other in an n-dimensional space with respect to the plane of the first two factors will artificially

appear close to each other on the projection. A projection on another plane such as that of the 1st and 3rd factor will usually resolve this kind of doubt immediately.

Now the proximity of a group of sample points and variable points is really what helps to identify groups and characterize them. In the study of igneous rocks where the variables are chemical elements, this arrangement will reflect differentiation for instance and help in naming the various groups of points according to their position relative to the poles.

Another useful graphical representation is obtained by linking the center of our plane to each of the variable points and remembering that each of these arrows represents the strength and direction according to which the different elements pull away from the center and differ-

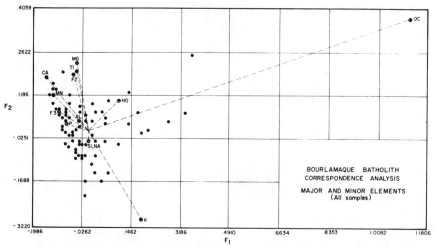

FIG. 2. Bourlamaque batholith Correspondence Analysis major and minor elements (all samples).

entiate individuals from the average point. This representation is equivalent to the plot of loadings used in Q–mode analysis; we will systematically plot it for each projection we make, see Figs. 2, 3, 4, 5, 6, and 8.

Application to the Bourlamaque Batholith

Geological setting

The chemical data are from the Bourlamaque batholith, a metadiorite body within the eastern end of the Noranda–Val d'Or volcanic belt, which in turn lies along the southern margin of the Superior Province of the Canadian Shield. The batholith has a surface area of 65 mi² (186 km²) and outcrops immediately north of the town of Val d'Or, which is in northwestern Quebec, 260 mi (415 km) northwest of Montreal.

This intrusion has been the object of field, petrographic, and geochemical studies (Campiglio 1973), the results of which will be presented in a forthcoming paper. For the purpose of the present study, it is worth pointing out that we had rather sound data on the geological history of the batholith *before* doing any interpretation. So, by referring to available geological data, we were able to check step by step the interpretation suggested by the proposed Correspondence Analysis method.

Petrographic studies of the batholith show that its primary igneous mineralogy has nearly everywhere been replaced by secondary metamorphic assemblages where chlorites, epidotes; paragonite (?), leucoxene, quartz, and acidic plagioclase are ubiquitous, while carbonates, although often present, are rather erratically distributed. Scattered primary minerals such as pyroxene, amphibole, and intermediate plagioclase are unusual. The above assemblages account for a low-grade metamorphism, locally patchy in its expression. On the other hand, from the geochemical behavior of the elements analyzed it appears that, on the whole and with only a few exceptions, the batholith has retained its primary chemistry during metamorphism.

Correspondence Analysis of Bourlamaque Batholith

Introduction

Raw material used consists of 75 chemical analyses for 22 elements. In the present study ʻCorrespondence Analysis shows that the first two factors are always able to explain at least 70% of the total variation. For this reason and also because local variations of batholith chemistry are possible, we will generally deal only with the projection plane having as axes

TABLE 1.   Correspondence analysis on major and minor elements (all samples), loadings of the first five factors with percentage of the total variance explained

	F1	F2	F3	F4	F5
CA	−0.1359	0.1631	−0.0191	−0.0413	−0.0576
NA	0.0107	−0.0485	0.0068	0.0749	−0.0263
K	0.1934	−0.3050	0.2834	−0.2078	−0.0334
MG	−0.0348	0.2118	0.0272	0.0298	0.0401
SI	0.0115	−0.0479	−0.0054	0.0032	0.0045
F3	−0.0977	0.0488	−0.0742	−0.0699	0.0686
F2	−0.0549	0.1755	0.0985	0.0363	0.0341
TI	−0.0429	0.1846	0.0328	0.0164	0.0726
MN	−0.1191	0.1053	0.0268	−0.0022	0.0094
HO	0.1148	0.0923	−0.0371	−0.0582	0.0705
AL	−0.0262	0.0208	0.0014	−0.0013	−0.0242
P	−0.0772	0.0169	0.1296	−0.0327	0.0703
OC	1.1179	0.3721	−0.0401	−0.0053	−0.0597
% expl.	46	34	5	4	3

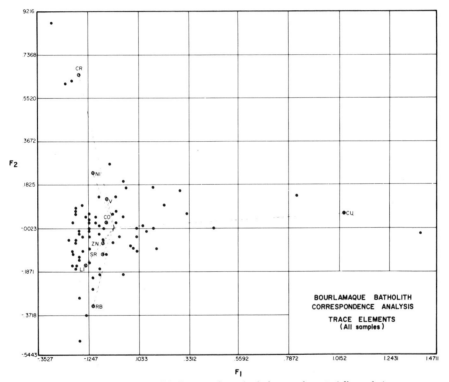

FIG. 3.   Bourlamaque batholith Correspondence Analysis trace elements (all samples).

TABLE 2. Correspondence analysis on trace elements (all samples), loadings of the first five factors with percentage of the total variance explained

	F1	F2	F3	F4	F5
CU	1.0389	0.0433	−0.0555	−0.0099	−0.0012
ZN	−0.0557	−0.0799	−0.0051	−0.0394	0.1437
LI	−0.1223	−0.1808	−0.1984	0.0191	0.0405
SR	−0.0556	−0.1245	0.0191	−0.0786	−0.0510
RB	−0.0991	−0.3497	−0.5002	0.3137	−0.0209
NI	−0.0913	0.2171	−0.0331	−0.0792	0.1260
CO	−0.0341	0.0148	0.0695	0.0169	0.0982
CR	−0.1561	0.6414	−0.2592	−0.0721	−0.0610
V	−0.0380	0.1059	0.1514	0.1384	−0.0179
% expl.	45	28	13	7	3

TABLE 3. Correspondence analysis on major, minor, and trace elements (all samples), loadings of the first five factors with percentage of the total variance explained.

	F1	F2	F3	F4	F5
CA	0.0300	0.0033	0.1198	0.0092	0.0454
NA	0.0691	−0.0914	−0.0830	−0.0102	0.0090
K	0.1113	−0.3120	−0.3944	0.2289	0.0034
MG	0.0487	0.1128	0.0502	0.0306	0.0271
SI	0.0699	−0.0943	−0.0800	0.0160	0.0040
F3	0.0263	−0.0514	0.0615	−0.0106	0.0087
F2	0.0232	0.0509	0.0467	0.0685	0.0949
TI	0.0137	0.0868	0.0598	0.0459	0.0201
MN	0.0448	−0.0217	0.0728	0.0076	0.0956
HO	0.0683	−0.0233	−0.0365	0.0814	−0.0560
AL	0.0542	−0.0700	−0.0186	0.0183	0.0158
P	0.0024	−0.0714	−0.0400	−0.0147	0.0882
OC	0.3123	0.2386	−0.4236	0.4013	−0.5037
CU	−1.0618	0.0271	−0.0652	−0.0090	−0.0031
ZN	0.0467	−0.0672	0.0115	−0.0484	0.1411
LI	0.1194	−0.1709	−0.1824	0.0095	0.0327
SR	0.0471	−0.1110	0.0342	−0.0814	−0.0548
RB	0.1017	−0 3496	−0.4816	0.2899	−0.0031
NI	0.0773	0.2311	−0.0270	−0.0874	0.1229
CO	0.0235	0.0269	0.0756	0.0142	0.0958
CR	0.1364	0.6543	−0.2728	−0.0819	−0.0555
V	0.0238	0.1240	0.1535	0.1400	−0.0178
% expl.	44	27	13	7	3

the first two factors. Higher order factors accounting usually for less than 10% of total variation are sometimes difficult to interpret and often geologically meaningless. On the other hand, loadings of the variable elements for the first two factors provided the key to the geological interpretation in this study.

Tables 1, 2, 3, 5, and 6 show for each of the Correspondence Analyses performed on the Bourlamaque data the loadings of the first five factors as well as the percentage of the total variance explained by each of them.

Notations used in these tables and in the figures are self-explanatory except for $F3(Fe_2O_3)$, $F2(FeO)$, $HO(H_2O^+)$, and $OC(CO_2)$.

Major and minor elements

Correspondence Analysis with major and minor elements on the bulk of the samples (Fig. 2) locates a plane explaining 80% of the total variation by its two first factors (Table 1). The first factor (46%) is characterized by CO_2 and, to a lesser degree, by H_2O. From a

306

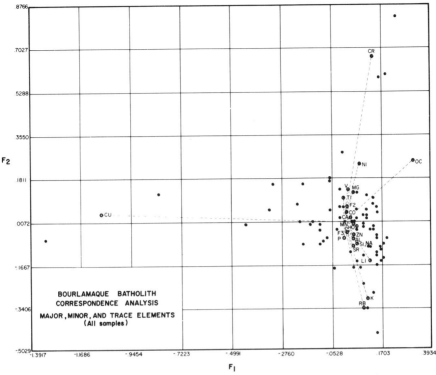

Fig. 4. Bourlamaque batholith Correspondence Analysis major, minor, and trace elements (all samples).

geological point of view the weight of K on this axis is rather fictitious, as a mere axes rotation would restore the true geological relationship. This rotation being visually obvious, we did not find it necessary in this case to perform an analytical rotation, which would not add any information and which would require more computation. The second factor (34%) is the result of the interaction of sialic and femic elements. Elements of the second factor are arranged on Fig. 2 in a geologically meaningful trend going from Mg, Ti, Fe^{2+}, Ca, and Mn in the upper part, through Fe^{3+}, Al, P, Si, and Na towards K in the lower part. The same figure gives information on the behavior of the samples. The mathematical formulation we developed in the first part, makes it possible to show on the same diagram elements and

samples together, a very useful feature of the method. The sample trend, more scattered if compared with the element trend (element location is the result of an average) matches fairly well the pattern outlined by the elements: Thus the sample points representing mafic, intermediate and silicic rocks of the batholith are encountered in that order as we cross the figure going to the bottom. Sample points

TABLE 4. Comparison between analyses with and without high chromium samples. Figures mean the percentage of the total variance explained by the first two factors

Analysis carried out with	All samples	High chromium samples excluded
Trace elements	74	80
Major + trace elements	70	78

TABLE 5. Correspondence analysis on trace elements (high chromium samples excluded), loadings of the first five factors with percentage of the total variance explained

	F1	F2	F3	F4	F5
CU	1.0373	0.0704	−0.0126	−0.0028	−0.0049
ZN	−0.0733	0.0139	−0.0062	−0.1044	−0.0819
LI	−0.1452	0.2597	0.0178	−0.0339	−0.1411
SR	−0.0783	0.0378	−0.1022	0.0252	0.0257
RB	−0.1484	0.6120	0.2811	0.0365	0.0707
NI	−0.0518	−0.1308	0.0362	−0.1842	0.0156
CO	−0.0371	−0.0944	0.0627	−0.0337	−0.0517
CR	−0.0180	−0.1428	0.1168	−0.1408	0.1582
V	−0.0281	−0.1547	0.0956	0.0867	−0.0177
% expl.	59	21	9	5	3

TABLE 6. Correspondence analysis on major, minor, and trace elements (high chromium samples excluded), loadings of the first five factors with percentage of the total variance explained

	F1	F2	F3	F4	F5
CA	0.0370	−0.1224	0.0259	0.0024	−0.0020
NA	0.0818	0.1009	0.0167	−0.0373	0.0228
K	0.1535	0.5022	0.1892	0.0176	0.0162
MG	0.0328	−0.1245	0.0832	0.0189	−0.0110
SI	0.0841	0.1004	0.0346	−0.0105	0.0392
F3	0.0398	−0.0353	−0.0151	0.0022	0.0302
F2	0.0187	−0.0878	0.1216	−0.0186	−0.0578
TI	0.0023	−0.0889	0.0526	0.0002	−0.0196
MN	0.0543	−0.0714	0.0291	−0.0327	−0.0607
HO	0.0697	0.0399	0.0997	0.0908	0.0384
AL	0.0669	0.0339	0.0375	−0.0017	0.0199
P	0.0129	0.0764	−0.0275	−0.0777	−0.1080
OC	0.2649	0.3890	0.4408	0.5619	0.1471
CU	−1.0617	0.0833	−0.0115	−0.0014	−0.0032
ZN	0.0607	−0.0020	−0.0217	−0.1056	−0.0922
LI	0.1384	0.2444	0.0016	−0.0368	−0.1216
SR	0.0658	0.0224	−0.1101	0.0306	0.0204
RB	0.1467	0.6030	0.2450	0.0221	0.0255
NI	0.0376	−0.1431	0.0249	−0.1810	0.0133
CO	0.0241	−0.1028	0.0584	−0.0372	−0.0479
CR	0.0039	−0.1467	0.1121	−0.1424	0.1615
V	0.0134	−0.1630	0.0960	0.0826	−0.0233
% expl.	57	21	8	5	3

representing the more altered rocks fall outside the above roughly oval-shaped trend because they have been pulled towards the CO_2 pole (1st factor).

Trace elements

Correspondence Analysis performed again on the bulk of the samples but with trace elements only (Fig. 3 and Table 2) is able to explain 74% of the total variance by the usual two first axes. Now the first factor (45%) is dominated by Cu against, again as second factor (29%), the other elements arranged as the major elements *i.e.* sialic vs. femic. So we find, in the order Rb, Li, Sr, and Zn against Co, Cu, V, Ni, and Cr. The first factor Cu, by pulling away the Cu–rich samples, acts here exactly as CO_2 did previously. Another interesting feature worth pointing out is the location of chromium as well as a few samples closely associated with it, well outside the cloud of the remaining elements and samples. This behavior will be discussed later in the light of the results given by the analysis in which major and trace elements were dealt with together. An objection could be raised concerning the meaning of copper, a key element in our interpretation. From Fig. 3 it could be inferred that copper location is essentially due to the presence of only two Cu–rich samples. This was checked by an analysis performed without the two above samples and this had *no effect* on Cu behavior.

Major, minor, and trace elements together

Obviously, the picture offered by the latter analysis (Fig. 4 and Table 3), where major and trace elements are simultaneously taken into consideration, does not add too much information to what we already know, but it has had the advantage of summarizing the results. Correspondence Analysis shows in fact its remarkable synthetic power providing a comprehensive picture of the whole system without a reduction of its analytical strength. Now the projection plane explains 70% of the total variance, 43% of which is explained by the first factor and 27% by the second. Cu and CO_2 pulling in opposite ways make up the first factor. The opposite sign of these elements is an interesting relationship confirmed by a rough negative correlation found in geochemical studies. The second factor is the usual pattern of sialic vs. femic elements. It remains to explain the particular locations of the few samples characterized by chromium. It is evident from Fig. 4 that the trend of ferromagnesian elements presents a gap at the level of chromium. In this case, according to the given rules of interpretation, some samples characterized by this element should belong to a family somewhat different than the rest of the batholith. Information is here simultaneously provided by both variables: The nature of the anomaly is concisely indicated by the variable element (chromium), while samples close to this element show the reader which samples are involved in the anomalous process. Returning to our problem we notice that these samples are spatially related and belong to an area that geochemical studies have revealed to be anomalous in magnesium and nickel and especially enriched in chromium.

Removal of the contamination

Considering the above feature peculiar to the area and due to a special geological process (contamination), we repeated the analysis without the contaminated rocks in order to see what role, if any, was played by these samples. The results of the new situation are the disappearance of the chromium gap (geologically significant), and the improvement of the power of the method. Details are shown in Tables 4, 5, 6 and in Figs. 5 and 6.

On the other hand the geological pattern offered by the previous analyses and in particular the meaning of the factors does not change.

Identification of factors

We are able to clearly identify on all diagrams the second factor (interaction of sialic and femic elements) as having been caused by magmatic differentiation. The first factor (Cu, CO_2, and H_2O) is also clearly identified as a product of metamorphism when the diagrams for major and trace elements are considered separately. However, due to the lack of a strict correlation between Cu and CO_2, when we consider the diagram for major, minor, and trace elements together, we cannot expect CO_2 and Cu to both contribute to a single factor only.

An illustration of the magmatic differentiation process can be seen on Fig. 7, where we projected on the second axis, all the elements except Cu, CO_2, and H_2O, which have been seen on previous diagrams to be orthogonal to that factor. The distance of Li, K, and Rb from the bulk of the other elements is quite probably due to the small number of differentiated rocks compared with the whole of the batholith, which is rather homogeneously intermediate in composition. This can also be interpreted as a measure of the chemical homogeneity of the batholith.

Removal of the first factor

Having noticed that the simpler the problem the more powerful the method becomes, we discarded from the subsequent analysis the elements considered as the expression of metamorphism *i.e.* Cu, CO_2, and H_2O, in order to dig further into our problem. This can be achieved by a new analysis without the above

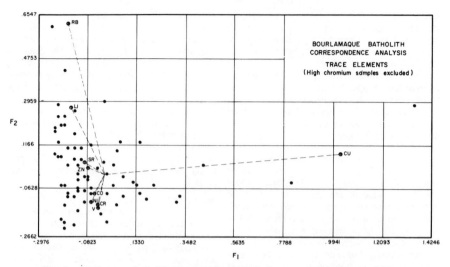

FIG. 5. Bourlamaque batholith Correspondence Analysis trace elements (high chromium samples excluded).

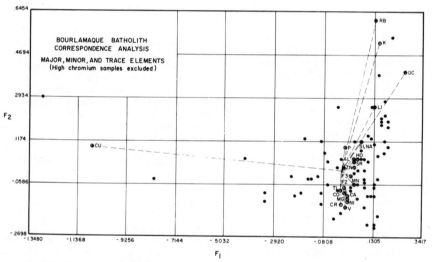

FIG. 6. Bourlamaque batholith Correspondence Analysis major, minor, and trace elements (high chromium samples excluded).

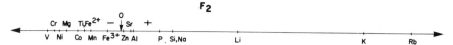

FIG. 7. Projection on second factor of major, minor, and trace elements (from Fig. 6; elements typical of first axis excluded).

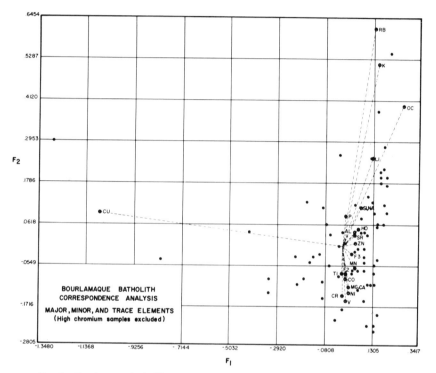

FIG. 8. Bourlamaque batholith Correspondence Analysis major, minor, and trace elements (high chromium samples excluded).

elements, but a simpler alternative is to make use of the previous analyses. As a matter of fact if we look at the plane determined by the second and the third factor, we automatically eliminate the first factor (metamorphism in our case). This is obvious for a mathematician; however we find it instructive to do the comparative study for the benefit of geologists less acquainted with orthogonality and multi-dimensional spaces. The near identity of the two planes so obtained is shown in Table 7, where the loadings of the factors that determine the planes are reported. The former plane, with factors 1 and 2 as coordinates, results from an analysis without the elements of metamorphism while the latter is the plane defined by factors 2 and 3 in the same body analysis, but with the elements of metamorphism (first factor) still present. The factors shown in Table 7 refer to an analysis carried out without the high chromium samples, since we noticed from Table 3 that the presence of these contaminated samples could completely hide the significance of the third factor.

Examining the plane of the second and third factor (Fig. 8) beside the magmatic differentiation (2nd factor) we identify within the third factor only strontium as a geologically significant feature. Strontium is in fact the only element (the weight of Fe^{3+}, P, and Zn being negligible), which now stands against all the others together. Spatially related high-strontium samples had indeed been outlined by the geochemical study and once again Correspondence Analysis is able to pinpoint this fact. Figure 8 shows, even with some scattering at least partially due to metamorphic effects, two subparallel trends of samples characterized by different concentrations of strontium corresponding to different areas of the batholith.

Conclusions

We have shown in this study that the method of Correspondence Analysis by considering at the same time samples and elements can help in extracting at minimum cost a maximum of information from a set of data. In this case we had previous knowledge of the batholith, but it is hoped that such an analysis could also be of major interest in cases where geological information is not already available. Also, as in standard factorial analysis, isofactor maps can be produced and striking results obtained; or in the case where factors are not clearly understandable, a rotation could be performed, although so far we have not found this necessary. A manual for the program is presently being prepared and it is hoped that it will be available shortly.

TABLE 7. Comparison of plane defined by factors 1 and 2 with Cu, CO_2, and H_2O removed and by factors 2 and 3 keeping all the elements

	F1	F2	F2	F3
CA	−0.1179	0.0233	−0.1224	0.0259
NA	0.1085	0.0201	0.1009	0.0167
K	0.5086	0.1972	0.5022	0.1892
MG	−0.1226	0.0778	−0.1245	0.0832
SI	0.1086	0.0354	0.1004	0.0346
F3	−0.0309	−0.0154	−0.0353	−0.0151
F2	−0.0905	0.1207	−0.0878	0.1216
TI	−0.0949	0.0524	−0.0889	0.0526
MN	−0.0665	0.0302	−0.0714	0.0291
HO			0.0399	0.0997
AL	0.0398	0.0368	0.0339	0.0375
P	0.0711	−0.0126	0.0764	−0.0275
OC			0.3890	0.4408
CU			0.0833	−0.0115
ZN	0.0032	−0.0138	−0.0020	−0.0217
LI	0.2594	0.0074	0.2444	0.0016
SR	0.0313	−0.1104	0.0224	−0.1101
RB	0.6047	0.2565	0.6030	0.2450
NI	−0.1372	0.0327	−0.1431	0.0249
CO	−0.1020	0.0588	−0.1028	0.0584
CR	−0.1526	0.1185	−0.1467	0.1121
V	−0.1652	0.0871	−0.1630	0.0960

Acknowledgments

All the programming was carried out by Yves Beauchemin and the help of Peter Dowd during the final edition of the paper was greatly appreciated. The field work for this study was financed by the Quebec Department of Natural Resources, the laboratory studies were supported by the NRC Grant A-5535 and statistical work was made possible by NRC Grant A-7035.

BENZECRI, J. P. 1970. Distance distributionnelle et métrique du chideux en analyse factorielle des correspondances. Laboratoire de Statistique Mathématique, Université de Paris 6. 3e édition.

CAMPIGLIO, C. 1973. Etude géochimique du batholite de Bourlamaque Ph.D. thesis, Ecole Polytechnique, Université de Montréal (in preparation).

CAZES, P. 1970. Application de l'analyse de données au traitement des problèmes géologiques. Thèse de 3e cycle, Paris.

CAZES, P., SOLETY, P., and VUILLAUME, Y. 1970. Exemple de traitement statistique de données hydrochimiques. Bulletin du BRGM, no. 4, pp. 75–90.

KLOVAN, J. E. and IMBRIE, J. 1971. An algorithm and Fortran IV Program for large scale Q–mode analysis. Mathematical Geology 3, pp. 61–67.

LEBART, L. and FENELON, J. P. 1971. Statistique et Informatique Appliquées, Dunod, Paris.

LE MAITRE, R. W. 1968. Chemical variation within and between volcanic rock series–a statistical approach. J. Petrol. **9**, pp. 220–252.

McCAMMON, R. B. 1968. Multiple component analysis and its application in classification of environments A.A.P.G. Bull. **52**, pp. 2178–2196.

PARKS, J. M. 1969. Multivariate Facies Maps. *In*: Symposium on computer applications in petroleum exploration. Kansas Geological Survey Computer Contribution 40.

POTENZA, R. 1972. L'analisi fattoriale come mezzo di studio dei precessi metasomatici. Rend. Soc. It. Mineralogia e Petrologia. XXVIII, pp. 545–558.

313

REFERENCES

Agterberg, F. P. 1964. Statistical techniques for geological data. *Tectonophys.* **1** (3):233-255.

Agterberg, F. P. 1967. Computer techniques in geology. *Earth Sci. Rev.* **3**(1):44-77.

Agterberg, F. P. 1974. *Geomathematics, Developments in Geomathematics 1.* Amsterdam: Elsevier, 596 pp.

Agterberg, F. P., and Robinson, S. C. 1972. Mathematical problems of geology. *Int. Stat. Inst. Bull.* **44**(1):567-596.

Ahrens, L. H. 1953. A fundamental law of geochemistry. *Nature* **172**(4390):1148.

Ahrens, L. H. 1954a. The lognormal distribution of elements (a fundamental law of geochemistry and its subsidiary). *Geochim. Cosmochim. Acta* **5**(2):49-73.

Ahrens, L. H. 1954b. The lognormal distribution of the elements (2). *Geochim. Cosmochim. Acta* **6**(213):119-120.

Ahrens, L. H. 1957. Lognormal-type distributions (3). *Geochim. Cosmochim. Acta* **11**(4):205-212.

Allen, P. 1944. Statistics in sedimentary petrology. *Nature* **153**(3872):71-74 (see Paper 6.)

Anderson, R. Y., and Koopmans, L. H. 1963. Harmonic analysis of varve time series. *J. Geophys. Res.* **68**(3):877-893.

Anderson, R. Y., and Koopmans, L. H. 1968. Statistical analysis of the Rita Blanca varve time series. In R. Y. Anderson and D. W. Kirkeland, eds. *Paleoecology of an early Pleistocene lake on the high plains of Mexico. Geol. Soc. Am. Mem.* **113**:59-75.

Anderson, T. W., and Goodman, L. A. 1957. Statistical inference about Markov chains. *Annals Math. Stat.* **28**:89-110.

Ayler, M. F. 1963. Statistical methods applied to mineral exploration. *Min. Cong. J.* **49**(11):41-45.

Bath, M. 1974. *Spectral Analysis in Geophysics, Developments in Solid Earth Geophysics.* Amsterdam: Elsevier, 563 pp.

Becker, R. M., and Hazen, S. W. 1961. Particle statistics of infinite populations as applied to mine sampling. *U. S. Bur. Mines Rept. Invest.* **5669,** 79 pp.

References

Benzecri, J. P. 1973. *L'Analyse des donnees, Tome II: L'Analyse des correspondances* Paris: Dunod, 620 pp.

Bingham, C. 1964. Distributions on the sphere and on the projective plane. Unpubl. doctoral dissertation, Yale Univ., New Haven, Conn., 140 pp.

Blackith, R. E., and Reyment, R. A. 1971. *Multivariate Morphometrics*. London: Academic Press, 435 pp. (See Paper 17.)

Blais, R. A., and Carlier, P. A. 1968. Applications of geostatistics in ore evaluation. *Can. Inst. Min. Met., Ore Reserve Estimation and Grade Control*, Spec. Vol. 9: 41-68.

Brinkmann, R. 1929. Statistisch-biostratigraphische untersuchungen on mitteljurassischen ammoniten, etc. *Abb. Ges. Wiss. Gottingen Math. Phys., K1. N.F.* 13(3):1-249.

Brower, J. C., and Veinus, J. 1974. The statistical zap versus the shotgun approach. *Math. Geol.* 6(4):311-332.

Burch, C. R. 1972. Tests of the normality of frequency distributions of Na, K, SiO_2 and Cl in the Dartmoor granite. *Trans. Roy. Geol. Soc. Cornwall* 20(3):179-198.

Burch, C. R., and Murgatroyd, P. N. 1971. Broken-line and complex frequency distributions. *Math. Geol.* 3(2):135-156.

Burma, B. H. 1949. Studies in quantitative paleontology, II: Multivariate analysis–a new analytical tool for paleontology and geology. *J. Paleo.* 23(1):95-103. (see Paper 7.)

Butler, J. C. 1975. Occurrence of negative open variances in tertiary systems. *Math. Geol.* 7(1):31-45.

Cattell, R. B. 1965. Factor analysis: An introduction to essentials. I: the purpose and underlying models. *Biometrics* 21:190-210. II: The role of factor analysis in research. *Biometrics* 21:405-435.

Chayes, F. 1949. On ratio correlation in petrography. *J. Geol.* 57(3):239-254.

Chayes, F. 1954. The lognormal distribution of the elements; a discussion. *Geochim. Cosmochim. Acta.* 6(2/3):119-120.

Chayes. F. 1956. *Petrographic modal analysis*. New York: John Wiley and Sons, 113 pp.

Chayes, F. 1960. On correlation between variables of constant sum. *J. Geophys Res.* 65(12):4185-4193.

Chayes, F. 1962. Numerical correlation and petrographic variation. *J. Geol.* 70(4): 440-452.

Chayes, F. 1964. Variance-covariance relations in some published Harker diagrams of volcanic suites. *J. Petrol.* 5(2):219-237.

Chayes, F. 1967. On the graphical appraisal of the strength of associations in petrographic variation diagrams. *Res. in Geochem.* (2):322-339.

Chayes, F. 1970. Effect of a single nonzero open variance on the simple closure test. In D. F. Merriam, ed. *Geostatistics.* New York: Plenum Press, pp. 11-12.

Chayes, F. 1971. *Ratio Correlation*. Chicago: Univ. Chicago Press, 99 pp.

Chayes, F. 1972. Effect of the proportion transformation on central tendency. *Math. Geol.* 4(3):269-270.

Chayes, F. 1975. A prior and experimental approximation of simple ratio correlations. In R. B. McCammon, ed., *Concepts in Geostatistics*. Berlin: Springer-Verlag, pp. 106-137.

Chayes, F. and Kruskal, W. 1966. An approximate statistical test for correlations between proportions. *J. Geol.* 74(5/2):692-702. (see Paper 15.)

Chenowith, P. A. 1952. Statistical methods applied to Trentonian stratigraphy in New York. *Bull. Geol. Soc. Am.* **63**(6):521–560.

Childs, D. 1970. *The Essentials of Factor Analysis.* London: Holt, Rinehart and Winston, 107 pp.

Chirvinsky (Tschirwinsky), P. N. 1909, Problème actuel de la pétrographie contemporaine lié à la question des méthodes de détermination quantitative de la composition minéralogique des roches (Russian with French summary). *Bull. Soc. Oural Sci. Natur.* **28**:9–46.

Chervinsky (Tschirwinsky), P. N. 1911. *Quantitative mineralogische und chemische Zusammensetzung der granite und greissen* (Russian with German summary). Moscow.

Claerbout, J. F. 1976. *Fundamentals of Geophysical Data Processing with Applications to Petroleum Prospecting.* New York: McGraw-Hill, 274 pp.

Clark, I., and Garnett, R. 1974. The detection of multiple mineralisation phases by statistical methods. *Trans. Inst. Min. Met.* **83**(809):A43–A52.

Clark, M. W. 1976. Some methods for statistical analysis of multimodal distributions and their application to grain-size data. *Math. Geol.* **8**(3):267–282.

Clarke, F. W. 1892. The relative abundance of the chemical elements. *Bull. Philos. Soc.* **11**:131–142.

Clarke, F. W. 1924. The data of geochemistry. *U.S. Geol. Surv. Bull.* (770).

Cole, A. J., ed. 1969. *Numerical Taxonomy.* London: Academic Press, 324 pp.

Cole, J. P., and King, C. A. M. 1969. *Quantitative Geography.* London: John Wiley and Sons, Ltd. 692 pp.

Collyer, P. L., and Merriam, D. F. 1973. An application of cluster analysis in mineral exploration. *Math. Geol.* **5**(3):213–224.

Cooley, W. W., and Lohnes, P. R. 1962. *Multivariate Procedures for the Behaviorial Sciences,* New York: John Wiley and Sons, 211 pp.

Cooley, W. W., and Lohnes, P. R. 1971. *Multivariate Data Analysis.* New York: John Wiley and Sons, 364 pp.

Court, A. 1952. Some new statistical techniques in geophysics. *Adv. in Geophys.* **1**: 75–85.

Craig, G. Y. 1974. A trifling investment of fact. In D. F. Merriman, ed., *The Impact of Quantification on Geology. Syracuse Univ. Geol. Contr.* **2**:5–18.

Cubitt, J. M. 1975. A regression technique for the analysis of shales by X-ray diffractometry. *J. Sed. Petrol.* **45**(2):546–553.

Curl, R. C. 1966. Caves as a measure of karst. *J. Geol.* **74**(5/2):798–830.

Dacey, M. F., and Krumbein, W. C. 1970. Markovian models in stratigraphic analysis. *Math. Geol.* **2**(2):175–191. (see Paper 20).

Dagbert, M., and David, M. 1974. Pattern recognition and geochemical data: An application to Monteregian Hills. *Can. J. Earth Sci.* **11**(11):1577–1585.

Dapples, E. C. 1975. Laws of distribution applied to sand sizes. In E. H. T. Whitten, ed., *Quantitative Studies in the Geological Sciences. Geol. Soc. Am. Mem.* **142**: 37–61.

David, M. 1977. Geostatistical Ore Reserve Estimation. *Developments in Geomathematics 2.* Amsterdam: Elsevier, 364 pp.

David, M., Campiglio, C., and Darling, R. 1974. Progress in R-mode and Q-mode analysis: Correspondence analysis and its application to the study of geological processes. *Can. J. Earth Sci.* **11**(1):131–146 (see Paper 23.)

Davis, J. C. 1970. Information contained in sediment size analysis. *Math. Geol.* **2** (2):105–112.

317

Davis, J. C. 1973. *Statistics and Data Analysis in Geology*. New York: John Wiley and Sons, 550 pp.

Davis, J. C., and Cocke, J. M. 1972. Interpretation of complex lithologic successions by substitutability analysis. In D. F. Merriam, ed., *Mathematical Models of Sedimentary Processes*. New York: Plenum Press, pp. 27–52. (see Paper 21.)

Davis, J. C., and McCullagh, M. J. 1975. *Display and analysis of spatial data*. New York: John Wiley and Sons, 378 pp.

Demirmen, F. 1969. Multivariate procedures and FORTRAN IV program for evaluation and improvement of classifications. *Computer Contributions* 31, 51 pp.

Denness, B., Cubitt, J. M., McCann, D., and McQuillan, R. 1978. Engineering evaluation of seabed sediments by cluster analysis. In D. F. Merriam, ed., *Recent Advances in Geomathematics*. Oxford: Pergammon Press, pp. 21–33.

De Wijs, H. J. 1951. Statistics of ore distributions, 1. *Geol. Mijnbouw* 13(11): 365–375.

De Wijs, H. J. 1953. Statistics of ore distributions, 2. *Geol. Mijnbouw* 15(1):12–24.

Drapeau, G. 1973. Factor analysis: How it copes with complex geological problems. *Math. Geol.* 5(4):351–364.

Dryden, A. L. 1931. Accuracy in percentage representation of heavy mineral frequencies. *Proc. Nat. Acad. Sci.* 17(5):233–238.

Dryden, A. L. 1935. A statistical method for the comparison of heavy mineral suites, *Am. J. Sci.* 29(173):393–408.

Durovic, S. 1959. Contribution to the lognormal distribution of elements. *Geochim. Cosmochim. Acta* 15(4):330–336.

Eisenhart, C. 1935. A test for significance of lithological variation, *J. Sed. Petrol* 5 (3):137–145. (see Paper 3.)

Ewing, C. J. C. 1931. A comparison of the methods of heavy mineral separation. *Geol. Mag.* 68(3):136–140.

Everitt, B. 1974. *Cluster Analysis*. London: Heinemann Educational Books, 122 pp.

Falconer, A. 1971. Uses of Q-mode factor analysis in the interpretation of glacial deposits. In E. Yatsu and A. Falconer, eds., *Research methods in pleistocene geomorphology*. 2nd Guelph Symposium on Geomorphology, pp. 148–185.

Fenner, P. 1969. *Models of Geologic Processes: An Introduction to Mathematical Geology*. Short Course Lecture Notes, Am. Geol. Inst., Washington, D.C., 359 pp.

Fenner, P. 1972. *Quantitative Geology*. Geol. Soc. Am. Spec. Pap. 146, 101 pp.

Fisher, R. A. 1953a. The expansion of statistics. *J. Roy. Stat. Soc., Ser. A.* 116(1): 1–6. (see Paper 1.)

Fisher, R. A. 1953b. Dispersion on a sphere. *Proc. Roy. Soc., Ser. A.* 217:295–305.

Fox, W. T. 1964. FORTRAN and FAP Program for Calculating and Plotting Time-Trend Curves Using an IBM 7090 or 7094/1401 Computer System. *Kansas Geol. Surv. Spec. Dist. Publ.* 12, 24 pp.

Fox, W. T. 1967. Simulation models of time-trend curves for paleoecologic interpretation. *Computer Contributions* 18:18–29.

Fox, W. T. 1975. Some practical aspects of time-series analysis. In R. B. McCammon, ed., *Concepts in Geostatistics*. Berlin:Springer-Verlag, pp. 70–89.

Fox, W. T., and Brown, J. A. 1965. The use of time-trend analysis for environmental interpretation of limestone. *J. Geol.* 73(3):510–518.

Friedman, G. M. 1962. On sorting, sorting coefficients and lognormality of the grain-size distribution of sandstones. *J. Geol.* 70(6):737–753.

Garrett, R. G., and Nichol, I. 1969. Factor analysis as an aid in the interpretation of regional geochemical stream sediment data. In F. C. Canney, ed., *Int. Geochem. Expl. Symp. Q.J. Colo. Sch. Mines*, **64**(1):245-264.

Gates, C. E., and Ethridge, F. G. 1972. A generalized set of discrete frequency distributions with FORTRAN program. *Math. Geol.* **4**(1):1-24.

Ghose, B. K. 1970. Statistical analysis of mixed fossil populations. *Math. Geol.* **2** (3):265-276.

Griffiths, J. C. 1960a. Frequency distributions in accessory mineral analysis. *J. Geol.* **68**(4):353-365.

Griffiths, J. C. 1960b. Aspects of measurement in the geosciences. *Mineral Ind. Bull.* **29**(4):1, 4-5.

Griffiths, J. C. 1962. Statistical methods in sedimentary petrography. In H. B. Milner ed., *Sedimentary petrography*, New York: Macmillan, pp. 565-617.

Griffiths, J. C. 1966. Exploration for natural resources. *Operations Res.* **14**(2):189-209.

Griffiths, J. C. 1967. *Scientific Method in the Analysis of Sediments*. New York: McGraw-Hill, 508 pp.

Griffiths, J. C. 1970. Current trends in geomathematics. *Earth Sci. Rev.* **6**(2):121-140.

Griffiths, J. C. 1971. Problems of sampling in geoscience. *Trans. Inst. Min. Met.* **80**: B346-356. (see Paper 18.)

Guarascio, M., David, M., and Huijbregts, Ch. 1976. *Advanced Geostatistics in the Mining Industry*. Dordrecht-Holland: D. Reidel Publishing Co., 461 pp.

Haas, A., and Viallix, J. R. 1975. Krigeage applied to geophysics. *Geophys. Prosp.* **24**:49-69.

Harbaugh, J. W., and Bonham-Carter, G. 1970. *Computer Simulation in Geology*. New York: John Wiley and Sons, 575 pp.

Harbaugh, J. W., and Demirmen, F. 1964. Application of Factor Analysis to Petrologic Variations of Americus Limestone (Lower Permian), Kansas and Oklahoma. *Kansas Geol. Surv. Spec. Dist. Publ.* 15, 41 pp.

Harbaugh, J. W., Doveton, J. H., and Davis, J. C. 1977. *Probability Methods in Oil Exploration*. New York: John Wiley and Sons, 269 pp.

Harbaugh, J. W., and Merriam, D. F. 1968. *Computer Applications in Stratigraphic Analysis*. New York: John Wiley and Sons, 282 pp.

Harker, A. 1909. *The Natural History of Igneous Rocks*, London: Methuen, 384 p.

Harman, H. H. 1967. *Factor Analysis*. Chicago: Univ. Chicago Press, 474 pp.

Harris, D. V. P. 1966. Factor analysis as a tool for quantitative studies in mineral exploration. *Mineral. Ind. Bull.* **2**:GG1-GG37.

Harvey, P. K., and Ferguson, C. C. 1976. On testing orientation data for goodness-of-fit to a von Mises distribution. *Computers and Geosciences* **2**(2):261-268.

Hawkins, D. M. 1972. On the choice of segments in piecewise approximation. *J. Inst. Math. Appl.* **9**(2):250-256.

Hawkins, D. M., and Merriam, D. F. 1973. Optimal zonation of digitized sequence data. *Math. Geol.* **5**(4):389-395.

Hawkins, D. M., and Merriam, D. F. 1974a. Segmentation of discrete sequences of geologic data. *Geol. Soc. Am. Mem.* **142**:311-315.

Hawkins, D. M., and Merriam, D. F. 1974b. Zonation of multivariate sequences of digitized geologic data. *Math. Geol.* **6**(3):263-269.

Hazen, S. W., Jr. 1967. Some statistical techniques for analyzing mine and mineral deposit samples and assay data. *U.S. Bur. Mines Bull.* **621**, 223 p.

Henley, S. 1976. The identification of discontinuities from areally distributed data. In D. F. Merriam, ed., *Quantitative techniques for the analysis of sediments*. Oxford: Pergammon, pp. 157-168.

Hersch, A. H. 1934. Evolutionary relative growth in the Titanotheres. *Am. Nat.* **68**: 537-561.

Hill, M. 1975. Correspondence analysis: A neglected multivariate method. *J. Roy. Stat. Soc., Ser. C* **23**:340-354.

Horton, R. E. 1945. Erosional development of streams and their drainage basins: Hydrophysical approach to quantitative morphology. *Bull. Geol. Soc. Am.* **56** (3):275-370.

Howarth, R. J. 1970. Principle components analysis of the geochemistry and mineralogy of the Portaskaig tillite and Kiltylanned schist (Dalradian) of Co. Donegal, Eire. *Math. Geol.* **2**(3):285-302.

Howarth, R. J. 1971a. An empirical discriminant method applied to sedimentary-rock classification from major-element geochemistry. *Math. Geol.* **3**(1):51-60. (see Paper 22.)

Howarth, R. J. 1971b. Empirical discriminant classification of regional stream-sediment geochemistry in Devon and east Cornwall. *Trans. Inst. Min. Met. B* **80**(777):142-149.

Howarth, R. J. 1973. FORTRAN IV programs for empirical discriminant classification of spatial data. *Geocom Bull.* **6**(1-2):1-31.

Howd, F. H. 1964. The taxonomy program—a computer technique for classifying geologic data. In *Computers in the Mineral Industries, Part A, Q.J. Colo. Sch. Mines* **59**(4):207-222.

Huijbregts, C., and Matheron, G. 1971. Universal Kriging (an optimal method for estimating and contouring in trend surface analysis). *Can. Inst. Min. Met., Decision Making in the Mineral Industry*, Spec. Vol. **12**:159-169.

Imbrie, J. 1963. *Factor and Vector Analysis Programs for Analyzing Geological Data.* Geography Branch, Tec. Rep. (6) ONR Task No. 389-135, Contract No. 1228 (26), 83 pp.

Imbrie, J., and Kipp, N. G. 1971. A new micropaleontological method for quantitative paleoclimatology: Application to a late Pleistocene Caribbean core. In K. K. Turekian, ed., *Lake Cenozoic Glacial Ages.* New Haven, Conn: Yale Univ. Press, pp. 71-181.

Imbrie, J., and Newall, N. 1964. *Approaches to Paleoecology.* New York: John Wiley and Sons, 432 pp.

Imbrie, J., and Purdy, E. G. 1962. Classification of modern Bahamian carbonate sediments. In *Classification of Carbonate Rocks: A Symposium. Am. Assoc. Petrol. Geol. Mem.* **1**:253-272. (see Paper 10.)

Imbrie, J., and Van Andel, T. H. 1964. Vector analysis of heavy mineral data. *Bull. Geol. Soc. Am.* **75**:1131-1155.

Ingerson, E. 1954. Geochemical work of the geochemistry and petrology branch, U.S. Geological Survey, *Geochim. Cosmochim. Acta* **5**(1):20-39.

Jaquet, J. M., Froidevaux, R., and Vernet, J. P. 1975. Comparison of automatic classification methods applied to lake geochemical samples. *Math. Geol.* **7**(3): 237-266.

Jardine, N., and Sibson, R. 1971. *Mathematical Taxonomy.* London: John Wiley and Sons, Ltd., 286 pp.

Jeran, T. W., and Mashey, J. R. 1970. A computer program for the stereographic analysis of coal fractures and cleats. *U.S. Bur. Mines Circ.* **8454**, 34 pp.

Jizba, Z. V. 1959. Frequency distribution of elements in rocks. *Geochim. Cosmochim. Acta* **16**(1/3):79-82.

Jones, T. A. 1968. Statistical analysis of orientation data. *J. Sed. Petrol.* **38**(1): 61-67.

Jöreskog, K. G., Klovan, J. E., and Reyment, R. A. 1976. *Geological Factor Analysis, Methods in Geomathematics,* vol. 1. Amsterdam: Elsevier, 180 pp.

Journel, A. G. 1973. Geostatistics and sequential exploration. *Min. Eng.* **25**(10):44-48.

Journel, A. G. 1975. Geological reconnaissance to exploitation—a decade of applied geostatistics. *Can. Inst. Min. Met. Bull.* **68**(758):75-84.

Joyce, A. S. 1973. Application of cluster analysis to detection of subtle variation in a granitic intrusion. *Chem. Geol.* **11**(4):297-306.

Kaesler, R. L. 1969. Numerical taxonomy in paleontology classification, ordination and reconstruction of phylogenies. *North Am. Paleo. Conv. Chicago, 1969, Proc. B,* pp. 84-101.

Kaesler, R. L., and McElroy, M. N. 1966. Classification of subsurface localities of the Reagan sandstone (Upper Cambrian) of central and northwest Kansas. In D. F. Merriam, ed., *Colloquium on classification procedures. Computer Contributions* **7**:42-47.

Kaesler, R. L., and Taylor, R. S. 1970. Cluster analysis and ordination in paleoecology of Ostracoda from the Green River Formation, (Eocene, USA). In H. J. Oertli, ed., *Paleoecologies Ostracodes Pau, 1970.* pp. 153-165.

Kittleman, L. R. 1964. Application of Rosin's distribution in size-frequency analysis of clastic rocks. *J. Sed. Petrol.* **34**(3):483-502.

Klovan, J. E. 1966. The use of factor analysis in determining depositional environments from grain-size distributions. *J. Sed. Petrol.* **36**(1):115-125.

Klovan, J. E. 1968. Selection of target areas by factor analysis. In *Proceedings of a Symposium on Decision Making in Mineral Exploration,* Vancouver, B. C., 1968, pp. 19-27.

Klovan, J. E. 1975. R-mode and Q-mode factor analysis. In R. B. McCammon, ed., *Concepts in Geostatistics.* Berlin: Springer-Verlag, pp. 21-69.

Koch, G. S., Jr., and Link, R. F. 1970. *Statistical Analysis of Geologic Data,* vol. 1. New York: John Wiley and Sons, 368 pp.

Koch, G. S., Jr., and Link, R. F. 1971. *Statistical Analysis of Geologic Data,* vol. 2. New York: John Wiley and Sons, 438 pp.

Kolmogorov, A. N. 1941. The lognormal law of distribution of particle sizes during crushing. *Dokl. Akad. Sci. USSR* **31**(2):99-101.

Kolmogorov, A. N. 1949. The solution of a problem in the theory of probability related to the mechanism of formation of sedimentary beds. *Dokl. Akad. Sci. USSR* **65**(6):793-796.

Korn, H. 1938. Schichtung und absolute zeit. *Neus Jahrb. Mineral. Abh. A* **74**(1):51-188.

Krige, D. G. 1951. A statistical approach to some basic valuation problems on the Witwatersand. *J.S. Afr. Soc. Chem. Met. Min.* **52b**:119-139.

Krige, D. G. 1960. On the departure of ore value distributions from the lognormal model in South African gold mines. *J.S. Afr. Inst. Min. Met.* **61**(4):231-244.

Krige, D. G. 1964. A brief review of developments in the application of mathematical statistics to ore valuation in the South African gold mining industry. *Q. J. Colo. Sch. Mines* **59**(4):785-794.

Krige, D. G. 1966. Two dimensional weighted moving average trend surfaces for ore valuation. *J.S. Afr. Inst. Min. Met.* **64**(1):13-79.

Krige, D. G. 1970. The role of mathematical statistics in improved ore valuation techniques in South African gold mines. In M. A. Romanova and O. V. Sarmanov, eds., *Topics in Mathematical Geology* New York: Consultants Bureau, 281 pp.

Krumbein, W. C. 1934. The probable error of sampling sediment for mechanical analysis. *Am. J. Sci.* **27**(159):204–214.

Krumbein, W. C. 1936a. The use of quartile measures in describing and comparing sediments. *Am. J. Sci.* **32**(188):98–111.

Krumbein, W. C. 1936b. Application of logarithmic moments to size frequency distribution of sediments. *J. Sed. Petrol.* **6**(1):35–47. (see Paper 4.)

Krumbein, W. C. 1937a. The sediments of Barataria Bay (La.). *J. Sed. Petrol.* **7**(1): 3–17.

Krumbein, W. C. 1937b. Sediments and exponential curves. *J. Geol.* **45**(6):577–601.

Krumbein, W. C. 1937c. Korngrösseneinteilungen und statistische Analyse. *Neues Jahrb. Mineral. Abh.A* **73**:137–150.

Krumbein, W. C. 1938a. Size frequency distributions of sediments and the normal phi curve. *J. Sed. Petrol.* **8**(3):84–90.

Krumbein, W. C. 1938b. Sampling, preparation for analysis, mechanical analysis and statistical analysis. In W. C. Krumbein and F. J. Pettijohn, eds. *Manual of Sedimentary Petrography.* New York: Appleton-Century, pp. 1–274.

Krumbein, W. C. 1939. Graphical presentation and statistical analysis of sedimentary data. In P. D. Trask, ed., *Recent Marine Sediments.* Tulsa: Am. Assoc. Pet. Geol., pp. 558–591.

Krumbein, W. C. 1954. Applications of statistical methods to sedimentary rocks. *Am. Stat. Assoc.* **49**(265):51–66.

Krumbein, W. C. 1955. Experimental design in the earth sciences. *Trans. Am. Geophys. Union* **36**(1):1–11. (see Paper 8.)

Krumbein, W. C. 1960. Some problems in applying statistics to geology. *Appl. Stat.* **9**(2):82–91.

Krumbein, W. C. 1969. The computer in geologic perspective. In D. F. Merriam, ed., *Computer Applications in the Earth Sciences.* New York: Plenum Press, pp. 251–275.

Krumbein, W. C. 1974. The pattern of quantification in geology. In D. F. Merriam, ed., *The Impact of Quantification on Geology. Syracuse Univ. Geol. Contr.* **2** 51–56.

Krumbein, W. C. 1975. Markov models in the earth sciences. In R. B. McCammon, ed., *Concepts in Geostatistics.* Berlin: Springer–Verlag, pp. 90–105.

Krumbein, W. C., and Dacey, M. F. 1969. Markov chains and embedded Markov chains in geology. *Math. Geol.* **1**(1):79–96.

Krumbein, W. C., and Graybill, F. A. 1965. *An Introduction to Statistical Models in Geology.* New York: McGraw-Hill, 475 pp.

Krumbein, W. C., and Imbrie, J. 1963. Stratigraphic factor maps. *Am. Assoc. Petrol. Geol. Bull.* **47**(4):698–701.

Krumbein, W. C., and Lieblein, J. 1956. Geological application of extreme-value methods to interpretation of cobbles and boulders in gravel deposits. *Trans. Am. Geophys. Union* **37**(3):313–319.

Krumbein, W. C., and Miller, R. L. 1953. Design of experiments for statistical analysis of geological data. *J. Geol.* **61**(6):510–532.

Krumbein, W. C., and Shreve, R. L. 1970. *Some Statistical Properties of Dendritic Stream Networks.* Tech. Rep. 13, ONR Task No. 389-150, Dept. Geol. Scis, Northwestern Univ. and Sp. Proj. Rept., NSF Grant 6A-1137, Dept. Geol., UCLA, 117 pp.

Krumbein, W. C., and Sloss, L. L. 1969. High speed digital computers in stratigraphic and facies analysis. *Am. Assoc. Petrol. Geol. Bull.* **42**(11):2650–2669.

Kupletsky, B. M. and Oknova, T. M. 1934. Quantitative mineralogical composition of nepheline rocks (Russian with English summary). *Trav. Inst. Petrog. Acad. Sci. USSR* **6**:17–28.

Lafitte, P. 1972. *Traite d'informatique geologique.* Paris: Masson, 624 pp.

Leitch, D. 1951. Biometrics and systematics in relation to paleontology. *Proc. Linn. Soc. London.* **162**(2):159–170.

Le Maitre, R. W. 1968. Chemical variations within and between volcanic rock series: A statistical approach. *J. Petrol.* **9**(2):220–252.

Link, R. F., and Koch, G. S., Jr. 1975. Some consequences of applying lognormal theory to pseudolognormal distributions. *Math. Geol.* **7**(2):117–128. (see Paper 14.)

Loewinson-Lessing, F. J. 1924. On limits and subdivisions of the family of andesites. *Izv. Geol. Kom.* **43**(6):723–735.

Loewinson-Lessing, F. J. 1925. On differentiation of basalts and andesites. *Izv. Geol. Kom.* **44**(4):411–423.

Loewinson-Lessing, F. J. 1930. On differentiation of liporite and dacite. *Compt. Rend. Acad. Sci. URSS, Ser. A.* **8**:179–184.

Loewinson-Lessing, F. J. 1933. Statistical characteristics of the chemistry of trachytes. *Izv. Acad. Sci. URSS* **7**(1):101–111.

Loewinson-Lessing, F. J. 1935. *Petrography.* Moscow: ONTI.

Loewinson-Lessing, F. J. 1936. *An Introduction to the History of Petrography.* Leningrad: ONTI, 398 pp.

Loudon, T. V. 1964. *Computer Analysis of Orientation Data in Structural Geology.* Tech. Rept. no. 13. Evanston, Ill.: Northwestern Univ., 130 pp.

Manson, V., and Imbrie, J. 1964. FORTRAN Program for Factor and Vector Analysis of Geologic Data Using an IBM 7090 or 7090/1401 Computer System. *Kansas Geol. Surv., Spec. Distr. Publ.* 13, 46 pp.

Mardia, K. V. 1972. *The statistics of directional data.* London: Academic Press, 357 pp.

Marsal, D. 1967. *Statistische methoden fur erdwissenschaftler.* Stuttgart: Schweizerbart, 152 pp.

Mason, B. 1962. *Meteorites.* New York: John Wiley and Sons, 274 pp.

Mather, P. M. 1973. Recent advances in factor analysis. In *Multivariate Analysis in Geography*, working paper set (1). Study Group in Quantitative Methods, Inst. Br. Geogr., pp. 45–47.

Mather, P. M. 1976. *Computational Methods of Multivariate Analysis in Physical Geography*, London: John Wiley and Sons, Ltd., 532 pp.

Matheron, G. 1962. *Traité de geostatistique appliquée*, Tomme 1. Paris: Editions Technip, 333 pp.

Matheron, G. 1963a. *Traité de geostatistique appliquée*, Tomme 2. Paris: Editions Technip, 172 pp.

Matheron, G. 1963b. Principles of geostatistics. *Econ. Geol.* **58**:1246–1266. (see Paper 19.)

Matheron, G. 1965. *Les variables régionalisées et leur estimation.* Paris: Edition Masson, 306 pp.

Matheron, G. 1967a. *Elements pour une theorie des mileux poreaux.* Paris: Edition Masson, 166 pp.

Matheron, G. 1967b. Kriging or polynomial interpolation procedures? A contribution to polemics in mathematical geology. *Can. Inst. Min. Met. Bull.* **60**(665): 1041–1045.

Matheron, G. 1969. *Le krigeage universal, Les cahiers du centre de morphologie methematique de Fontainebleau,* Fasc. 1. Paris: l'Ecole Nationale Superieure des Mines de Paris, 83 pp.

Matheron, G. 1970. Random functions and their applications in geology. In D. F. Merriam, ed., *Geostatistics,* New York: Plenum Press, pp. 79–87.

Matheron, G. 1971. *The Theory of Regionalised Variables and Its Applications, Les cahiers du centre du morphologie mathematique de Fontainebleau,* Fasc. 5. Paris: L'ecole Nationale Superieure des Mines de Paris, 211 pp.

Matheron, G. 1975. *Random Sets and Integral Geometry.* New York: John Wiley and Sons, 261 pp.

McCammon, R. B. 1966. Principle components analysis and its application to large scale correlation studies. *J. Geol.* 74(5/2):721–733.

McCammon, R. B. 1968. Multiple component analysis and its application in classification of environments. *Am. Assoc. Petrol. Geol. Bull.* 52(11/1):2178–2196.

McCammon, R. B. 1969. Aspects of classification. In P. Fenner, ed., *An Introduction to Mathematical Geology: Models of Geologic Processes.* Am. Geol. Inst., R-C1-Rm-C41.

McCammon, R. B. 1970. Component estimation under uncertainty. In D. F. Merriam, ed., *Geostatistics.* New York: Plenum Press, pp. 45–61.

McCammon, R. B. 1975. *Concepts in Geostatistics.* Berlin: Springer-Verlag, 168 pp.

McCammon, R. B., and Wenniger, G. 1970. The dendrograph. *Computer Contributions* 48, 28 pp.

Melguen, M. 1973. Correspondence analysis for recognition of facies in homogeneous sediments off an Iranian river mouth. In B. H. Purser, ed., *The Persian Gulf.* Berlin: Springer-Verlag, pp. 99–113.

Melguen, M. 1974. Facies analysis by "correspondence analysis": Numerous advantages of this new statistical technique. *Mar. Geol.* 17(3):165–182.

Merriam, D. F. 1969. *Computer Applications in the Earth Sciences.* New York: Plenum Press, 281 pp.

Merriam, D. F. 1970. *Geostatistics.* New York: Plenum Press, 177 pp.

Merriam D. F. 1972. *Mathematical Models for Sedimentary Processes.* New York: Plenum Press, 271 pp.

Merriam, D. F. 1974. *The impact of quantification on geology.* Syracuse Univ. Geol. Contr. 2, 104 pp.

Merriam, D. F. 1976a. Mathematical geology. *Geotimes* 21(1):27.

Merriam, D. F. 1976b. *Random Processes in Geology.* Berlin: Springer-Verlag. 168 pp.

Merriam, D. F. 1976c. *Quantitative Techniques for the Analysis of Sediments.* New York: Pergamon Press, 178 pp.

Merriam, D. F. 1978. *Recent Advances in Geomathematics.* Oxford: Pergammon Press, 233 pp.

Merriam, D. F. In prep. *Computer Fundamentals for Geologists.* Darmouth, N. H.: COMPUTE.

Meyer, S. L. 1975. *Data Analysis for Scientists and Engineers.* New York: John Wiley and Sons, 513 pp.

Middleton, G. V. 1962. On sorting, sorting coefficients and the lognormality of the grain-size distribution of sandstone: A discussion. *J. Geol.* 70(6):754–756.

Middleton, G. V. 1964. Statistical studies on scapolites. *Can. J. Earth Sci.* 1:23–24.

Miesch, A. T. 1969. Critical review of some multivariate procedures in the analysis of geochemical data. *Math. Geol.* 1(2):171–184. (see Paper 16.)

Miesch, A. T., Chao, E. C. T., and Cuttitta, F. 1966. Multivariate analysis of geochemical data on tektites. *J. Geol.* 74(5):673–691.

Miller, R. L. 1953. Introduction to special issues on statistics in geology. *J. Geol.* **61**(6):479–481.

Miller, R. L., and Goldberg, E. D. 1955. The normal distribution in geochemistry. *Geochim. Cosmochim. Acta* **8**(1/2):53–62.

Miller, R. L., and Kahn, J. S. 1962. *Statistical Analysis in the Geological Sciences.* New York: John Wiley and Sons, 483 pp.

Morisawa, M. 1971. *Quantitative geomorphology: some aspects and applications.* Binghamton, New York: State Univ. of New York, 315 pp.

Mundry, E. 1972. On the resolution of mixed frequency distributions into normal components. *Math. Geol.* **4**(1):55–60.

Nichol, I., Garrett, R. G., and Webb, J. S. 1966. Automatic data plotting and statistical interpretations of geochemical data. In E. M. Cameron, ed., *Proc. Symp. Geochemical Prospecting, Ottawa.* Dept. Mines. Tech. Surv., Geol. Surv. Can. Pap. 66–54, pp. 195–210.

Niggli, P. 1923. Anwendungen der mathematischen statistik auf probleme der mineralogie und petrologie. *Neues Jahrb. Mineral.* **48**:167–222.

Niggli, P. 1935. Die charakterisierung der klatischen sedimente nach der Komzusammensetzung. *Schweiz. Min. Petrol. Mitt.* **15**(1):31–38.

Niggli, P., De Quervain, F., and Winterhalter, R. U. 1930. *Chemismus schweizerischer gesteine.* Bern. 389 pp.

Olea, R. A. 1975. *Optimum Mapping Techniques Using Regionalized Variable Theory.* Lawrence: Kansas Geol. Surv., 137 pp.

Ondrick, C. W., and Griffiths, J. C. 1969. FORTRAN IV computer program for fitting observed count data to discrete distribution models of binomial, Poisson, and negative binomial. *Computer Contributions* **44**. 20 pp.

Otto, G. H., 1937, The use of statistical methods in effecting improvements in a Jones sample splitter. *J. Sed. Petrol.* **7**(3):110–132.

Parks, J. M. 1966. Cluster analysis applied to multivariate geologic data. *J. Geol.* **74**(5/2): 703–715. (see Paper 12.)

Parks, J. M. 1970. FORTRAN IV program for Q-mode cluster analysis on distance function with printed dendrogram. *Computer Contributions* **46**:32 pp.

Parks, J. M. 1972. Cluster and factor analysis in classification and correlation of paleoenvironmental data. *Int. Stat. Inst. Bull.* **44**(1):539–551.

Pearson, H. S. 1928. Chinese fossil Suidae. *Paleontologia Sinica* V, fasc. 5, Ser. C., pp. 1–75.

Pearson, K. 1895. Contribution to the mathematical theory of evolution, 2: Skew variation in homogeneous materials. *Phil. Trans. Roy. Soc. A* **186**:343–414.

Pincus, H. J. 1951. Statistical methods applied to the study of rock fractures. *Bull. Geol. Soc. Am.* **62**(2):81–130.

Pincus, H. J. 1952. Some methods for operating on orientation data, *Bull. Geol. Soc. Am.* **63**(3):431–434.

Pincus, H. J. 1953. The analysis of aggregates of orientation data in the Earth Sciences. *J. Geol.* **61**(6):482–509.

Potenza, R. 1973. A geomathematical investigation of syntexis in a gabbroic formation. *Math. Geol.* **5**(4):321–339.

Potter, P. E. and Blakely, R. F. 1968. Random processes and lithologic transitions. *J. Geol.* **76**(2):154–170.

Prentice, J. E. 1949. The statistical method in paleontology. *Brit. Sci. News.* **3**(25): 17–19.

Purdy, E. G. 1963. Recent calcium carbonate facies of the Great Bahama Bank, 1: Petrography and reaction groups. *J. Geol.* **71**(3):334–355.

Rao, C. P., Mann, C. J., and Carozzi, A. V. 1973. Factor analysis testing of micro-facies and interpreted environmental factors in Ste. Genevieve Limestone (Mississippian) Illinois. *J. Geol.* **81**(1):65–80.

Rao, C. R. 1952. *Advanced Statistical Methods in Biometric Research.* New York: John Wiley and Sons, 390 pp.

Razumovsky, N. K. 1940. Nature of the distribution of the metal contents in ore deposits. *Compt. Rendu. Acad. Sci. URSS* **28**(9):815–817.

Razumovsky, N. K. 1941. On the Role of the Logarithmically Normal Law of Frequency Distribution in Petrology and Geochemistry. *Comptes Rend. (Dokl.) de l' Academ. Sci. de'l URSS* **33**(1):48–49. (see Paper 5.)

Razumovsky, N. K. 1948. The lognormal distribution law of matter and its special features. *Zap. Leningr. gornogo. Inst.* **20**:105–121.

Read, W. A., and Dean, J. M. 1972. Principal component analysis of lithological variables from some Namurian (E_2) parallic sediments in Central Scotland. *Bull. Geol. Surv. Gr. Brit.* **40**:83–99.

Reyer, E. 1877. Beitrage zur fysik der eruptionen und der eruptivgesteine. Vienna. 225 pp.

Reyment, R. A. 1961. Quadrivariate principal components analysis of Globigerina yeguaensis. *Stockholm Contr. Geol.* **8**:17–26.

Reyment, R. A. 1963. Multivariate analytical treatment of quantitative species associations: An example of Paleoecology. *J. Anim. Ecol.* **32**:535–547. (see Paper 11.)

Reyment, R. A. 1970. Symposium on biometrical methods in paleontology. *Bull. Geol. Inst. Univ. Uppsala, New Ser.* **2**, 89 pp.

Reyment, R. A. 1971. *Introduction to Quantitative Paleoecology.* Amsterdam: Elsevier, 226 pp.

Reyment, R. A. 1974. The age of zap. In D. F. Merriam, ed., *The Impact of Quantification on Geology Syracuse Univ. Geol. Contr.* **2**:pp 19–26.

Rhodes, J. M. 1969. The application of cluster and discriminant analysis in mapping granite intrusions. *Lithos* **2**:223–237.

Richardson, W. A. 1923. The frequency distribution of igneous rocks. *Mineral. Mag.* **20**(100):1–19 (see Paper 2.)

Richardson, W. A., and Sneesby, G. 1922. The frequency distribution of igneous rocks. *Mineral. Mag.* **19**(97):303–313.

Robb, R. C. 1935. A study of mutations in evolution, Part I: Evolution in the equine skull. *J. Genet.* **31**(1):39–46.

Romanova, M. A. 1957. Geologia verchnei tchasti krapnotsvetnych otlozhenii poluostrova tcheleken. *Trudy Obsbch. Estest. Leningr. Univ.* **49**(2):116–126.

Romanova, M. A., and Sarmanov, O. V. 1970. *Topics in Mathematical Geology.* New York: Consultants Bureau, 281 pp.

Royle, A. G., and Hosgit, E. 1974. Local estimation of sand and gravel reserves by geostatistical methods. *Trans. Inst. Min. Met.* **80**(809):53–62.

Saager, R., and Sinclair, A. J. 1974. Factor analysis of stream sediment geochemical data from the Mount Nansen area. Yukon Territory, Canada. *Mineral. Deposita (Berl.)* **9**:243–252.

Saha, A. K., Bhattacharyya, C., and Lakshmipathy, S. 1974. Some problems of interpreting the correlations between the modal variables in granitic rocks. *Math. Geol.* **6**(3):245–258.

Sarmonov, O. V. 1961. False correlations between random variables. *Trudy Mat. Inst. Akad. Nauk SSSR* **64**:174–184.

Sarmanov, O. V., and Vistelius, A. B. 1969. Correlation between percentage values. *Dokl. Akad. Nauk SSSR* **126**(1):22–25.

Saxena, S. K., and Walter, F. S. 1974. A statistical-chemical and thermodynamic approach to lunar mineralogy. *Geochim. Cosmochim. Acta* **38**(1):79-95.

Scheidegger, A. E. 1960. Mathematical methods in geology. *Am. J. Sci.* **258**(3): 218-221.

Schmid, K. 1934. Biometrische untersuchungen au foraminiferen aus d. phacen von ceram. *Ecologae Geol. Helvetiae* **27**(1):46-128.

Schuenemeyer, J. H., Koch, G. S., Jr., and Link, R. F. 1972. Computer program to analyze directional data, based on the methods of Fisher and Watson. *Math. Geol.* **4**(3):177-202.

Schwarzacher, W. 1964. An application of statistical time-series analysis of a limestone-shale sequence. *J. Geol.* **72**(2):195-213.

Schwarzacher, W. 1975. *Sedimentation Models and Quantitative Stratigraphy: Developments in Sedimentology 19.* Amsterdam: Elsevier, 382 pp.

Sepkoski, J. J., Jr. 1974. Quantified coefficients of association and measurement of similarity. *Math. Geol.* **6**(2):135-153.

Sharapov, I. 1971. *Applications of Mathematical Statistics in Geology.* Moscow: Nedra, 260 pp.

Shaw, D. M. 1961. Element distribution laws in geochemistry. *Geochim. Cosmochim. Acta* **23**(112):116-134.

Shaw, D. M., and Bankier, J. D. 1954. Statistical methods applied to geochemistry. *Geochim. Cosmochim. Acta.* **5**(3):111-123.

Sichel, H. S. 1947. An experimental and theoretical investigation of bias error in mine sampling with special reference to narrow gold reefs. *Bull. Inst. Min. Met.* **483**:1-41.

Sichel, H. S. 1952. New methods in the statistical evaluation of mine sampling data. *Trans. Inst. Min. Met.* **61**(6):261-288.

Sichel, H. S. 1973. Statistical valuation of diamondiferous deposits. *Proc. 10th. Int. Symp. Appl. Comput. Mineral. Ind., J.S. Afr. Inst. Min. Met.* **73**(7):235-243.

Simpson, G. G., and Roe, A. 1960. *Quantitative Zoology.* New York: McGraw-Hill, 440 pp.

Size, W. B. 1973. Interpretation of factor analysis on modal data from the Red Hill syenitic complex. *Math. Geol.* **5**(2):191-199.

Smart, J. S. 1976. Joint distribution functions for link lengths and drainage areas. In D. F. Merriam, ed., *Random Processes in Geology* Berlin: Springer-Verlag, pp. 112-123.

Smith, F. G. 1966. *Geological Data Processing Using FORTRAN IV,* New York: Harper and Row, 284 pp.

Sneath, P. H. A., and Sokal, R. R. 1973. *Numerical taxonomy.* San Francisco: W. H. Freeman and Co., 573 pp.

Sokal, R. R., and Rohlf, F. J. 1970. Biometry: *The Principles and Practice of Statistics in Biological Research.* San Francisco: W. H. Freeman, 776 pp.

Sokal, R. R., and Rohlf, F. J. 1973. *Introduction to Statistics.* San Francisco: W. H. Freeman, 368 pp.

Sokal, R. R., and Sneath, P. H. A. 1963. *Principles of Numerical Taxonomy.* San Francisco: W. H. Freeman, 359 pp.

Sorby, H. C. 1908. On the application of quantitative methods to the study of the structure and history of rocks. *Q. J. Geol. Soc. London* **64**(2):171-233.

Steinmetz, R. 1962. Analyses of vectorial data. *J. Sed. Petrol.* **32**(4):801-812.

Strahler, A. H. 1952. Dynamic basis of geomorphology. *Bull. Geol. Soc. Am.* **63** (8):923-938.

Strahler, A. N. 1954. Statistical analysis in geomorphic research. *J. Geol.* **62**(1): 1-25.

Strahler, A. N. 1956. Quantitative slope analysis. *Bull. Geol. Soc. Am.* **69**(3):279–300.

Strahler, A. N. 1964. Quantitative geology of drainage basins and channel networks. In V. T. Chow, ed., *Handbook of Applied Hydrology.* New York: McGraw-Hill, pp. 39–76.

Strahler, A. N. 1968. Quantitative geomorphology. In R. W. Fairbridge, ed., *Encyclopedia of Geomorphology.* New York: Reinhold, pp. 898–912.

Symons, F., and De Meuter, F. 1974. Foraminiferal associations of the mid-Tertiary Edegem Sands at Terhagen, Belgium. *Math. Geol.* **6**(1):1–16.

Teil, H. 1975. Correspondence factor analysis: An outline of its method. *Math. Geol.* **7**(1):3–12.

Teil, H. 1976. The use of correspondence analysis in the metallogenic study of ultrabasic and basic complexes. *Math. Geol.* **8**(6):669–682.

Teil, H., and Cheminee, J. L. 1975. Application of correspondence factor analysis to the major and trace elements in the Erta Ale Chain (Afar, Ethiopia). *Math. Geol.* **7**(1):13–30.

Thiergartner, H. 1968. *Grandprobleme der statistischen behandlung geochemischer daten.* Leipzig: Deutscher verlag Fur Grundstoffindustrie, 99 pp.

Till, R. 1974. *Statistical Methods for the Earth Scientist: An introduction.* New York: Halsted Press, John Wiley and Sons, 154 pp.

Till, R., and Colley, H. 1973. Thoughts on the use of principal components analysis in petrogenic problems. *Math. Geol.* **5**(4):341–350.

Tryon, R. C., and Bailey, D. E. 1970. *Cluster Analysis.* New York: McGraw-Hill, 347 pp.

Tukey, J. W. 1970. Some further inputs. In D. F. Merriam, ed., *Geostatistics.* New York: Plenum Press, pp. 163–174.

Twomey, S. 1977. *Introduction to the Mathematics of Inversion in Remote Sensing and Indirect Measurements, Developments in Geomathematics 3.* Amsterdam: Elsevier, 243 pp.

Udden, J. A. 1898. *The Mechanical Composition of Wind Deposits.* Rock Island, Ill.: Augustana Library, Publication No. 1.

Udden, J. A. 1914. Mechanical composition of clastic sediments. *Bull. Geol. Soc. Am.* **25**:655–744.

Vistelius, A. B. 1949. K voprosu o mechanizme sloeobrazovania, *Dokl. Akad. Nauk SSSR* **65**(2):191–194.

Vistelius, A. B. 1952. Kirmak inskaija svita vostotchnogo Azerbaijana. *Dokl. Akad. Nauk Azerb. SSR* **8**(1):17–23.

Vistelius, A. B. 1956. The problem of studying correlations in mineralogy and petrography. *Zap. Vses. Mineral. Obshch.* **85**(1):58–74.

Vistelius, A. B. 1957. Regionalnaija litostratigrafija i uslovija formirovanija produktivnoi tolstchi ijugo vostochnogo kavkaza. *Trudy Obsbch. Estest. Lening. Univ.* **69**(2):126–150.

Vistelius, A. B. 1960. The skew frequency distributions and the fundamental law of the geochemical process. *J. Geol.* **68**(1):1–22.

Vistelius, A. B. 1961. Sedimentation time-trend functions and their application for correlation of sedimentary deposits. *J. Geol.* **69**(6):703–728. (see Paper 9.)

Vistelius, A. B. 1962. Problems of mathematical geology: I. History of the question. *Geol. Geofiz.* (12), p. 3–9.

Vistelius, A. B. 1963. Problems in mathematical geology. II: models and processes in paragenetic analysis. *Geol. Geofiz* (7), p. 3–16. III: Random processes. *Geol. Geofiz.* (12), p. 3–10.

Vistelius, A. B. 1967. *Studies in Mathematical Geology.* New York: Consultants Bureau, 294 pp.

Vistelius, A. B. 1968. Mathematical geology: A report of progress. *Geocom. Bull.* **1**(8):229–264.

Vistelius, A. B. 1969. Preface. *Math. Geol.* 1(1):1–2.

Vistelius, A. B. 1976a. Mathematical geology and the development of the geological sciences. *Math. Geol.* **8**(1):3–8.

Vistelius, A. B. 1976b. Mathematical geology and the progress of the geological sciences. *J. Geol.* **84**(6):629–651.

Vistelius, A. B., and Sarmanov, O. V. 1947. Stochastic basis of a geologically important probability distribution. *Dokl. Akad. Nauk Sci. SSSR* **58**(4):631–634.

Vistelius, A. B., and Sarmanov, O. V. 1961. On the correlation between percentage values: Major component correlation in ferromagnesium micas. *J. Geol.* **69**(2): 145–153.

Vogt, J. 1921–1923. The physical chemistry of the crystallisation and magmatic differentiation of igneous rocks. *J. Geol.* **29**: 318–350, 426–443, 515–539, 627–649; **30**:611–630, 659–672; **31**:233–252, 407–419.

Waitr, I., and Stenzel, P. 1974. Application of factor analysis to classification of engineering-geological environments. *Math. Geol.* **6**(1):17–33.

Washington, H. S. 1920. Chemistry of the earth's crust. *J. Franklin Inst.* **190**:757–815.

Washington, H. S. 1925. The chemical composition of the earth. *Am. J. Sci.* **9**: 352–378.

Watermeyer, G. A. 1919. Application of the theory of probability in the determination of ore reserves. *J. S. Afr. Soc. Chem. Met. Min.* **19**.

Watson, G. S. 1966. The statistics of orientation data. *J. Geol.* **74**(5/2):786–797. (see Paper 13.)

Webb, W. M., and Briggs, L. I. 1966. The use of principal components analysis to screen mineralogical data. *J. Geol.* **74**(5/2):716–720.

Webster, R. 1973. Automatic soil boundary location from transect data. *Math. Geol.* **5**(2):27–37.

Weiss, A., ed., 1969. *A Decade of Digital Computing in the Mineral Industry—A Review of the State-of-the-Art,* New York: Am. Inst. Min. Metall. Petrol. Engrs., 952 pp.

Weiss, M. P., Edwards, W. R., Norman, C. E., and Sharp, E. R. 1965. *The American Upper Ordovician Standard, VII. Stratigraphy and petrology of the Cynthiana and Eden Formations of the Ohio Valley.* Geol. Soc. Am. Spec. Pap. 81, 76 pp.

Wentworth, C. K. 1922. A scale of grade and class terms for clastic sedimentology. *J. Geol.* **30**(5):377–392.

Wentworth, C. K. 1929. Method of computing mechanical composition types in sediments. *Bull. Geol. Soc. Am.* **40**(4):771–790.

Whitten, E. H. T. 1964. Process and response models in geology. *Bull. Geol. Soc. Am.* **75**(5):455–464.

Whitten, E. H. T., ed. 1975. *Quantitative Studies in the Geological Sciences. Geol. Soc. Am. Mem.* **142**, 406 pp.

Wilks, S. S. 1963. Statistical analysis in geology. In T. W. Donnelly, ed., *The Earth Sciences: Problems and Progress in Current Research.* Chicago: Univ. of Chicago Press, pp. 105–136.

Yevjevich, V. M. 1972. *Probability and Statistics in Hydrology.* Fort Collins, Colo: Water Resources Publications, 302 pp.

References

Yule, G. U., and Kendall, M. G. 1953. *An introduction to the theory of statistics.* London: Griffin, 701 pp.

Zodrow, E. L. 1975. Closure Correlation. *Computer Applications* **2**(1 and 2):267–274.

Zodrow, E. L. 1976. Empirical behaviour of Chayes' Null Model. *Math. Geol.* **8**(1): 37–42.

Zodrow, E. L., and Sutterlin, P. G. 1971. Toward a definition of a mineral sample in geology. *Math. Geol.* **3**(3):313–316.

AUTHOR CITATION INDEX

SUBJECT INDEX

About the Editors

JOHN M. CUBITT is presently employed by Syracuse University in the Department of Geology and is a faculty member of the Institute for Energy Research. At Syracuse, he teaches aspects of computer applications in geology, mineral and energy resources, and general geology. He has had ten years of experience working with statistics in geology including two years of research and application in the Computer Unit of the Institute of Geological Sciences, London, England.

Dr. Cubitt received a B.Sc. (Hons.) from Leicester University in 1970 and a Ph.D. from the same institution in 1975. He is known for his study of the geochemistry and mineralogy of Pennsylvanian and Permian shales in Kansas, his geological and geotechnical mapping of the seafloor and was a member of the 4-man team (including Dr. Henley) who developed the data-base management and analysis system, G-EXEC. As director of an IGCP project, he organizes and conducts research on computer correlation. Currently, he is also regional editor of *Computers & Geosciences.*

STEPHEN HENLEY gained 1st class Honors in geology at Nottingham in 1967, and went on in 1970 to take a Ph.D. for work on geology and geochemistry in the Devonian of southwest England. In the course of this work, he developed an interest in statistics and computer methods, an interest that deepened in the succeeding two years when he took over responsibility for computing in the geological branch of the Australian Bureau of Mineral Resources, Canberra. In 1973, he joined the Computer Unit of the Institute of Geological Sciences, and has worked on a number of different projects in the London, Edinburgh, and Keyworth offices. Currently, Dr. Henley's principal interests are in providing computer support (both at home and overseas) for geological-aid projects in developing countries and in developing a general system for computer modelling integrated with a geological data-base management system, (G-EXEC).

A